D1279619

REMOTE SENSING

REMOTE SENSING
With Special Reference to Agriculture and Forestry

Committee on Remote Sensing for Agricultural Purposes
Agricultural Board
National Research Council

NATIONAL ACADEMY OF SCIENCES
Washington, D. C. | 1970

This study was supported by the
National Aeronautics and Space Administration
and by the
United States Department of Agriculture

Standard Book Number 309-01723-8

Available from:

Printing and Publishing Office
National Academy of Sciences
2101 Constitution Avenue
Washington, D. C. 20418

Library of Congress Catalog Card Number 77-600961

We dedicate this book to the memory of
HARRY J. KEEGAN
Spectrophotometrist, National Bureau of Standards (1921–1966)
Professor, Clemson University (1966–1968)

Preface

In this report, we attempt to develop concepts, value judgments, and background on potential uses of remote sensing in agriculture and forestry and to give a technical appraisal of state-of-the-art sensors and discrimination techniques.

For resources to be managed wisely, there must be accurate and timely information, and remote sensing from aerospace platforms can provide quantitative data from which large amounts of needed information can be extracted. Much of this can be made available not only to responsible officials but also to the public at large.

We hope that this report will aid communication among physical scientists, data-processing specialists, agricultural scientists, and foresters. Technical and mathematical details are minimal; only those essential for accuracy and clarity are included. Some original research is reported, and much of the information is derived from reports of limited distribution.

Although the book emphasizes applications of remote sensing to agriculture and forestry, it has implications for all earth resource fields. Geologists, for example, are concerned with plants and soils as clues to geological composition and formation. Hydrologists wish to know the extent and amount of rainfall, depth and density of snowpack, and levels of rivers, lakes, and irrigation reservoirs. Oceanographers are interested in chlorophyll content and surface temperatures of ocean fish-ing areas. And geographers need continually updated information on dynamic use of land, especially for urbanization planning.

We acknowledge with gratitude the support and encouragement received from Dale Jenkins of the National Aeronautics and Space Administration and from Howard B. Sprague, formerly Executive Secretary of the Agricultural Board. We also wish to thank Mrs. Hilda Taft, Willow Run Laboratories, The University of Michigan, for her help in preparing the final manuscript.

J. R. SHAY, *Chairman*
Committee on Remote Sensing
for Agricultural Purposes

Contents

I

Uses, Potentialities, and Needs in Agriculture and Forestry

PERCY R. LUNEY and HENRY W. DILL, JR. Economic Research Service, U.S. Department of Agriculture, Washington, D. C.

INTRODUCTION

"Remote sensing" is the term currently used by a number of scientists for the study of remote objects (earth, lunar, and planetary surfaces and atmospheres, stellar and galactic phenomena, etc.) from great distances. Broadly defined and as used here, remote sensing denotes the joint effects of employing modern sensors, data-processing equipment, information theory and processing methodology, communications theory and devices, space and airborne vehicles, and large-systems theory and practice for the purposes of carrying out aerial or space surveys of the earth's surface. It is this view of remote sensing that appears to warrant the current enthusiasm about it.[1]

Remote sensing has the potential for revolutionizing the detection and characterization of many agricultural and forestry phenomena. Recent studies indicate that remote-sensing techniques can be used in the ultraviolet, visible, infrared, and microwave regions of the electromagnetic spectrum to collect data that give a measure of the reflectance, emittance, dielectric constant, surface geometry, and equivalent blackbody temperature of plants, soils, and water. With a minimum amount of ground sampling, these data will permit (a) identification and area measurements of the major agricultural crop types; (b) mapping of soil and water temperatures; (c) mapping of surface water, including snowpack; (d) mapping of disease and insect invasion; (e) mapping of gross forest types; (f) mapping of forest-fire boundaries; (g) assessment of crop and timber-stand vigor; (h) determination of soil characteristics and soil moisture condition; (i) delineation of rangeland productivity; (j) mapping of areas of high potential forest-fire hazard; and (k) mapping of major soil boundaries.

Effective management of our agricultural and forestry resources calls for large masses of current information. Much of the existing information is inadequate, but much that is needed has never even been collected, and the job of collecting it is time-consuming and extremely expensive. Although remote sensing cannot acquire all types of information needed, it does appear capable of doing much of the job satisfactorily. It has the potential for providing accurate and timely information on which to base important economic decisions.

Land, with all its various features (soils, vegetation, geology, topography, and surface and subsurface water), constitutes the resource base for both agriculture and forestry. The physical phenomena of these two sectors are, however, so different that the following discussion treats them separately.

Agriculture

Description of the Agricultural Sector

As a basis for discussion, the term "agriculture" is here defined to include the activities of planning, producing, and marketing crops and livestock, and miscellaneous by-products derived from plant and animal sources. The primary objectives of modern agriculture are to cultivate the soil in such a manner that it will produce more abundantly and, at the same time, to protect it from deterioration and misuse. In countries with modern agricultural programs, colleges and government agencies attempt to increase output by disseminating knowledge of improved agricultural practices, such as rotation of crops, fertilization, pest control, irrigation, erosion control, reclamation of land, and plant and animal breeding.

The world agricultural enterprise is vast; it comprises some 10 billion acres of arable land, land under tree crops, meadow, and pasture. Brown[2] has divided world agricultural production into seven geographic regions. In Figure 1, these regions are indicated, and the acreage of arable and tree-crop land, the permanent meadow and pasture, and the grain included in arable land and land under tree crops are noted for each region. The grain crops (wheat, rice, corn, millet and

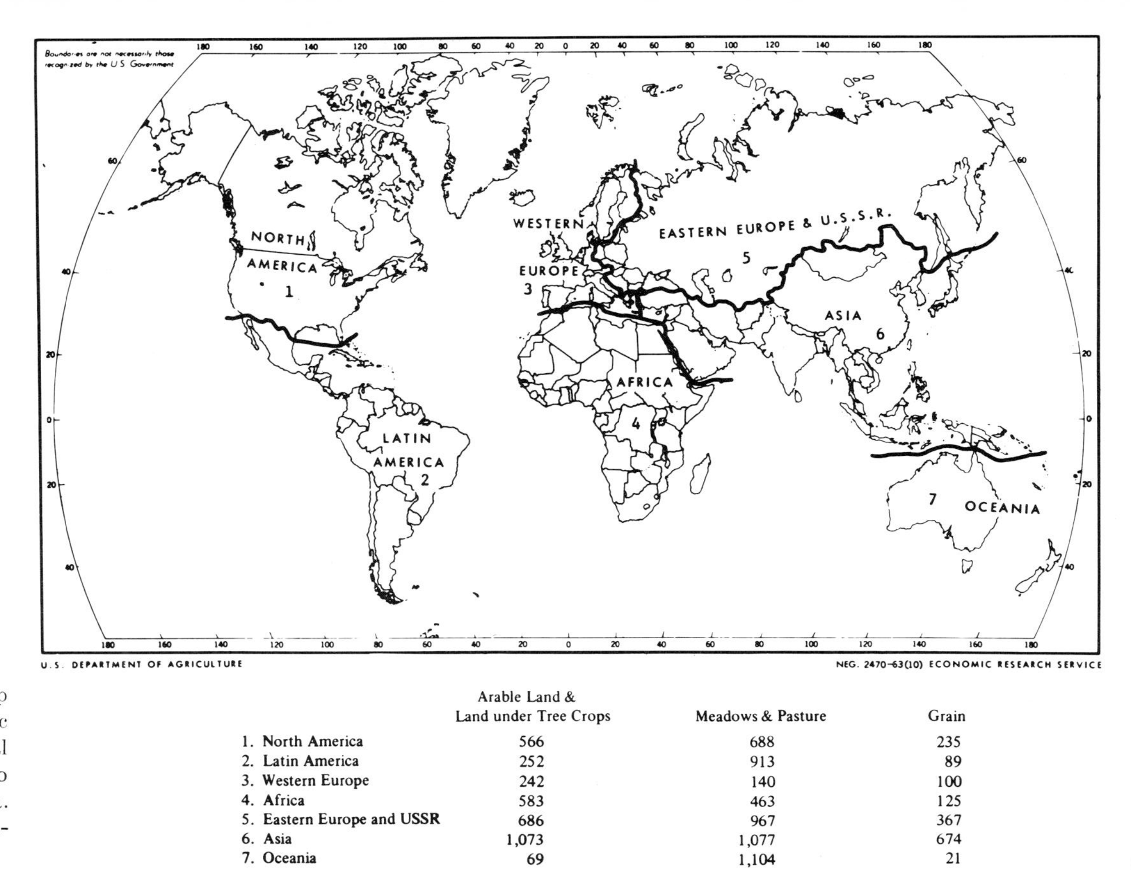

FIGURE 1 World map showing seven geographic regions of agricultural production, with a key to land use in each area. (Data expressed in millions of acres.)

	Arable Land & Land under Tree Crops	Meadows & Pasture	Grain
1. North America	566	688	235
2. Latin America	252	913	89
3. Western Europe	242	140	100
4. Africa	583	463	125
5. Eastern Europe and USSR	686	967	367
6. Asia	1,073	1,077	674
7. Oceania	69	1,104	21

sorghum, barley, oats, and rye) occupy about 1.7 billion acres. Significant acreages of grains are grown in all seven of the geographic regions, ranging from 21 million acres in the Oceania region of Australia and New Zealand to 674 million acres in Asia. Within the seven regions, the best and most extensive areas of cropland are in the plains. Plains, however, occupy only about 30 percent of the land.[3] Hilly lands of fairly low altitude (600–3,000 ft) are next in importance. Locally, terraces and valleys in mountainous terrain may be farmed intensively to meet the food needs of the native people.

Agricultural crop production is also diverse. The principal crops of the world and their acreages are listed in Table 1. About 96 percent of the crop area is occupied by annual crops—those planted and harvested within a year. The other 4 percent is occupied by perennial plants such as fruits, sugarcane, beverage crops, and rubber—all of which are important export crops. Crop production is a year-round operation. The planting dates of certain crops in the Northern Hemisphere coincide with the harvest periods of these crops in areas of the Southern Hemisphere. Corn, for example, is planted in May and harvested in November in the Corn Belt of the United States; in central Brazil corn is planted in November and harvested in May. In regions of mild climate, two or three successive crops of the same species may be grown. Dates of planting and harvesting of a specific crop within each hemisphere are dependent upon latitude or elevation. The advance of wheat harvest northward from early June in Texas to late August in the Canadian provinces is a well-known example.

Agricultural crop production in a given region is variable from year to year. The yield variations emphasize the fact that crop production is primarily a biological system, with its operations carried out in open-air environments that are subject to little control by the farmer. As in any biological system, the crop organisms are subject to devastation by climatic extremes such as heat, cold, drought, hail, and flooding, and by such biological pests as weeds, insects, fungi, bacteria, viruses, and nematodes. Climatic extremes often occur with little forewarning, and damage can be quite extensive. Damage from hail, frosts, and floods is well known. Also, insect and disease agents can multiply rapidly when the environment is favorable and cause dramatically severe losses. Wheat rusts and leaf-spot diseases may cause a yield reduction of 50 percent or more when extended cool wet periods occur after flowering. However, the onset of dry weather during this period stops progress of these diseases, and further yield reductions do not occur. Even so, foliage destruction may be sufficient to incur yield losses of 15–20 percent over extensive acreages.

TABLE 1 Harvested Area of Principal Crops of the World, 1964[4]

Crop	Area (1,000 acres)	Share of Total Cultivated Area (%)
Grains	1,689,979	72.8
Wheat	529,802	22.8
Rice	306,168	13.2
Corn	278,492	12.0
Millet and sorghum	255,016	11.0
Barley	174,459	7.5
Oats	76,110	3.3
Rye	69,932	3.0
Oilseeds	165,345	7.1
Soybean	70,994	3.1
Peanut	42,384	1.8
Rapeseed	19,522	0.8
Sunflower	18,113	0.8
Sesame	14,332	0.6
Roots and Tubers	123,061	5.3
Potatoes	60,542	2.6
Sweet potatoes and yams	40,279	1.7
Cassava	22,240	1.0
Pulses	94,643	4.1
Fibers	98,375	4.2
Cotton	85,500	3.6
Flax	4,992	0.2
Jute	6,153	0.3
Hemp	1,730	0.1
Fruits and Vegetables	84,463[a]	3.6
Sugar	44,348	1.9
Sugarcane	23,690	1.0
Sugar beet	20,658	0.9
Beverage Crops	16,492	0.7
Coffee[b]	11,562	0.5
Cocoa[b]	1,774	0.1
Tea	3,156	0.1
Tobacco	9,464	0.4
Rubber	5,585	0.2
Total	2,322,291[c]	100.0[d]

[a]Figure from Brown.[2]
[b]For major producing countries.
[c]This total refers to the area of the main crops only.
[d]Because of rounding, totals do not add to 100.

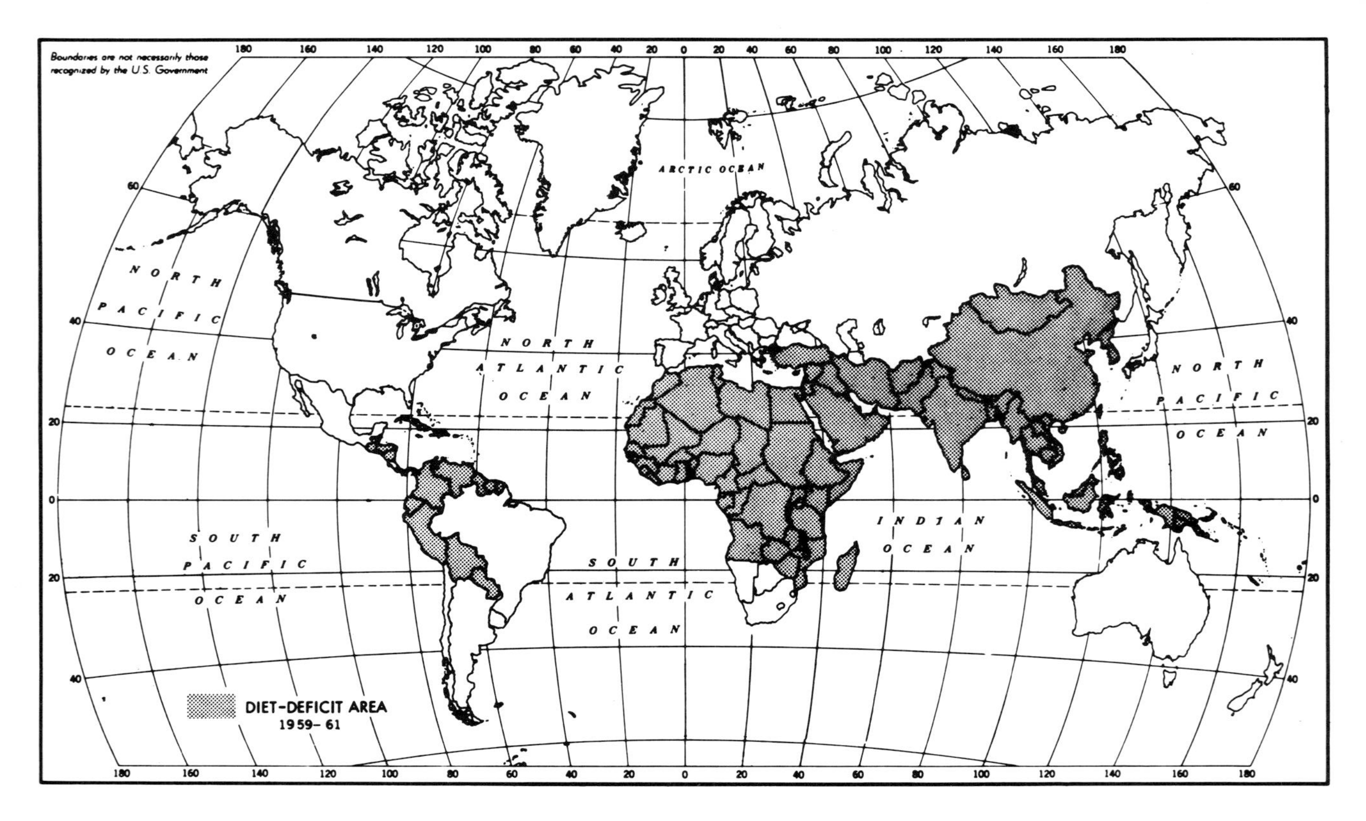

FIGURE 2 Map showing diet-deficit areas of the world, 1959–1961.

Because of the size, remoteness, diversity, variability, and vulnerability of the world's crops, accurate information on production during and following the crop season is limited to the crops grown in the more advanced countries. Even for these, the need for increased accuracy, timeliness, and frequency and degree of detail of crop information is continually growing.

One promising means of meeting the current and future needs for crop information is through remote sensing by a system of operational observation satellites. Such satellites, equipped with multiband sensors, offer the potential of providing macroscopic surveys of the earth on a synoptic basis and detailed observations of selected areas. Ideally, these satellites would be programmed to provide required agricultural data routinely, much as the Tiros and Nimbus satellites of today provide weather information.

World Food–Population Disparity

When the characteristics of crop production in agriculture are viewed in light of the objective of eliminating human starvation, the magnitude of the task ahead is impressive. The world food–population disparity has been growing since World War II. Figure 2 shows the diet-deficit areas of the world, where the problem is complicated by rapidly growing populations. The gains in food production are literally being eaten up as fast as they are made. The result is actual reduction in *per capita* food production in many already hungry nations. The Food and Agriculture Organization of the United Nations estimates that the areas of the world that have lagged behind in *per capita* food production contain two thirds of the world's population. By the year 2000, it is estimated that the world's population will be over 6 billion; today it stands at about 3 billion. Most of this increase will occur in Asia, Africa, and Latin America.[5] Within the next 30 years, the challenge that faces mankind is to double the entire food-production capability that has been achieved since time began.

A more optimistic note, however, is that many areas of the world have untapped agricultural resources. As depicted in Figure 3, only about 10 percent of the world's land is cultivated; an additional 21 percent is potentially cultivable. Although many of these areas are plagued with inhospitable climates, difficult access, debilitating diseases, and other obstacles, there is evidence that they may be cultivated when sufficient need, ingenuity, or investment is brought to bear.

In the face of the world problem of our times—to manipulate the world's resources and environments to gain a better life—there is need

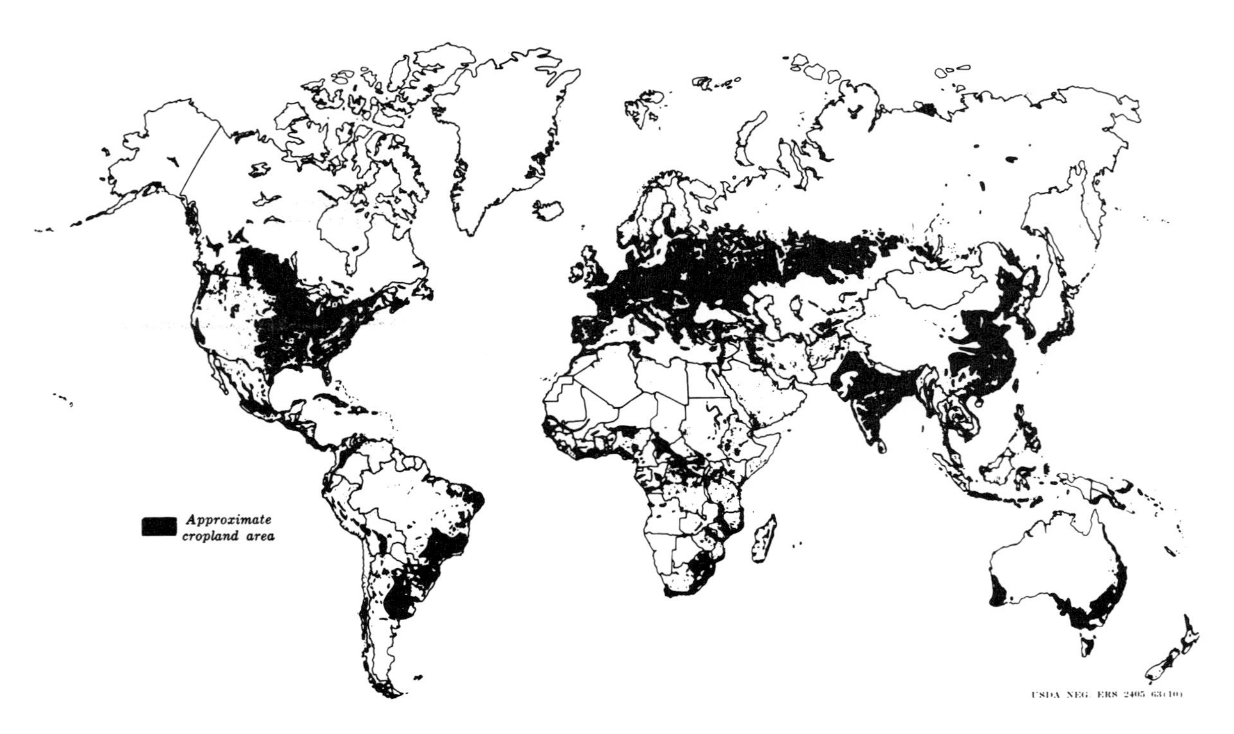

FIGURE 3 Map indicating approximate cropland area of the world, i. e., arable land including fallow and tree and bush crops. Partly because sufficiently detailed data on land use are not available for some countries and partly because the map is small, the shaded portions include scattered areas of land not used for crops, and the unshaded portions include scattered cropland areas.

to accelerate the preparation of natural-resource inventories to assist national and international groups. If man is to plan and operate an efficient and highly industrialized society, he needs more accurate, timely, and detailed information on the use of the terrain and its potential. Initial inventorying and mapping of many areas will lead to agricultural and geological resource interpretations to assist in the search for unused but potentially arable areas, construction materials, water, fuel, and other elements. Synoptic data on slopes, vegetation, floods, lithology, and depth of bedrock would provide the basic parameters for estimating arable land resources.

The use of remote-sensing techniques applied to a global program of assessing unused but potentially arable land resources could hasten the attainment of a better balance between food requirements and food production for the world. In most countries, and especially in the developing countries, there are few reliable statistics on unused but potentially arable land. Synoptic surveys of agricultural land from space altitudes should permit the identification of present use of the land and show population-settlement patterns and road and transportation networks. These surveys should also permit the identification of land characteristics, such as major soil types, drainage, and topographical relief patterns, as a basis for evaluating the best potential use of the land.

Specific Uses of Remote Sensors for Taking Agricultural Data

Remote sensors are currently used in many phases of agriculture. The sensor most employed is the aerial camera, which provides black-and-white panchromatic, black-and-white infrared, Ektachrome Infrared, and color photographs.* A recent study of the use of airphotos in a sample of countries throughout the world indicates that they are being used to an increasing extent, particularly in the larger countries and in the developing areas, to aid in agricultural planning, development, and administration.[6] The reasons generally given for their use include the need (a) to gather up-to-date information, (b) to survey large areas, (c) to accelerate development of agriculture, and (d) to provide for the most efficient utilization of trained scientists, a scarce commodity. An additional reason is that much of the present agricultural development is on a project basis, involving a natural land unit such as a river floodplain or watershed rather than a political unit (e.g., country, district, or province). The data, however, are collected by political units.

*For a discussion of types of film, see Chapter 2.

Since many projects cut across political boundaries, air surveys of the natural unit must be carried out if data on both sides of a political boundary are to be obtained.

The use of multiband sensing in providing agricultural data would appear to be most appropriate in the categories discussed below, where airphotos are now being used.

Land-Use Inventories Up-to-date information on major land use, acreage, and distribution of crops is a basic need in agriculture throughout the world. In both the developed and the developing countries, these data are essential for efficient management of agricultural resources. The historical continuity in the collection, compilation, and interpretation of data on use of land in the United States has proved invaluable in the study of present land-resource problems. Planning for future growth is greatly enhanced because reasonably uniform information exists about major land use since about 1920. Information on the present extent, location, and productivity of land used for different purposes is needed for analysis. The present concern with area redevelopment is an example of the need for careful examination of regional difference in utilization of resources. Competition for use of land is currently attracting much attention. Urban development, need for more recreation areas (particularly near large centers of urban population), and the preservation of wildlife habitats are matters of great interest to those concerned with use of land resources.

Historically, land-use data have been compiled by census interview or by field mapping. In some countries, major classes of land use are mapped in the field and published, usually on a small scale. For most countries, land-use inventory data consist of data compiled by census from personal interview, mail questionnaire, study of sample areas, or some combinations of these means. Generally, these methods take a relatively long time because the number of trained scientists is limited or because the number of less-well-trained people required is large. These factors account for the relatively long intervals between census projects in many countries.

Pressing needs for land-use inventory in many of the developed countries and in some of the developing countries have already prompted the use of remote sensing (black-and-white airphotos from recent overflights) to provide up-to-date information.

Airphotos are used to make maps designed to show distribution of major land use and specific crops in many countries. Measurements from these airphoto maps provide data on acreage of cropland, pasture, and other categories of agricultural use. In the more developed areas, airphotos are used frequently to provide up-to-date information on land use where agricultural changes are taking place.[7] By comparison of recent airphotos with older coverage (Figure 4), such changes as urban development, reversion of farmland to forest, and development of new farmland by clearing, draining, or irrigation can be ascertained.

Other remote-sensing technology offers considerable potential for obtaining significantly improved rural and urban land-use data. The synoptic coverage afforded by orbital height will materially reduce data-handling problems, and most area coverage will be continuous. A single image will provide an effective mosaic that would normally require many man-hours to assemble from conventional contact prints. Other advantages of an orbiting spacecraft are repeated coverage, reduced data-acquisition time, reduced cost (in light of repeated coverage) and, in some cases, higher quality data.

Soil Surveys Information about soils is basic for efficient sustained agricultural production. Soil characteristics, although they can be modified to some extent, generally determine the type of crop that can be grown and the production potential of that crop. The need for classification and mapping as a base for optimum utilization emphasizes the requirement for increasing knowledge about soils. Information on soils is particularly important in an area in which new land is being brought into production. In many areas, pressing agricultural development took place without the essential data on soils needed for assessing the potential for successful development. There are many examples of failures in draining or irrigating land not suitable for drainage and irrigation. In other situations, land suitable for forest was cleared, put into agricultural production, and then found to be too erodible for crop production.[8]

Before 1930, soil surveys in the United States took a long time and required many trained scientists working in the field both to classify the soil and to make the base maps. Need for some means of obtaining soil information more quickly and efficiently prompted the use of airphotos for field-survey work.[9]

Although present-day soil surveyors have become adjusted to discerning nuances in the gray scale on conventional aerial panchromatic photographs to differentiate soil boundaries, these photographs fall short of the ideal surface-mapping base for generalizing about slight differences in topographic, geologic, and ecologic features. By utilizing data from a multiband remote-sensing system, soil mapping may be quicker and more efficient than with panchromatic airphotos alone. Not only would these data provide improved coverage, but also additional information

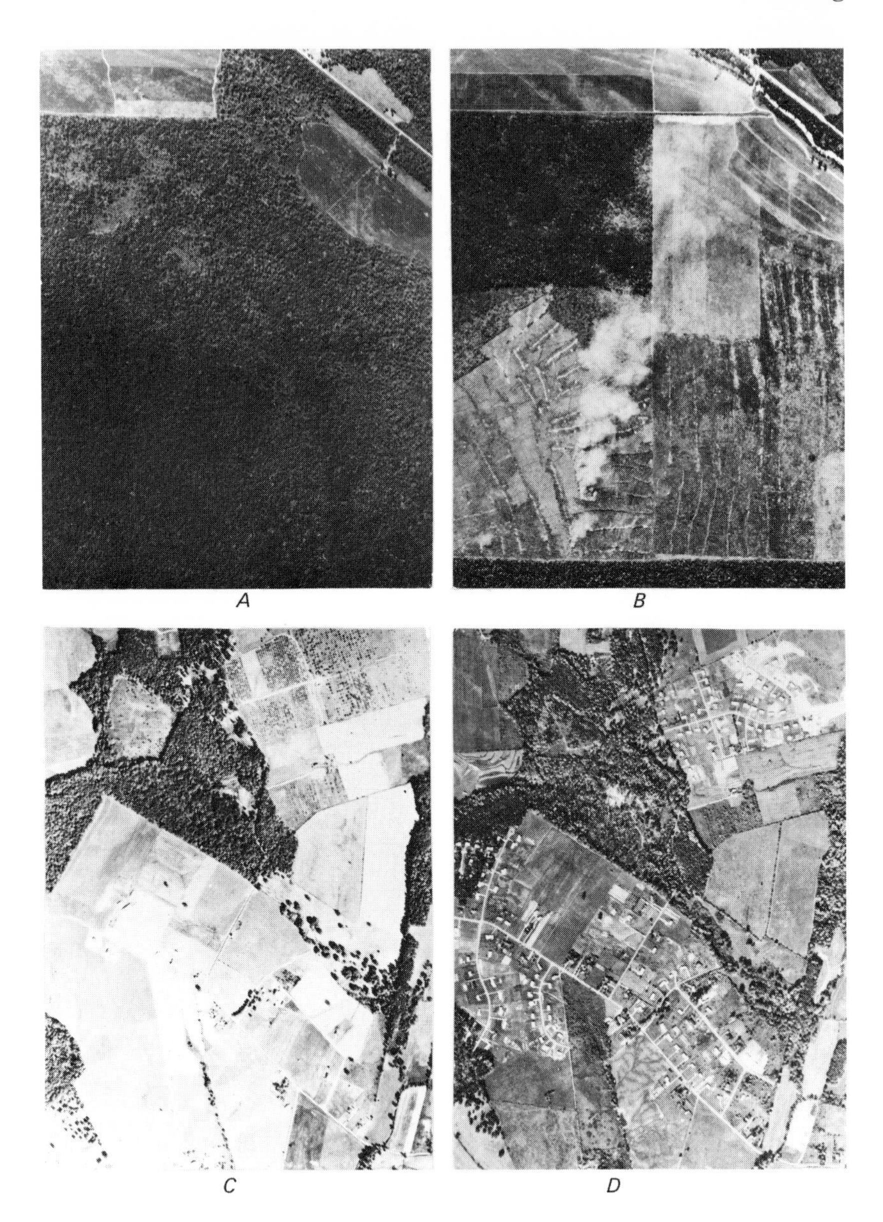

FIGURE 4 Airphotos showing changes in the use of agricultural lands in the United States. *A* (1956) and *B* (1964): new farmland being developed by clearing and draining forestland in Madison Parish, Louisiana. The smoke in *B* is from burning windrows of trees pushed over and piled with a bulldozer. *C* (1951) and *D* (1963): farmland shifting to residential use in Montgomery County, Maryland.

about the soil landscape should require taking fewer soil samples and would thus provide an expanded soil-mapping program with little or no increase in manpower requirements.

Crop Condition Estimates and Yield Forecasting Data on crop yields as well as forecasts during the growing season are of vital importance to agriculture. These data affect all phases of agricultural production, as well as the processing, storage, and disposal of agricultural products. In the United States, forecasting and estimating yields has been carried on for a long period, and yield and production forecasting of nationally important crops is done at monthly intervals, often beginning two months or longer before harvest.[10] Statistical methods are discussed in Chapter 8.

The main types of data obtained are: (1) data for making forecasts of crop production during the growing season and (2) annual estimates of crop production. The forecast is a statement of the most likely yield or production on the basis of known facts on a given date. It assumes that weather conditions and damage from insects or other pests during the growing season will be about the same as the average of previous years when the reported condition on the given data was similar to the present reported condition. These forecasts are generally based on grower appraisals of crop condition and expected yield, with persistent bias removed by charting. These forecasts have been quite satisfactory over the years, as shown in the case of wheat for 1965 in Table 2. Occasionally, however, in seasons when unusually large changes in crop conditions are encountered, they may not be fully reflected in the subjective appraisals.

As part of a program for improving crop-estimating methods, exploratory work was initiated to develop procedures for basing forecasts and estimates on objective counts and measurements. In keeping with the bench-mark data provided by the enumerative surveys, objective estimates of yield of comparable precision were contemplated, and forecasting counts and measurements that would be independent of judgment appraisals were sought. Forecasting yield before the crop is mature, and even before the fruit has been set by the plant, was recognized as being more difficult. The necessity for breaking yield into components that are predictable was evident, so the factors of plant numbers, fruit numbers, and fruit development in conjunction with plant maturity were studied. Forecasting models for cotton, corn, wheat, and soybeans have been developed and are being used.

Forecasts and end-of-year estimates of yield per acre for major crops in many states are based also on objective yield-survey indications.

TABLE 2 Winter Wheat—Yield Forecasts per Acre Compared with Final Estimate, 1965[11]

State	May 1		June 1		July 1		
	Forecast (*bu*)	Percentage of Final Estimate	Forecast (*bu*)	Percentage of Final Estimate	Forecast (*bu*)	Percentage of Final Estimate	Final Estimate (*bu*)
Ohio	32	100	31	97	31	97	32
Indiana	35	103	35	103	35	103	34
Illinois	37	103	37	103	38	106	36
Michigan	37	112	36	109	34	103	33
Missouri	32	114	30	107	30	107	28
Nebraska	20	100	22	110	23	115	20
Kansas	20	92	24	100	26	108	24
Oklahoma	26	93	28	100	29	104	28
Texas	20	91	22	100	22	100	22
South Dakota	19	112	19	112	25	147	17
Montana	26	90	24	83	28	97	29
Idaho	37	88	38	90	40	95	42
Colorado	10	62	13	81	16	100	16
Washington	41	98	41	98	42	100	42
Oregon	36	92	36	92	39	100	39

In the objective yield survey, during the growing season and after harvest, trained enumerators visit selected sample fields that are chosen with known probability to get counts and measurement of crop-growth characteristics on small plots. From these samples, probable and final yield estimates are derived with a computable degree of probability and are available to the Crop Reporting Board of the U.S. Department of Agriculture when it makes yield estimates.

Although in the United States forecasting and estimating yields of major crops has been carried on for a hundred years, present methods produce results that are often at variance with the real situation. The use of multiband sensing techniques in conjunction with repetitive overflights to provide data on crops more rapidly may reduce the lag between observation and data reduction by at least 50 percent and improve accuracy. Further, a remote-sensing system will permit quick response to any major disaster that calls for an adjustment in the forecast.

Acreage Estimates of Specific Crops Acreage estimates are an essential part of agricultural programs throughout the world because they provide basic data for research, program planning, and administration. Surveys on intentions to plant and estimates of acreage help farmers plan their plantings, serve as direct measures of land utilization, and are prime indicators of the future demand for various farm supplies and labor. The successive data needed annually include prospective plantings, actual plantings, acreage for harvest, and actual harvested acreage.

In the United States, acreage estimates are generally based on two types of information: (1) absolute-acreage data or bench-mark data from complete enumeration, and (2) acreage indications of change in crops from year to year obtained by questionnaire (either mail or personal enumeration) from samples of farmers. Complete enumeration is the ideal method, but the time, personnel, and expense involved have limited this procedure, for example, to five-year intervals. In addition, the time required to process the data precludes its use for current-year forecasts. The agricultural census data in the United States do not become available until a year or more after they are taken and are therefore used for bench marks for future years and for "truing-up" historical estimates.

To overcome the problems of timing, personnel, and expense of complete enumeration, remote sensing is being used in some countries to obtain crop-acreage data. Black-and-white airphotos are used for both actual field surveys and as inputs for stereoscopic interpretation. Crop-acreage data are compiled by measurement of the areas identified on the airphotos, which also record the distribution pattern and location. Although the data so obtained from airphoto interpretation have great value, it is difficult or impossible to identify many of the major crops by this means, particularly the cereals, as well as to distinguish between corn, potatoes, and soybeans at all stages in the growing season.

Use of remote multiband sensing appears to hold considerable potential in providing current data on crop acreages. Research at Purdue University has demonstrated the feasibility of in-flight recording of spectral signatures of several crop species, including wheat, oats, corn, soybeans, alfalfa, red clover, and rye.[12] These data may be digitized and computer-processed into a map-type display from which both the species and the acreage can be derived.

Disaster Location and Mapping Locating and mapping natural disasters present problems in management of agricultural resources. Floods, hurricanes, fires, and severe epidemics of insects and disease occur rapidly; many disrupt normal channels of communication and present complex problems in assessing damages and planning programs of salvage and restoration to normal conditions. In several situations, use of aerial survey and interpretation of airphotos has provided data that could be used as a basis for action while the event was taking place, as well as providing information on areas potentially subject to damage.[13]

During floods, airphotos may indicate areas from which people and livestock should be rescued before being isolated or swept away by flood crests. Airphotos taken while floods are receding can provide data on areas that have been inundated. In these areas the direct damage to crops and the degree of silting or flood scour can be shown. Damage to farm buildings, roads, and other facilities can also be determined. Although airphotos can provide a great deal of information, they are generally not available for analysis immediately following natural disasters. Since most disasters occur swiftly, time does not permit quick coverage by aircraft. In addition, weather conditions accompanying most natural disasters prohibit aerial photography. Nevertheless, disaster applications as weather conditions permit are highly rewarding uses of remote sensing.

Detection of Crop Insect Pests and Diseases Detection and control of crop disease and plant insect-pest infestation is an important phase of agricultural management throughout the world. Plant disease and insect pests not only have devastating effects on crop yield but also reduce the quality and market value of products. Damage may begin at the time the crop is planted, continue throughout its growing period, and persist after harvest, when products are transported and placed in storage. Unless pests are prevented or controlled, there can be no sustained improvement in agricultural productivity.

The economic benefits of early detection of disease or infestation of crops are tantamount to increasing agricultural productivity. Even in developed countries, losses in crop value caused by insects and diseases are high. In the United States, the average annual losses caused by plant diseases are estimated at about $3.7 billion.[14] Average annual losses from crop insect pests amount to $3.8 billion. Even larger losses occur elsewhere, particularly in the developing countries. No total estimate is available for the world, but a few examples from surveys by the UN Food and Agriculture Organization follow. In the United Arab Republic, it is estimated that the annual loss to major crops from disease and insects amounts to $168 million. The desert locust, which invades the northern half of Africa and the Near East to India and Pakistan, caused losses totaling $42 million during the nine-year period 1949–1957, in spite of intensive and costly control programs.

Control measures have improved greatly in recent years, but costs of application are high, and many countries cannot afford to establish a plant-protection service to survey the distribution and prevalence of disease and insect pests. Again, the lack of good communication to transmit information and provide assistance to farmers hampers control efforts in many areas. The need for early detection and control of diseases or pests in new environments cannot be overemphasized. This is illustrated by the spectacular occurrence of the maize rust in Africa and the Far East. The rust apparently never caused much loss in its native area (probably North or South America), but when it was found in Sierra Leone in 1949, its destructiveness caused great alarm. From Sierra Leone it spread rapidly across central Africa, in several years reaching Kenya and Rhodesia; finally it crossed the Indian Ocean to Southeast Asia. Its effect on maize was so devastating that the west African countries had to import maize to make up for the losses that resulted.[15]

Up to 1965 very little use had been made of multiband sensing to overcome the problems of survey and detection of crop disease and insect infestation. Since that time black-and-white infrared and Ektachrome Infrared photographs have been used experimentally to detect plant diseases. Other experiments indicate that certain plant diseases can be detected by a remote sensor before they can be detected through visible means. Although precise signatures for crop disease have not yet been established, the detected loss in vigor calls attention to the possibility that a pathogen is present.

Currently, the U.S. Agricultural Research Service is spending about $3 million annually in survey work to detect plant diseases and insect outbreaks.* A remote-sensing system could provide for earlier detection of the presence of diseases and insects than is possible by current methods.

Wildlife Management Wildlife is more difficult to inventory than domestic farm animals. Indirect estimates of migrating secretive animals are necessarily inaccurate. Lack of accurate counts handicaps scientific study and resources management. The U.S. Fish and Wildlife Service takes an annual census of many kinds of animals, ranging from antelope to waterfowl, and from sea lions to salmon. In the past, wildlife censuses have been taken by estimating the numbers or by direct counting from a photograph taken from the ground. In recent years, aerial photography has been used on an experimental basis for taking these censuses. According to evaluations by the Fish and Wildlife Service, these experiments have revealed a considerable potential for the use of aerial photography (remote-sensing data) in wildlife management. For example, remote-sensing data could be used at frequent intervals to monitor wildlife conditions.

Water-Supply Information and Management Freshwater in lakes and streams, and underground is a moving resource required by man and animals for their existence and for the production of food and fiber. The current problem of freshwater is one of distribution in space and time. Gross maldistributions of freshwater are manifested in floods and droughts.

The tasks of water-resource management are to find freshwater at low cost, to forecast the future supply, and to control the location, quantity, quality, and timing of that supply. Measurable hydrologic phenomena consist of two kinds: (1) worldwide or regional conditions that change rather slowly and (2) those rapidly fluctuating processes that govern the short-term water balance of the earth. Measurement of the first kind, which may vary daily or weekly, is useful for prediction of hydrologic conditions up to several months ahead. Measurable elements in this group are the quantity and distribution of snow and ice in terrestrial snowpacks and extent and depth of frozen soil and

*Personal communication. 1967. U.S. Department of Agriculture, Agricultural Research Service.

of unfrozen soil over permafrost. The rapidly fluctuating conditions that can change within hours are the conditions that affect water balance at the earth's surface, variations in soil moisture, and surface temperature of lakes and streams.

Snow-depth measurements are important to irrigated areas in the western part of the United States and to some unirrigated areas in the northeastern part of the country where planning for water use in the summer is becoming more dependent on snow cover. For the irrigated regions of the country, an estimate of equivalent water volume from snow cover is already a major part of the water-supply prediction problem. In the western states, measurement of the snow cover, which is a good indication of later soil-moisture condition, assists greatly in assessing the coming season's crop-growing conditions and in forecasting and estimating the wheat crop yield. The Soil Conservation Service of the Department of Agriculture issues a water-supply forecast annually about mid-April for the western states.

The quantity and distribution of water stored in the mountain snowpack is determined by snow surveyors, who measure the water content and estimate the acre-feet of runoff from each mountain watershed. The snow-survey courses are invariably located in remote mountain areas inaccessible by ground transportation. In the early days of snow surveys, several days would be necessary to survey one course. Today, both helicopters and aerial surveying are used to speed the process. Helicopters are used mainly to transport personnel to and from the courses. Aerial photography, used since the early 1950's, appears to be cheaper and easier. In the aerial surveys, markers bearing a ladderlike series of crossbars are placed adjacent to a regular snow course. Overflight photographs of the markers are compared with form drawings to obtain snow depth. For each marker, the snow density is estimated from previous ground measurements. By this method, several basins, each with many snow courses, can be photographed in one day, with resultant savings.

Utilization of groundwater and surface water for agricultural purposes in the 21 humid-area states has been increasing in recent years. This water is used for the irrigation of crops whose per-acre value is high. In a research project by the Indiana Department of Conservation in cooperation with Purdue University, a procedure has been developed for using aerial photography to locate groundwater supplies.[16] This research indicates a strong relationship between areas of groundwater-bearing formations and terrain characteristics, such as the pattern of each "wrinkle" of the earth, the shapes of gullies, the patterns of streams and their tributaries, the tonality of the photograph of an area, and other recognizable features, many of which can be evaluated from aerial photography.

Irrigation Management Forecasts of available water resources have substantial economic value to irrigation farmers who must plan the coming season's crops, to power companies whose schedules of water release from reservoirs become increasingly efficient with better knowledge, and to many other water users. The value of good forecasts increases as freshwater becomes scarce and as floodplain occupancy becomes more intensive.

A remote-sensing system can materially assist in the maintenance and operation of existing irrigation systems. About one half of the water is lost between the dam head of a canal-system input and the crop in the field, most of it apparently by leakage from the canals and ditches. Remote sensing of water and wet–dry boundaries would appear to be an ideal method of monitoring leaks and breaks as change in wetlands or vegetation cover adjacent to canals is noted.

Remote-sensing techniques could also be used to advise farmers on irrigation timing and water quantity. For a known soil and drainage condition, it appears probable that certain remote sensors will be able to detect the changes in surface moisture with enough confidence that water-table depths can be deduced.[17] If this information is available, farmers can avoid overapplication or underapplication of irrigation water. If the water table is too low, plants suffer from lack of moisture. If the water table is too high, irrigation water is lost through evaporation at the surface with the coincidental deposits of dissolved salts in the soil, a condition that, if permitted to continue, renders the soil totally unfit for agriculture.

Weather Forecasting Of the many factors affecting the success or failure of agricultural production, none plays a more decisive role than weather. Weather manifests its influence in agricultural production through its effects on the soil (plant growth, development, yield, and composition) and on practically every phase of animal growth and production.

In the more developed countries, farming has become heavily dependent upon machines and chemicals. These, along with irrigation and improved cultural practices, have given some measure of "weatherproofing" to agricultural production. But each farming operation that has to be repeated and each washed-off chemical application that has to be replaced because of adverse weather conditions adds to the production cost. Each seed that is planted at the wrong time and rots in a cold wet soil adds to farm inefficiency.

Weather affects farm animals directly through its influence on appetite and general well-being and indirectly through its influence on the incidence of disease, parasites, and food supply of all kinds. Weather also plays a major role at calving, farrowing, and lambing time, and frequently determines the number of surviving young.

Thus, throughout the year, farmers must make many decisions on the basis of weather forecasts. Good forecasts enable the farmer both to avoid losses caused by growing certain crops under unfavorable conditions and to exploit favorable conditions when they occur. If adverse weather is predicted, a farmer can weigh the cost of protection against the probable weather-caused loss. Some enterprises can be protected against the effects of adverse weather, but such protective measures may be expensive. Crops may be protected from drought by irrigation and by harvesting in advance of predicted severe weather. Livestock can be moved from the open range to shelters in advance of blizzards or other severe weather. Fruit trees may be protected from frost by orchard heaters. The employment of such protective measures can be justified only on the basis of increased profits or decreased losses.

In the United States, direct weather-caused losses in crop output alone are estimated to be between $1 billion and $2 billion per year. Advance weather information could reduce or eliminate much of this. Even a 5 percent reduction in weather damage to crops would yield benefits between $50 million and $100 million per year.

In recent years the U.S. Department of Commerce has become increasingly aware of agriculture's need for specialized weather forecasts. In 12 areas in which there are valuable concentrations of weather-sensitive crops, that department's Weather Bureau provides specialized forecasts and advisories. The areas are: (1) New Jersey; (2) an area including northern Virginia, northeastern West Virginia, western and central Maryland, and south central Pennsylvania; (3) South Carolina; (4) a tri-state area including southern Georgia, southern Alabama, and northwestern Florida; (5) the Mid-South, including northern Mississippi, western Tennessee, Arkansas, northeastern Louisiana, and the Boot Heel of Missouri; (6) the Lower Rio Grande Valley of Texas; (7) western Lower Michigan; (8) southern Idaho; (9) Oregon; (10) the High Plains area of Texas; (11) Indiana; and (12) Kentucky.

The forecasts are divided into several different categories, depending on valid time targets:

1. A 1–6-hour forecast using radar data (reasonably accurate).
2. The 2-day general agricultural weather forecast, with an outlook for the third day (not as accurate as item 1 above).
3. The 5-day outlook (not as accurate as item 2 above).
4. The 30-day outlook (the least accurate).

In addition, specialized agricultural advisories (state and federal), designed with the immediate problems of the local area in mind, are prepared in support of such critical operations as pest control, irrigation, harvesting, and defoliation. Throughout the remainder of the United States the agricultural managers must, with few exceptions, plan and carry out their operations on the basis of forecasts and advisories that are designed to serve the general public.[18]

If improved weather forecasting, e.g., a 5–7-day forecast, were available and used, agricultural managers might be able to avoid such costly repetitions as relocating livestock, reseeding, refertilizing, and respraying for pests. Such potential benefits are recognized by the agribusiness community. On a worldwide scale, accurate forecast of rainfall would aid in planning future food requirements; a crop loss in one hemisphere might be compensated for by additional plantings in the other hemisphere.

The use of satellite-borne remote sensors in the Tiros and Nimbus programs to observe the earth's weather is perhaps the earliest and best-known application of remote sensing from orbital altitudes to earth-related problems. Recent investigations of the National Aeronautics and Space Administration and the Environmental Science Services Administration have demonstrated that useful weather data may be obtained by visual and infrared imaging of the cloud patterns. Use of the collected qualitative data by the meteorologist has significantly improved the accuracy of the day-to-day forecasting. Also, recent advances in improving the computational stability of numerical models of the atmospheric circulation, along with expected improvements in speed of electronic computers, should make it possible in the next 5–10 years for meteorologists to forecast large-scale weather (large storms) up to two weeks in advance.*

Rangeland Management Information on condition of rangeland and forecasts of future condition are important items in the economics of

*Personal communication. 1968. Environmental Science Services Administration, Washington, D. C.

many parts of the world. In the United States, the reported condition of range feed, cattle and calves, and sheep and lambs in 17 states (the western states, North Dakota, South Dakota, Nebraska, Kansas, Oklahoma, and Texas) is issued each month. These reports include descriptive comments for each state on range-feed conditions, feed prospects, moisture conditions, livestock conditions, and other matters.

In addition, special reports are made on the capacity of the several wheat-pasture areas of Kansas, Oklahoma, and Texas. In the wheat-pasture area, data are collected showing: (1) percent of seeded wheat with sufficient growth to pasture; (2) percent of seeded wheat being pastured; and (3) areas of wheat pasture required to carry a 400-lb calf through the fall and winter. The relative condition of top growth available for grazing is indicated on a map that shows the rating of winter wheat pasture as excellent, good, fair, or poor. Data for this report are provided by wheat growers, ranchers, and farmers.

The carrying capacity of rangeland often may be increased two- or threefold by more intensive management. Because of the extensive nature of rangelands, remote sensing appears to be the most efficient and economical means for providing accurate and up-to-date inventories of the plant communities and continuing appraisals of range conditions.

Color-infrared photography and infrared scanning are among the most promising sensing means for these inventories. Soil and foliage temperatures appear to be related to transpiration rates, soil moisture, relative humidity, thermal capacity of the soil and vegetation, etc. When these parameters are coupled with the loss of reflected infrared in the vegetation, certain predictive capabilities for range assessment appear.

The feasibility of using remote sensing (e.g., Gemini IV color photography) to redefine an existing vegetation-resource map of an arid region has already been demonstrated by Poulton, as shown in Figure 5 and Plate 1.[19] Thus it appears that good-quality color photographs taken from space platforms and interpreted in conjunction with adequate ground study can provide much of the data needed to improve small-scale vegetation-resource maps in arid and semiarid regions. The data that can be extracted from synoptic small-scale space-platform photographs will be especially useful in broad county, state, and national planning and in natural resource policy formulation.

Livestock Surveys Use of airphotos in making livestock inventories is still in the research stage, and although this practice appears prom-

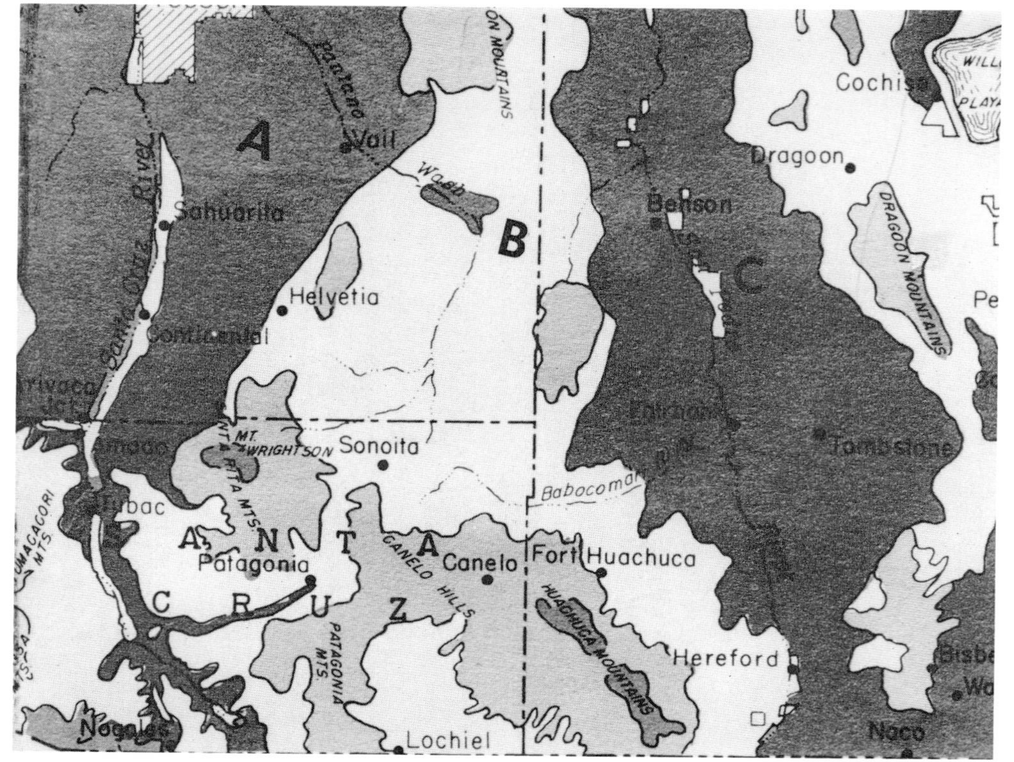

FIGURE 5 A portion of a vegetation-resource map of Arizona showing the major vegetation-resource areas of the region. *A* and *C* = desert scrub; *B* = desert grassland. This map represents the best that can be done at this scale from information presently available and shows approximately the same area as that covered by the Gemini IV frame reproduced in Plate 1. (*Adapted from Humphrey*[19].)

ising, it has not resulted in substantial modification of present inventory methods. The inventory data desired present problems for remote sensing because they include counts in each field by kind of animal (e.g., sheep, cattle), use (e.g., dairy, beef), breed, sex, age, and vigor. It is expected that airphotos at scales smaller than 1 : 12,000 will be useful for livestock surveys when keys and other aids are developed and when special photography is obtained (see Chapter 4).

Forestry

Description of the Forestry Sector

Forestry, in general, is the management of forests and forestlands for wood, forage, water, wildlife, and recreation. Because the major forest product is wood, modern forestry is concerned chiefly with timber management (especially reforestation), with maintenance and improvement of the existing forest stands, and with fire control. Wood is one of the world's principal natural resources; it is renewable, and nearly all countries possess it or have the potential to create it. The necessity for government supervision has long been recognized and is employed in varying degrees in all countries. In forested areas the basic rule of management is to cut each year a volume of timber no greater than the volume of wood that grew during that year on standing trees.

Forestlands, as distinguished from agricultural lands, are defined by the UN Food and Agriculture Organization as "all lands bearing vegetative associations dominated by trees of any size, exploited or not, capable of producing wood or of exerting an influence on the local climate or on the water regime."[20] Lands from which forests have recently been clear-cut or burned out but that will be reforested in the near future are included. Brushland, groups of trees outside the forest, or trees along roads, on agricultural lands (such as orchards), and in parks are not included.

Forests of one type or another cover nearly a third of the world's land area, as shown in Figure 6. They range from scrub and thorn equatorial swamp to mountain rainforest, and from homogeneous plantations to luxuriant jungles. The forests are distributed very unevenly throughout the world, and their usefulness varies from providing raw material for a growing variety of industries to yielding poles and firewood for domestic consumption, and from regulating the water regime to protecting wildlife and satisfying recreational demands.

Little is known quantitatively about the world's forests. Over large parts of the world, forests have not been surveyed, and all that can be

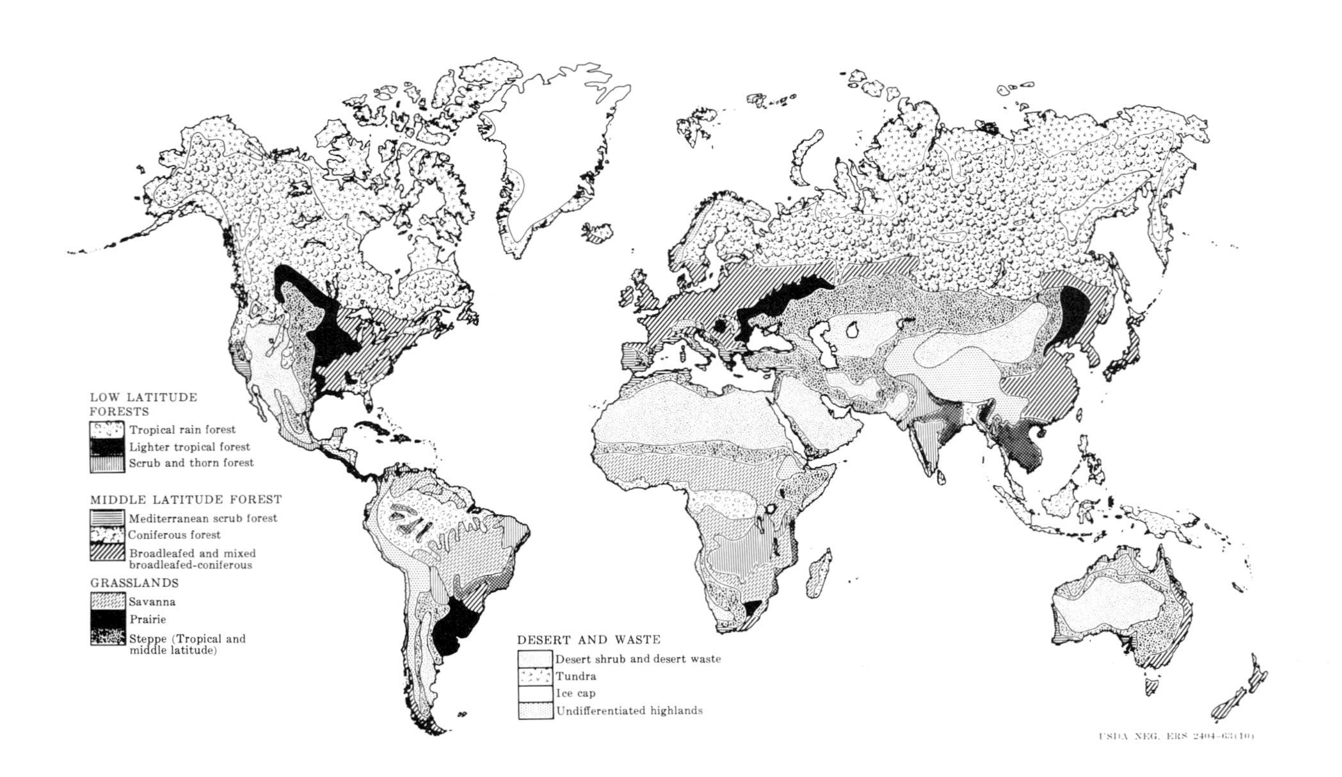

FIGURE 6 Natural vegetation map of the world.

reported on a national or regional scale is the area of the forests and major forest conditions. Generally speaking, it is only for the temperate forests that useful estimates can be made of the aggregate volume of growing stock and of the total annual volume increment of the forests.[21]

A world estimate of forestland by regions, as reported by the UN Food and Agriculture Organization, is shown in Table 3.

According to the 1963 World Forest Inventory, about one third of the world's forest is coniferous. It is located mainly in the Temperate Zone of the Northern Hemisphere, with 45 percent being located in the USSR and 36 percent in North America. The remaining two thirds is broad-leaved and is mainly in South America and Africa, which continents have about 75 percent of the world's broad-leaved stand.

The distribution, magnitude, and condition of forests are related to the historical development of the regions containing them. In Europe, the USSR, North America, and Japan, the regions where most wood is consumed, the forests are predominantly coniferous, and the mixed forests of these regions have a large coniferous content.

In the tropics, the greater part of the volume of wood is found concentrated in rainforests (which contain the largest volume of the world's broad-leaved woods). These forests have been little used or managed for the production of wood, mainly because they are concentrated in regions of relatively sparse population at an early stage of development. Examples are the Amazon basin in South America, West Central Africa, and Southeast Asia. These forests usually contain a

TABLE 3 Land Categories[20]

Region	Forestland (millions of acres)	Total Land Area (millions of acres)	Forestland as Percentage of Total Land Area
North America	1,853	4,633	40
Central America	188	672	28
South America	2,199	4,349	51
Africa	1,754	7,339	24
Europe	356	1,164	31
USSR	2,249	5,298	42
Asia	1,359	6,672	20
Pacific area	235	2,081	11
World	10,193	32,208	32

wide variety of broad-leaved species, ranging from the extremely hard, heavy, and slow-growing type to the very soft, light, and quick-growing type. The UN Food and Agriculture Organization has estimated that there are some 2,100 million acres of tropical forests carrying roughly 4,414 billion cubic feet (gross volume) of wood. This is about the same as the volume on the much larger areas of north-temperate forests, but relatively little of this volume is of usable sizes, and even less is in commercially accepted species.

The remaining forests consist mainly of dry open woodlands such as savanna. These forests occupy wide areas of South America, Southern Asia, and Africa. They contain very low volumes per acre, mostly in small trees of species with little commercial value.

It is estimated that about 40 percent of the world's forest is unproductive because of ecological conditions favoring growth of unwanted cover (tundra, maquis, chaparral, etc.) and also because of low economic productivity and prohibitively high transportation cost. Most of this vast vegetated unproductive area is considered wildland or undeveloped land, although some has suitable soil for growing food crops.

As indicated in the above discussion, forests constitute a principal form of land use, and as such have a number of highly important functions, including protection and recreation, as well as wood production. The demand for wood is rising and, as a result, there is a severe strain on the capacity of many of the traditional sources of supply. For example, because the world's softwood resources are largely unmanaged, they are proving inadequate to meet current requirements without continued overcutting.

Population pressures are forcing forest production off more accessible sites to make room for urban growth and additional cropland. Although forests are a renewable resource, the land base on which they must be produced is vulnerable. As this land base must meet increasing requirements, information about its total area and distribution becomes increasingly important.

Sustained use of forest resources includes two main objectives: effective management for production and preservation of the resource by adequate controls. Effective management of forest resources requires good inventory data. No accurate inventory of the forest resources of the world has ever been completed. Attempts have been made, but time lapse between start and finish has made it impossible to compare or collate the data in a single time frame. The UN Food and Agriculture Organization is continually attempting to evaluate the world forest resource. It publishes periodic estimates, but high costs, poor records, and refusal of cooperation by some governments have left many gaps.

Specific Uses of Remote Sensors for Taking Forestry Inventory Data

The most pressing need in forestry is a worldwide inventory to round out our knowledge concerning (1) forest location and distribution (mapping), (2) tree size, (3) tree quality, (4) tree growth rate, and (5) site quality. To meet this need, remote-sensing methods have begun to be employed in the United States and in other parts of the world. To date, the chief instrument has been the camera; panchromatic, infrared, and Ektachrome Infrared airphotos have been taken for use in field mapping and stereoscopic interpretation. Airphotos save time and money and permit coverage of large, and often inaccessible, areas. In addition, airphotos facilitate maximum use of trained personnel, who are scarce.

The following paragraphs outline a few of the possibilities for utilizing remote sensing to obtain information in the five categories named above. For details and examples, see Chapter 4.

Mapping Information on the location and distribution of forests is a basic need in management, particularly information about the distribution of species within an area. Aerial photographing followed by stereoscopic interpretation, usually with some field checking, is the most common method, and has been used throughout the world. However, even use of airphotos does not eliminate the problem of costs and time, and many areas remain unsurveyed.

Aerial cameras or other remote sensors mounted in an orbiting spacecraft could record the forest cover of the earth in a short enough time to allow for completion of a world vegetation map. Ground resolution of 50–100 ft would be adequate in most cases, and species determination would be facilitated if simultaneous panchromatic color and color-infrared imagery were obtained.

Tree-Size Measurement A conventional forest inventory includes the following items relevant to tree size: species, diameter (of bole or crown), density (number of trees per acre or degree of crown closure), height (merchantable or total), and age of the stand. Determination of tree height can be carried out by stereoscopic analysis of airphotos. Tree height controls the amount and type of product that can be cut from a forest.

A great deal of information can be obtained on the above items from airphotos. They appear to provide the most-needed data, especially measurement of tree height by differential parallax measurement on stereoscopic pairs of airphotos. However, the cost of aerial coverage is high. Multiband sensing for forest inventory is under study and appears to have some potential for adding to the precision of surveys.

Tree Quality The quality of concern is the number of crooked, damaged, or diseased trees that will affect yield. In the past, it has been necessary to get this information by ground surveys. Whether multiband sensing has any practical potential for differentiating between sound and unsound trees remains to be researched.

Tree Growth Rate Growth rate, a valuable management parameter, is estimated by associating measurements of volume at different points in time with tree-size determination. It is usually based on whole stands rather than on individual trees and has been obtained by calculations from ground measurements. It is highly possible that series photographs or serial information taken by sensors other than cameras at predetermined intervals will provide more accurate growth-rate data.

Site Quality Determination of site quality is suggested by tree size, volume per unit area, soil type, and other factors, but no consistently applicable index of quality has been standardized. Foresters generally use total height at a designated age as a "site index," but such indices are applicable only to single species for a limited geographic area. To date, site quality has also been determined by ground measurements alone. Further study is needed to determine whether remote sensing, either directly, or indirectly by calculation, can give meaningful site-quality information.

Specific Uses of Remote Sensors for Taking Forest-Fire Detection Data

Fire detection and control is a vital aspect of forest management, and any improvement in methods of early detection and surveillance of going fires would provide large benefits. Losses from fire are significant throughout the world. Conservative estimates for annual loss in the United States range from $300 million to $500 million. For the rest of the world, with much less effective fire-detection systems, it is estimated that the annual loss amounts to at least $400 million.[22]

Fire detection poses distinct types of problems: (1) surveillance or mapping of large fires or several active fires close enough together to

coalesce and create a large fire; (2) detection of small (diameter of 20 ft) fires; and (3) detection of "sleeper" (incipient) fires that smoulder inside the bark or roots of a tree after a lightning strike.
der inside the bark or roots of a tree after a lightning strike.

A recently developed airborne thermal-infrared scanner is now being instantaneous and precise mapping of fire lines day or night, and through dense smoke. When fully operational, this system will save millions of dollars annually in control costs and resource values and should reduce the toll of human injury and death. A refinement of this system that is now being studied would detect and pinpoint very small fires automatically.

Detection of fires from orbiting spacecraft containing infrared scanners or passive microwave sensors has been proposed. The primary advantage of an orbiting sensing system would be the large-area synoptic view. Remote sensors, particularly infrared scanners, would achieve better resolution from aircraft altitudes than from orbit. In addition, the fixed orbital path of the satellite renders the fire a target of opportunity. It is obvious that a geosynchronous orbiting system would be necessary for continuous monitoring, and eventually, sensors may be developed that could detect large fires in remote areas.

Specific Uses of Remote Sensors for Taking Insect-Pest and Disease-Detection Data in Forestry

Pest-caused losses to forest resources and forest products far exceed those from fire both in the United States and throughout the world. Although figures for U.S. losses are the principal ones available, world losses from insects and disease are believed comparable. In just one and one half years (October 1962 to April 1964), the southern pine beetle killed 50 percent of the pine timber on 2.5 million acres in Honduras, Central America.[23] Before the epidemic subsided, about 8.7 billion board-feet of pine was killed over an area of 4.2 million acres. In the United States, the annual insect-caused mortality of standing timber is high; for example, the monetary loss was estimated to be $579 million in 1965. In addition to the direct loss of standing timber, related losses from reduced growth, lowered quality of timber, and impaired forest reproduction are difficult to estimate.

These disastrous losses not only demonstrate the dynamic nature of the increase in the beetle population, but also point to the importance of early detection of outbreaks.

Annual losses from disease are estimated to be about $82.6 million. Of this amount, about $53 million represents growth loss and lower quality, and $29 million, loss of standing timber. For example, heart rots are of particular importance, since they attack all species of timber and cause heavy losses from cull and lowered quality of remaining wood.

In addition to the direct loss of standing timber, related losses from reduced growth, lowered quality of timber, and impaired forest reproduction are difficult to estimate. Color transparencies and Ektachrome Infrared airphotos have been used successfully in the United States to estimate tree losses and to locate infestations for salvage and control operations. It is possible that mass killing of coniferous timber, such as occurred in Honduras during 1962–1964, could be monitored successfully by color photography from satellites.

Summary

The earth's resources are being depleted by exponential population growth and the increasing level of *per capita* consumption. Effective management calls for large masses of current information on existing agricultural and forestry resources. The information generally available in the United States is, in many respects, far from adequate. In other countries, particularly the developing nations, the available information is inadequate or nonexistent, and the job of collecting such information from "on-the-ground" observations would be time-consuming and expensive. Fortunately, many of the data needs are amenable to aerial-survey techniques. Although aerial surveys cannot acquire all types of information needed, they are capable of doing much of the job in a satisfactory manner.

Today, aerial photographs (see Chapter 2) are indispensable in making land-use inventories, soil surveys, acreage allotments, farm-conservation plans, disaster evaluations, assessment of water supplies, and forest maps. They are invaluable in forest-fire detection and have a host of other uses.

Recent advances in remote-sensing techniques, particularly multiband sensing (see Chapter 3), may stimulate an increase in the use of aerial surveys and facilitate the extrapolation of aerial-survey techniques to space. The use of spacecraft, with its capacity for repetitive global coverage and simultaneous large-area coverage, can provide data never before available.

Specific applications of remote sensing in taking agricultural and forestry data are discussed in Chapters 4, 6, and 7. Chapter 9 is concerned with future research and development.

REFERENCES

1. Holter, M. R. 1967. Tools for the Future. 13th Ann. Meeting Amer. Astron. Soc., Dallas, Texas.
2. Brown, L. R. 1963. Man, land and food. Foreign Agr. Econ. Rep. No. 11. Dep. Agr., Washington, D. C. p. 16–23.
3. Van Royen, W. 1954. The agricultural resources of the world. Prentice-Hall, New York. p. 1–7.
4. Food and Agriculture Organization of the United Nations. 1966. World crop statistics, 1948–64. Rome.
5. Food and Agriculture Organization of the United Nations. 1962. Population and food supply. Freedom from hunger campaign, Basic Study No. 7. Rome. p. 1–10.
6. Dill, H. W., Jr. 1967. Worldwide use of airphotos in agriculture. Agr. Handbook No. 344. Econ. Research Service, Dep. Agr., Washington, D. C. p. 1–11.
7. Dill, H. W., Jr. 1959. Use of the comparison method in agricultural airphoto interpretation. Photogram. Eng. 25:44–49.
8. Christian, C. S. 1963. Survey and assessment of land resources. Report to United Nations conference on the application of science and technology for the benefit of the less developed areas. Geneva.
9. U.S. Soil Conservation Service. 1966. Aerial photo interpretation in classifying and mapping soils. Agr. Handbook No. 294. Dep. Agr., Washington, D. C., p. 1–20.
10. U.S. Statistical Reporting Service. 1964. Scope—Methods. Statistical Reporting Service Misc. Pub. No. 967. Dep. Agr., Washington, D. C. p. 2–31.
11. U.S. Statistical Reporting Service. 1965. Crop Production (monthly issues), Crop Reporting Board. Dep. Agr., Washington, D. C.
12. Purdue University. 1967. Remote multispectral sensing in agriculture. Res. Bull. No. 832. Laboratory for Agricultural Remote Sensing. Vol. II, Agr. Exp. Sta., Lafayette, Ind. p. 44–51.
13. American Society of Photogrammetry. 1960. Manual of photographic interpretation. George Banta Publishing Co., Menasha, Wis. p. 615–624.
14. U.S. Agricultural Research Service. 1965. Losses in agriculture. Handbook No. 291. Dep. Agr., Washington, D. C. p. 3–54.
15. United Nations Conference on the Application of Science and Technology for the Benefit of Less Developed Areas. 1963. Summation of papers and discussion on agriculture. Geneva.
16. Howe, Robert H. L. 1958. Procedures of applying air photo interpretation in the location of ground water. Photogram. Eng., 24:(1) 35–49.
17. Myers, V. I. *et al.* 1967. Spectral sensing in agriculture. Ann. Rep. on NASA Contract R–09–038–002. Fruit, Vegetable, Soil and Water Research Laboratory, Agr. Res. Service, Dep. Agr., Washington, D. C., p. 6-1–6-5.
18. Environmental Science Services Administration. 1967. Federal plan for a national agriultural weather service. Dep. Commerce, Washington, D. C., p. 1–22.
19. Humphrey, R. R. 1963. Arizona natural vegetation. Univ. of Arizona Bull. A-45 (map). Agr. Exp. Sta. and Coop. Ext. Service, Tucson, Ariz.

20. Food and Agriculture Organization of the United Nations. 1963. World forest inventory. Rome. p. 15–24.
21. Food and Agriculture Organization of the United Nations. 1966. Unasylva 20 (1–2):80–81; 46–51.
22. University of Michigan. 1966. Peaceful uses of earth-observation spacecraft. Vol. II. Survey of applications and benefits. Inst. Sci. Tech. Rep. 7219-1-F(II). p. 63–69.
23. Beal, J. A., W. H. Bennett, and D. E. Ketcham. 1964. Beetle explosion in Honduras. Amer. For. Nov.:31–33.

2

Imaging with Photographic Sensors

Robert C. Heller U.S. Department of Agriculture, Forest Service, Berkeley, California

Introduction

This chapter deals mainly with aerial photographic developments since 1960 and those physical and technical factors that appear relevant to photography as a means of remote sensing. For illustrations of particular applications see Chapters 4 and 6.

The present review of photographic films, techniques, or past applications is not exhaustive; these subjects have been covered extensively in the literature. The *Manual of Photographic Interpretation,*[1] published in 1960 by the American Society of Photogrammetry, has comprehensive coverage of the many uses to which aerial photography is being applied. Chapters 2, 3, 5, and 11 in that volume are directed particularly to agriculture and forestry and might profitably be reviewed by the reader.

Early balloon experimenters first pointed out the usefulness of coupling the camera with the airborne platform. With the development of stable aircraft, improvements in aerial photography and its application to military and civil uses have advanced to a highly sophisticated state. From all indications, photography will continue to be one of the most valuable and useful sensors for agriculturalists and foresters, whether it is used on the ground, from aircraft, or from orbital satellites.

Factors Affecting Quality of Aerial Imagery

The factors that have a direct bearing on the quality of photographic imagery from aerial cameras include: (1) physical factors, such as ground luminance and reflectance, atmospheric scattering of light, angle of sun, and spectral quality of sunlight; (2) film, emulsion, and filter properties; (3) camera and equipment factors, such as lens characteristics, image motion-compensation devices, shutter mechanisms, and multispectral techniques. The right combination will produce high-quality photographs that resolve fine detail and provide good tonal or color differentiation. If any one of these factors is not optimal, the imagery will be degraded. Agricultural and forestry photography could be upgraded markedly by use of modern precision aerial cameras.

Physical Factors

Illumination and Reflectance Illumination can be defined as light (luminous flux) falling on a given area. For purposes of aerial photography, it is made up of sunlight penetrating the atmosphere and of skylight that results from atmospheric scattering and reflection.

The luminous reflectance from ground targets varies immensely. Carman and Carruthers[2] measured reflectance from ground objects and found, for example, that black asphalt reflects only 2 percent, timberland 3 percent, open grassland 6 percent, concrete 36 percent, and snow 80 percent. Ground-contrast ratios can go as high as 1,000 : 1 in bright sunlight when an asphalt road adjoins an area of snow; however, these contrasts do not hold for high-altitude photography. Seldom does the contrast ratio exceed 10 : 1 over cities and towns; and for very-high-altitude photography, it would more likely be 5 : 1. Agricultural and forest lands have relatively low reflectance, as indicated above—particularly when the vegetation completely covers the soil. At medium scales of photography (1 : 20,000), bare soil, irrigation ditches, field boundaries, rock outcrops, crop foliage, and water surfaces are typical objects that furnish contrast in agricultural and forest photographs. These contrasts or tonal differences permit the photo interpreter to make judgments about the vegetation he is viewing.

Scattering of Light Light is scattered both as it passes through the atmosphere and as it is reflected from the earth's surface. The amount of scattering depends on both the number of gas molecules and the number of larger particles (Mie particles) present, the latter consisting of dust, water, smoke, etc. The gas molecules are suspended in the air to altitudes of 30,000 ft, but the Mie particles seldom rise above 15,000 ft. The gas molecules scatter light at a rate inversely proportional to the fourth power of the wavelength (Rayleigh effect; see Chapter 3), which affects sensing of the ultraviolet and blue portions of the spectrum more than sensing of the longer wavelengths (red and near infrared). This explains why most panchromatic aerial photography is exposed through a filter such as the minus blue (Wratten No. 12), or amber (Wratten No. 15), to reduce the fogging effects that haze causes on aerial exposures. Because the Mie particles are made up of many differing elements and concentrations, it is difficult to estimate their total effect on the aerial photograph. High concentrations of Mie particles, as around cities during inversion conditions, will, however, degrade the imagery taken through them.

Angle of the Sun The angle at which the sun strikes the earth's surface affects not only the quantity (lumens) of light being reflected to the aerial camera but also the spectral quality.

According to charts of natural illumination plotted by Brown[3] in clear weather, illumination drops off as latitude increases north or south from the equator, by season of the year, and by hours before or after local apparent noon. For example, at lat. 40°N, illumination at noon can drop from 11,000 fc in June to 5,000 fc in December. A similar dropoff occurs in June, 4 hours before or after local apparent noon. Because many agricultural or forest lands are poor reflectors, they should normally be photographed within 2 hours of local apparent noon.

Specht *et al.*[4] discuss the effect that sun angle has on aerial camera exposure. From microdensitometer measurements on aerial panchromatic film of rural and urban scenes, they computed values of minimum apparent scene luminance as influenced by solar altitude. Their data show that the greatest dropoff in luminance takes place below 20° solar altitude. These findings were incorporated into a revised (1966) *Kodak Aerial Exposure Computer,* which takes solar altitude into account.

Spectral Sensitivity The sun's energy and the spectral sensitivity of the human eye peak in the green portion of the visible spectrum. Skylight is essentially blue. Daylight appears to be white but varies with sun angle. At low angles, the blue portion of sunlight is almost completely scattered because of the extreme depth of atmosphere that the light must penetrate; this scattering makes the sun appear reddish in early morning and late afternoon. For both optimum illumination and complete spectral distribution, air photographs, whether panchromatic,

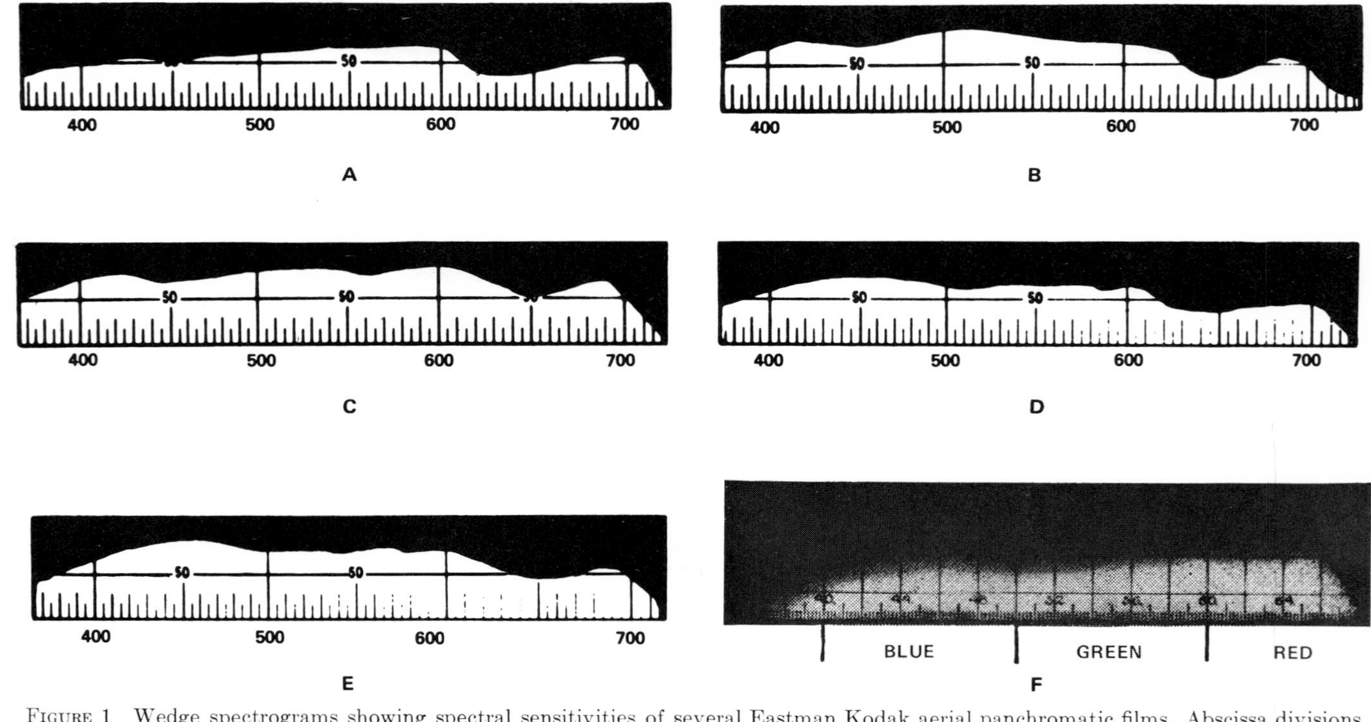

FIGURE 1 Wedge spectrograms showing spectral sensitivities of several Eastman Kodak aerial panchromatic films. Abscissa divisions are in millimicrons.

A. High-definition aerial film type 3404.
B. Special high-definition aerial film type SO-243.
C. Special fine-grain aerial film type SO-190.
D. Panatomic X aerial film type 3400.
E. Plus X aerial film type 3401.
F. Super XX aerial film type 5425.

infrared, or in color, should be exposed at midday. For more details see Condit and Gumm.[5]

To avoid specular reflection from water surfaces, the Coast and Geodetic Survey, U.S. Department of Commerce, is practicing an exception to midday exposure.[6] The bright sunspot reflected back to the camera during midday photography washes out needed underwater detail on aerial color transparencies. Consequently, on coastal water missions, color photography by the Coast and Geodetic Survey is exposed at low sun angles, with film having high emulsion speeds. Except in large areas flooded for rice production or areas adjacent to large bodies of water where vegetational detail is needed along shorelines, specular reflection is probably not a serious problem to agriculturalists or foresters.

Spectral energy of sunlight on a clear day approximates very closely that of a blackbody heated to 6,000° K and peaks in the visible portion of the spectrum.* Most aerial film is sensitive to wavelengths from 0.36 to 0.72 μ, with aerial infrared films extending the range to 0.9 μ (Figures 1 and 2), a range that coincides with the spectral energy peak of sunlight. The lower (blue) end of the film-sensitivity range is determined by the cutoff range of optical glass (0.36 μ), and the red or near-infrared end is determined by the spectral sensitivity of aerial film.

Film and Emulsion Properties

Panchromatic Film Most black-and-white aerial film ranges in spectral sensitivity from 0.36 to 0.72 μ. This is a red range extended beyond most panchromatic film used on the ground.

A comparison of several aerial films developed by Eastman Kodak Company demonstrates some of the properties of aerial film commercially available and those under development.

The accompanying wedge spectrograms (Figure 1) show that the high-definition (resolution) and fine-grain film is more sensitive to the red end of the spectrum and less to the blue than are Plus X or Super XX aerial film. The minus blue filter commonly used during aerial photography to cut off blue light below 0.5 μ reduces film sensitivity more on the faster film than it does on the fine-grain film. Increased red sensitivity reduces the filter factor required (see Table 1).

Plus X aerial film (3401) resolves fewer lines per millimeter than high-definition aerial film, but its relative sensitivity is 33 times as

*The reader should refer to Table 1 (Chapter 3) to compare the sensitivities of various sensors with respect to the regions of the electromagnetic spectrum.

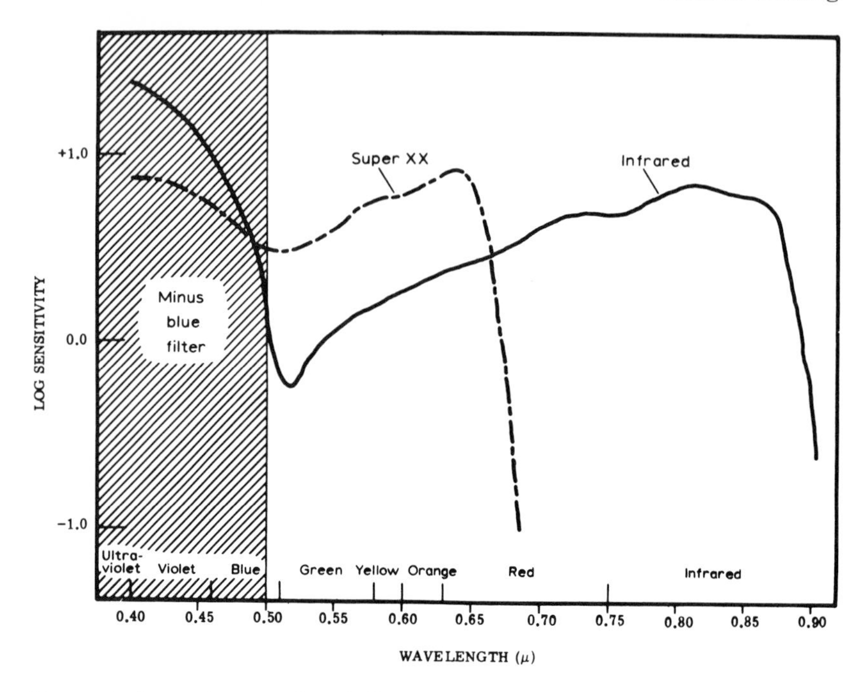

FIGURE 2 Spectral sensitivity of Kodak Infrared Aerographic film (Type 5424) and comparison with Kodak Super XX Aerographic film (Type 5425) at various wavelengths. Note the effect of the minus blue filter and the cutoff wavelengths for both films.

TABLE 1 Comparison of Aerial Panchromatic Films[a]

Name	Type	Resolution (line pairs/mm)[b]	Speed Relative to Type 3404	Granularity Values	Wratten No. 12 Filter Factor
High-definition aerial	3404	550	1	0.023	1.5
High-definition aerial	SO-243	440	1.2	0.016	1.6
Special fine-grain aerial	SO-190	180	3.7	—	1.5
Panatonic X aerial	3400	150	9.0	0.052	1.9
Plus X aerial	3401	100	33.0	0.088	1.7
Super XX aerial	5425	75	41.0	—	2.0

[a] Courtesy of Eastman Kodak Company.
[b] Average difference in target luminance 6.3 : 1

great (Table 1). For example, Plus X (3401) is normally exposed under aerial conditions at apertures of about *f*/5.6 to *f*/8 at 1/500 sec. To resolve images, high-definition aerial film type 3404 would require wider lens apertures and slower shutter speeds than are available (or desirable) on modern aerial cameras. Thus, in spite of the high definition possible with newer films like this, their slow speeds may preclude their use except under the most ideal conditions.

Infrared Film The only infrared aerial emulsion film available in the United States is manufactured by the Eastman Kodak Company. Spectral sensitivities of infrared film extend from 0.36 μ to 0.9 μ (Figure 2). Infrared film is less sensitive to the green part of the spectrum than panchromatic film, but its sensitivity extends beyond the red into the reflective portion of the infrared. Unless filters that cut out the blue or blue-green wavelengths are used in conjunction with infrared film, there is little advantage in using it instead of panchromatic film. Resolution capabilities of infrared film are lower than those of Super XX aerial film—about 55 line pairs/mm at 6.3 : 1 target luminance.

When infrared film is used with either a Wratten 25A (red) or 89B (dark red) filter, broad-leaved vegetation (angiosperms) appears lighter in tone, whereas water, streams, and moist soil appear darker in tone on positive prints. Broad-leaved plants that are diseased or losing vigor show up to be darker than their healthier neighbors. These are some of the uses that appeal to agriculturalists and foresters. They are illustrated in Chapters 4 and 6 of this book and in Chapter 2 of the *Manual of Photographic Interpretation*.[1]

Because the blue end of the spectrum is effectively cut off by filters, infrared film has superior haze-cutting capabilities. One drawback, however, is the degradation of contrast in the prints, which show little tonal gradation because of poor shadow penetration. For this reason, cartographers avoid using infrared film because of the difficulty in seeing the ground in forested areas through their mapping instruments.

Granularity Granularity of photographic emulsions increases directly with increased film sensitivity or speed; high-speed films are more granular than slow-speed films. Brock *et al.*[7] suggest that granularity be measured with a microdensitometer equipped with a small aperture (24 μ). The difference between films is expressed in the size of the standard deviation about a specified density level (e.g., 1.0 above base fog). For example, Plus X has a standard deviation about 4 times greater than 3404, and this increased granularity is directly related to lower resolution (Table 1).

Acutance Acutance is a term used to define edge sharpness of images on film and paper. With microimage scanning instruments, this film property can be evaluated quantitatively, and sharpness does not have to be diagnosed by subjective visual perception.

Often two different films will resolve an equal number of lines per millimeter, but one film appears to record image detail—large and small—better than the other film. This apparent difference is measurable by exposing film to a knife edge laid in contact with the emulsion. After processing, the film is scanned with a microdensitometer, and the acutance is calculated from the density–distance curve obtained. Knife-edge traces of an ideal emulsion and of two different typical emulsions are shown schematically in Figure 3. Films exhibiting both high resolving power and acutance are desirable for vegetation analysis.

Color Film Vast improvements that have been made in both color and false-color aerial films since 1960 include better resolution, increased emulsion sensitivity, shorter processing times, coating on glass plates and polyester film base, and finally, improvement in the color-coupling

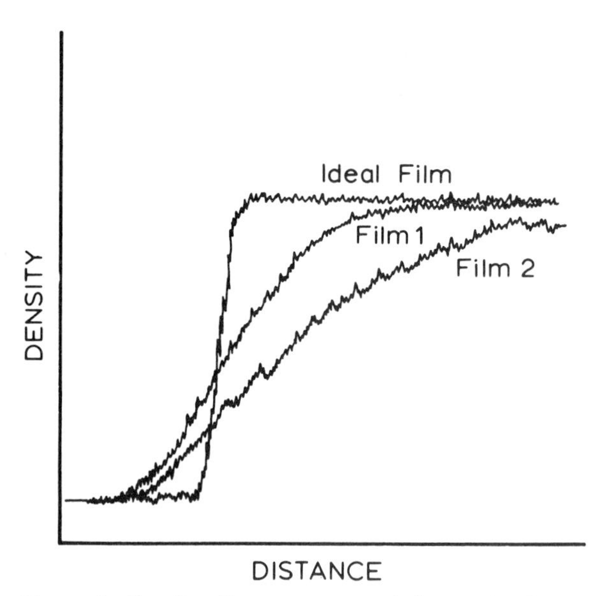

FIGURE 3 Density–distance trace made by a microdensitometer across a knife-edge exposure of a theoretically ideal film emulsion and of two different emulsions more representative of actual practice.

dyes for fidelity and stability. The reader is referred to a recent publication, *Manual of Color Aerial Photography*,[8] which is comprehensive in its treatment of color film and its use in the study of earth resources.

Color film is of three types: (1) color-positive, (2) color-infrared, and (3) color-negative. Color-positive film is sensitized to three primary colors: blue, green, and red. When exposed and processed, it produces transparencies that, when viewed by white light, appear similar to the original scene. Fritz[9] developed a table that shows clearly the principles of operation for color-positive and color-infrared film (Table 2).

Two film manufacturers produce most of the aerial color film in the United States: Eastman Kodak Company and General Aniline and Film Corporation (GAF). In addition, Agfa Aktiengesellschaft in Leverkusen, West Germany, supplies a small amount of the aerial color film used in the United States.

The spectral sensitivity of Anscochrome D/200 film (GAF) to light of equal intensity at all wavelengths is shown in Figure 4. Figure 5 shows the transmission curves of Anscochrome dyes. As in panchromatic film, the spectral sensitivity ranges from the cutoff by optical glass at 0.36 μ in the blue to 0.7 μ in the red. Kodak Ektachrome Aero film (E-3 process) exhibits very similar spectral sensitivities. The resolution qualities and recommended film speed for proper aerial exposure are shown in Table 3.

Gray-scale values or photographic tones provide the only measure-

TABLE 2 Principles of Operation for Normal Color Film (Type 8442) and Infrared-Sensitive Color Film (Type 8443)[9]

Normal color film				
Spectral region	Blue	Green	Red	Infrared
Normal sensitivities	Blue	Green	Red	—
Color of dye layers	Yellow	Magenta	Cyan	—
Resulting color in photographs	Blue	Green	Red	—
Infrared-sensitive color film[a]				
Kodak Ektachrome Infrared Aero (E-3 process)	Blue	Green	Red	Infrared
Sensitivities with Wratten No. 12 yellow filter	—	Green	Red	Infrared
Color of dye layers	—	Yellow	Magenta	Cyan
Resulting color in photographs	—	Blue	Green	Red

[a] Note that the blue-sensitive layer is masked by the yellow filter and that dye couplers (dye layers) are shifted one color band to the right and linked with the green-, red-, and infrared-sensitive layers.

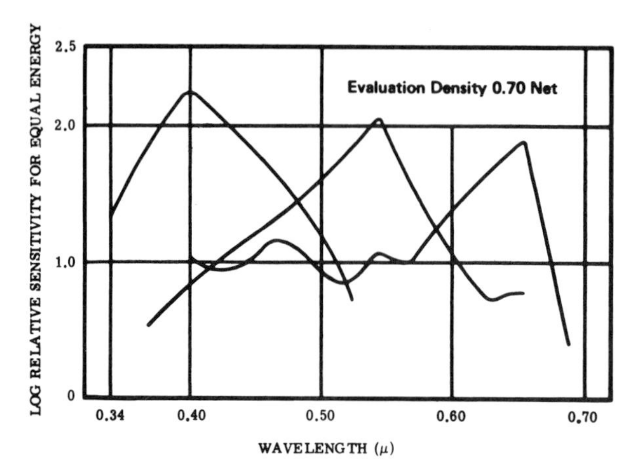

FIGURE 4 Spectral sensitivity curves for Anscochrome D/200 film.

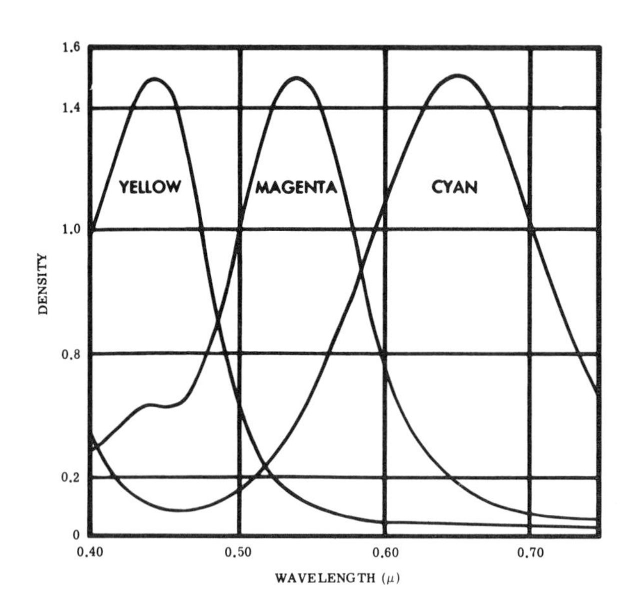

FIGURE 5 Spectral transmission curves for Anscochrome dyes.

TABLE 3 Resolving Power and Speeds of Modern Color Film

Film	Resolving Power (line pairs/mm) Targets with		Film Speed, ASA
	High Contrast	Low Contrast	
Anscochrome D/500	125	40	500
Anscochrome D/200	125	40	160–200
Kodak Ektachrome Aero (E-3 process)	Not published		160[a]
Kodak Ektachrome Infrared Aero (E-3 process) with Wratten No. 12 filter	Not published		160[a]

[a] Determined from repeated flights over vegetation, with Weston Exposure Meter aimed vertically from plane. Eastman Kodak Company recommends use of Kodak Aerial Exposure Computer with the index set at 25 for Ektachrome Aero and at 10 for Ektachrome Infrared Aero.

ments available to separate objects on black-and-white films; color film offers the two additional dimensions of hue (color) and chroma (strength of color). These two dimensions can be quantified by use of a Munsell Color System[10] or by the International Commission on Illumination (ICI) system.

Photo interpreters are accustomed to identifying objects not only by shape and form but also by color. An agriculturalist or forester trained to recognize plants by morphological features also associates a color with a plant. Therefore, color in photographs gives him one more factor on which to base his judgment. More training is required for an interpreter to be able to recognize objects by tones (values) of gray than by the normal colors he associates with the object. For example, the accuracies of photo interpretation of tree species were 17 percent higher on large-scale color transparencies than on panchromatic prints at the same scale.[11]

According to Evans,[12] the human eye can separate more than 100 times more color combinations (hues, values, and chromas) than gray-scale values (ratio of 20,000 to 200). Color film can match a ground scene very closely, but colorimetric measurements on film do not necessarily match spectral measurements on the ground.[13] Although the eye can separate 20,000 color combinations, color film cannot discriminate to that degree. Nevertheless, color film has been very effective in separating rock types,[14] identifying tree species,[10] locating insect-damaged timber,[15,16] and locating underwater obstructions.[17] Generally speaking, it compares favorably with the other sensing media in the multispectral tool kit.

False-Color Film False-color film is so named because objects do not appear on the transparencies of prints in the same colors as they do in nature. In the United States, Kodak Ektachrome Infrared Aero film (E-3 process) is such a film. This film may be referred to by any of the following synonymous names elsewhere in this book: "Infrared Aero Ektachrome," "camouflage-detection film," "false-color film," "infrared-color film," and "color-infrared film." Its three layers are sensitized to three colors—infrared (to 0.9 μ), green, and red. The film is always used with a yellow filter (Wratten No. 12), which prevents blue light from exposing the film; thus, only reflected green, red, and infrared wavelengths reach the emulsion. One of the greatest assets of this film is its ability to penetrate haze. Fritz's chart[9] (Table 2) depicts the sensitivities of the emulsion, the color of the dye layers, and the resulting colors on the transparencies. The spectral-sensitivity curves and spectral densities of the dyes of Kodak Ektachrome Infrared Aero film (E-3 process) are shown in Figure 6. According to Tarkington and Sorem,[13] the present emulsion released in 1962 is a much-improved version, has better resolution, is three times as fast as the old emulsion, and has less granularity.

One poorly understood concept of subtractive reversal color film (both color and false-color) is that the dye responses when the film is processed are inversely proportional to the exposure of the respective layers (or wavelengths) (Figure 6). This fact explains why the cyan-coupled dye, which is linked to the infrared-sensitive layer, produces little or no cyan dye on the film when viewed over white light. Instead, the green- and red-sensitive layers that are coupled with the yellow and magenta dyes come through strongly, and this combination makes healthy vegetation, which is highly infrared-reflective, appear reddish.

Another way to understand subtractive color films is in relation to the colors absorbed or subtracted through the coupled complementary dyes in the processed film. In the above case, when color-infrared transparencies are viewed over white light, healthy vegetation appears red. This is because the large amounts of yellow and magenta dyes subtract blue and green light from white light, respectively. The lack of cyan dye permits red light to be transmitted to the eye. In addition, the reflectance of healthy vegetation in the green and red portions of the spectrum is considerably lower than the infrared reflectance (about

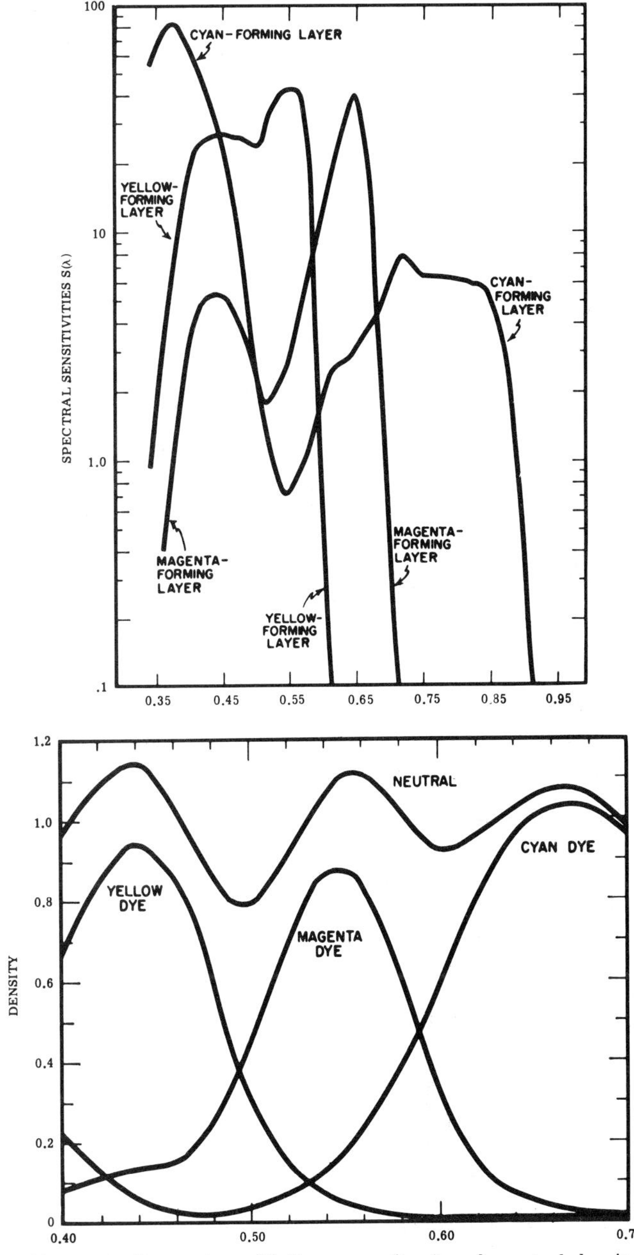

FIGURE 6 Spectral sensitivity curves (top) and spectral density curves (bottom) of Kodak Ektachrome Infrared Aero film (E-3 process). (*Courtesy Eastman Kodak Co.*)

5 times on spectrophotometer curves), which induces a strong yellow and magenta dye response because of the inverse relationship described above.

When vegetation begins to die, it loses infrared reflectance; this causes more of the cyan and less of the yellow and magenta to show. In addition to the near-infrared phenomenon, a stronger color contrast occurs on false-color film than on normal color film and frequently makes identification of diseased vegetation easier.

The predecessor to this film was known as camouflage-detection film. Developed during World War II, it was produced to detect painted targets that were camouflaged to look like vegetation. Because healthy vegetation is a much stronger reflector in the near infrared than in the green portion of the spectrum, vegetation appears in various hues of red on the film; objects painted green appear blue and can be distinguished immediately. As on black-and-white infrared film, when vegetation in the angiosperm group loses vigor through internal or external disease manifestations or soil salinity (Plate 19), it reflects less infrared light and first shows up as a darker red to black on infrared-color film (Plate 12). In cases of plants affected by moisture loss, infrared reflectance is reduced while the visible portion of the spectrum remains unchanged; here the infrared-color film shows stressed leaves as a lighter red to white color. In some cases, plants under stress show up on this film before symptoms of decadence or death are visible on the ground.

Plants in the gymnosperm group (pine, spruce, fir, etc.) display somewhat different color characteristics. Loss of infrared reflectance usually does not show up before visible signs of death occur. Dying needles of conifers, which appear yellow-green on color transparencies, assume a pink hue on Ektachrome Infrared transparencies; yellow trees appear whitish; and yellow-red and red trees show up yellow on the infrared-color film (Plate 2). Once a photo interpreter translates what the false colors represent in terms of vegetation and natural resources, he frequently finds this the fastest way to discriminate such things as conifers from hardwoods, dying from healthy broad-leaved plants (angiosperms), and presence of soil moisture from dry earth.

Russian film makers also produce both normal- and false-color films. Figure 7, copied from a translation by Mikhailov,[17] shows the sensitivity of the Soviet two-layer film called SN-2. Often referred to as "spectrozonal" film, this is a combination of one panchromatic and one infrachromatic layer on a common base. During developing, color dyes are introduced into both layers to produce images in various colors.

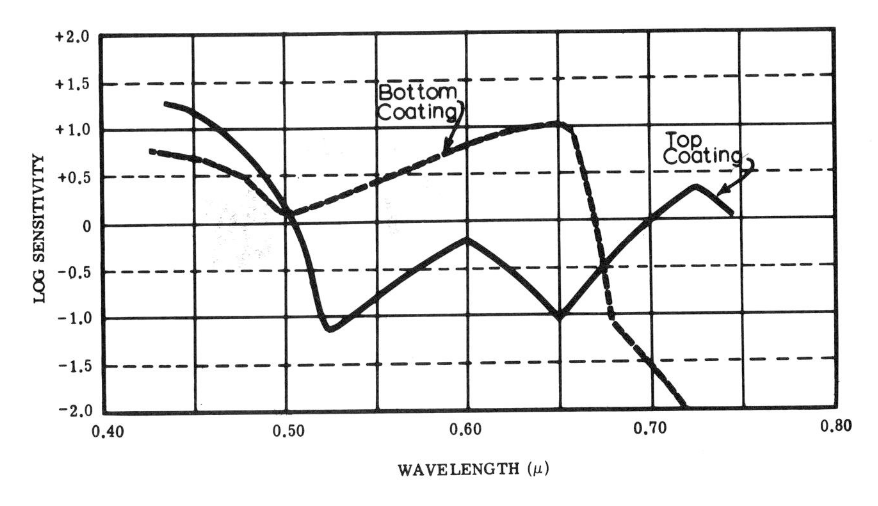

FIGURE 7 Spectral sensitivity of Russian spectrozonal aerial film (SN-2).

Because the dyes in spectrozonal film are different from the three dyes used in Ektachrome Infrared Aero film (E-3 process), vegetation and other natural features appear in colors different from those of the Kodak product. For example, coniferous species appear green, broad-leaf tree species, yellow and orange, and bare soil, blue on the SN-2 film, whereas the same objects on Ektachrome Infrared Aero film (E-3 process), appear reddish-brown, bright-red, and green to blue, respectively.

Mikhailov[16] reports that the SN-2 film is 50 to 100 percent faster than the Russian color film (CN) and that Russian analysts of terrain and vegetation have fully accepted spectrozonal film because the arbitrary colors make the identification of detail much easier. Furthermore, because the film is used with haze-cutting filters, spectrozonal pictures can be taken from higher altitudes than normal-color pictures.

Color-Negative Film Negative-type color film is produced by two manufacturers: Ektacolor Aero film by Eastman Kodak Company, and Agfacolor by Agfa Aktiengesellschaft in West Germany. These color films have the dye-coupler components incorporated in the emulsion layers at the time of manufacture. After color developing and bleaching away of the silver images, the remaining dye images are not only negative to the tone gradations of the subject (as with black-and-white emulsions) but also are complementary colors of the original scene. For example, a green tree would appear magenta on the negative film, but on Ektacolor Aero film the complementary colors are masked by an over-all amber gelatin layer so that the complementary colors are not easily distinguishable.

Theoretically, this kind of color material is exceptionally versatile—color prints, color transparencies, and black-and-white prints and transparencies all can be made from one negative. Unfortunately, there are several drawbacks to the use of color-negative film for aerial photography; poor resolution is the most serious. Furthermore, color prints made from negative aerial film are much inferior in resolution to color transparencies. In addition, Cooper and Smith[18] found that their color balance was not entirely satisfactory, particularly in the greens and blues, which are so critical in vegetation analysis. Manipulation of filters during exposure of prints may restore color balance somewhat if it is lacking in the original negative, but this is a questionable procedure for accurate vegetation identification. The film lends itself better to portrait or snapshot photography in which color balance can be altered in the darkroom to produce a pleasing picture rather than to provide accurate color representation needed in the analysis of natural resources.

Another disadvantage is a low ASA (American Standards Association) film-speed rating, 40, which reduces usefulness under low light conditions. In comparison with positive aerial film (ASA 160–200), negative aerial film requires about three times as much exposure. When the two are exposed under identical light conditions, negative aerial film requires a slower shutter speed—a factor that contributes to image movement and loss of resolution. Finally, only short lengths (40 ft) can be processed in wind–rewind developing equipment* because the developing solution becomes exhausted by longer lengths. (Short film lengths create space and handling problems in aircraft—and would be more serious in spacecraft.)

In March 1966, Eastman Kodak Company announced a new aerial color-negative product, Kodak Aero-Neg Color System (Kodak Ektachrome MS Aerographic, Estar Base, 2448). According to the manufacturer, this color negative has true complementary colors and all the theoretical advantages of the Ektacolor Aero film. It has better resolution, higher speed sensitivity, and is available in normal lengths.

In March 1967, General Aniline and Film Corporation reported the development of a new color film, Anscochrome D/500, having an ASA

*Morse B-5A or Zeiss wind-and-rewind processing equipment is commonly used for developing aerial films and can accommodate only limited quantities of solutions (5½ and 8 gallons, respectively).

rating of 500 and a capability of doubling the sensitivity to an ASA of 1,000 by developing the film longer in the first solution. This film can also be processed to either a positive or a negative transparency. In spring 1969, this company produced a color film without a blue-sensitive emulsion layer; it is designed to be exposed from high altitudes, and it responds to reflectivities in the green and red portions of the spectrum. Its usefulness for agricultural and forestry application has not yet been demonstrated.

Plate 3 is a display of four kinds of aerial film commonly used for resource photography. These aerial photographs were taken July 30, 1967, over color test panels set up on the University of California campus at Davis. The reader may compare two color emulsions, the infrared color and a color-negative film over the same site. The response of the hardwood shade trees, grass, and color panels on the infrared and color-negative films is particularly interesting and should be compared with the natural three-color films.

Physical Characteristics of Film How is film affected by low atmospheric pressure, radiation, relative humidity, processing, and temperature? What is its tear resistance and strength?

Harvey and Myskowski[19] report that if film is reconditioned in a normal atmposphere after being subjected to a low-pressure atmosphere, no loss of dimensional stability or serious effect (blistering or cracking) to the emulsion results.

Photographic emulsions are sensitive to both ultraviolet radiation and penetrating radiation from x rays, gamma rays, and high-energy particles. The earth's atmposphere provides some shielding from ultraviolet radiation for conventional aerial photography. Outside the atmosphere extra care is needed to prevent stray light from entering camera systems because the ultraviolet energy level is considerably higher there.

Alpha and beta particles also cause exposure of sensitized silver halide crystals; in fact, film aboard artificial earth satellites has traced their frequency and occurrence. Fortunately, though film is highly sensitive to these particles, it is protected easily by normal camera construction and by the thin metal surrounding aerial-film canisters.

In space, x rays and gamma rays are more penetrating than in the atmosphere, and the shielding required for film must probably be as effective as that required for human radiation protection. All forms of hard radiation produce similar effects on film: a decrease in film speed, loss of resolution, a decrease in gamma (loss of sensitivity), an increase in fog level, and an increase in granularity.

Dimensional stability of films has been improved greatly by the development of polyester materials to support the emulsion. For example, shrinkage resulting from processing is about one fifth as great on polyester-base film (0.02 percent) as on cellulose-triacetate-base film (0.10 percent). Relative humidity also has less effect on polyester-base film than on cellulose triacetate. This improved dimensional stability is important to photogrammetrists who are interested in keeping images in their proper spatial orientation. It will be doubly important when synoptic photographs are taken from space, so that displacement of objects is kept to a minimum.

Tear resistance, tensile strength, and break strength for polyester film bases are also greater than for the cellulose-triacetate bases. These are important qualities to consider for both aircraft and spacecraft photography when access to camera magazines is often difficult and the opportunity to capture imagery is lost because of delays in maintaining operational capability. Complete specifications and properties of aerial and polyester-base films are described in two pamphlets available from Eastman Kodak Company.[20,21]

Filters Used in Aerial Photography

Filters provide a means of selecting, amplifying, or eliminating portions of the electromagnetic spectrum in which photosensitive materials respond. Mention has been made of filters that reduce some of the degrading photographic effects of atmospheric scattering of light in the ultraviolet and blue end of the spectrum. The spectral-transmission curves of the haze-cutting filters most commonly used by commercial and map-making agencies when exposing aerial films are reproduced here by permission of Eastman Kodak Company.[22] The caption to each figure (Figures 8–13) describes the film used with each filter and the approximate filter factor. Note that the filter factor is often different for panchromatic and infrared emulsions.

Haze-cutting filters for color films are less dense than those for black-and-white films, so the proper color balance can be maintained. Should dense filters be used, an aerial color transparency will assume an over-all hue similar to that of the filter. There are several instances in which aerial transparencies taken from pressurized aircraft have had an over-all bluish cast. The discoloration was traced to bluish, optically flat glass used between the camera lens and ground targets to maintain aircraft pressurization. Excessive filtration of color film reduces the contrast of all other hues normally present and should be avoided if possible.

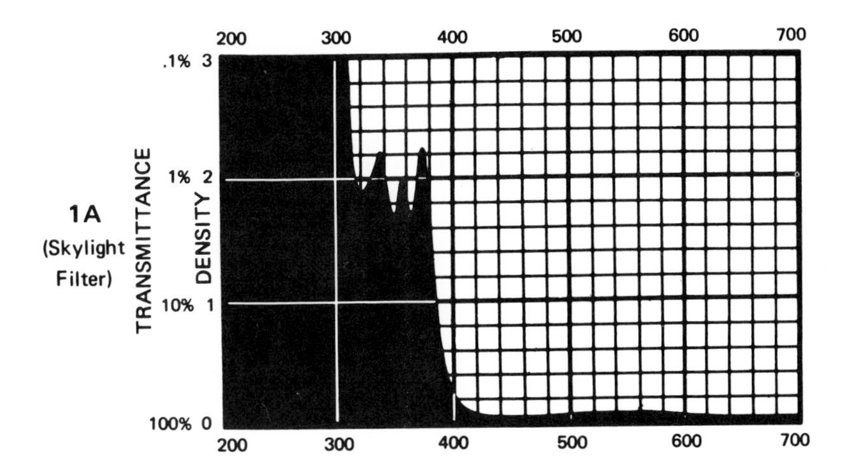

FIGURE 8 Wratten No. 1A filter (skylight). No reduction in exposure required. Used with Anscochrome color films (D/200, D/500, Ultraspeed, D/100) with light haze and altitudes to 10,000 ft. (*Courtesy Eastman Kodak Co.*)

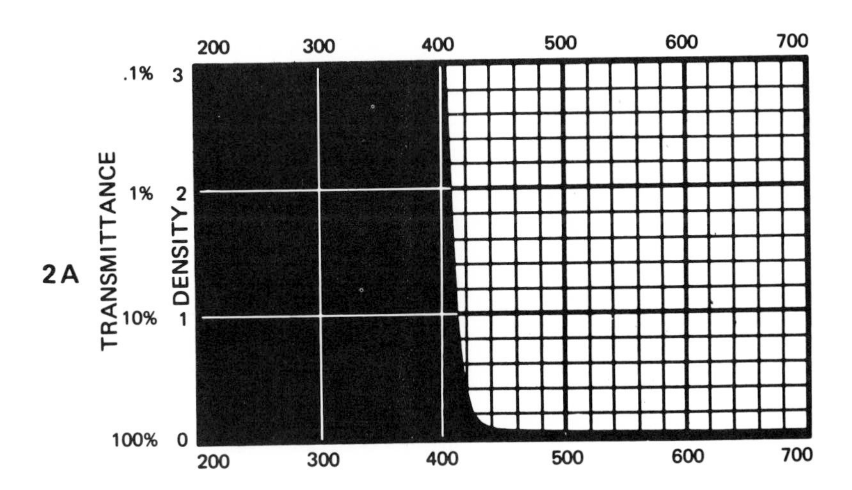

FIGURE 9 Wratten No. 2A filter. No reduction in exposure required. Used with Anscochrome color films (D/500, D/200, Ultraspeed, D/100) with moderate to heavy haze and altitudes above 10,000 ft. (*Courtesy Eastman Kodak Co.*)

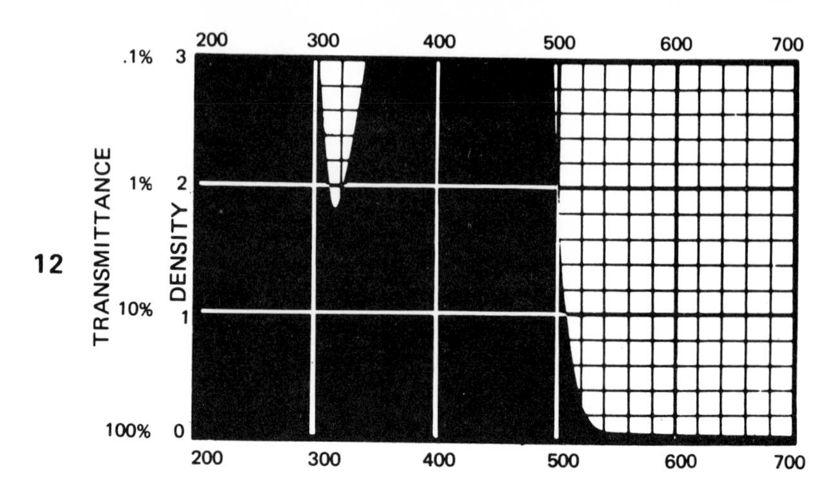

FIGURE 10 Wratten No. 12 filter (minus blue). Used with both panchromatic and infrared emulsions (including Ektachrome Infrared Aero) to reduce atmospheric scattering of blue light. Filter factor is 2.0 with panchromatic film, 1.5 with infrared. (*Courtesy Eastman Kodak Co.*)

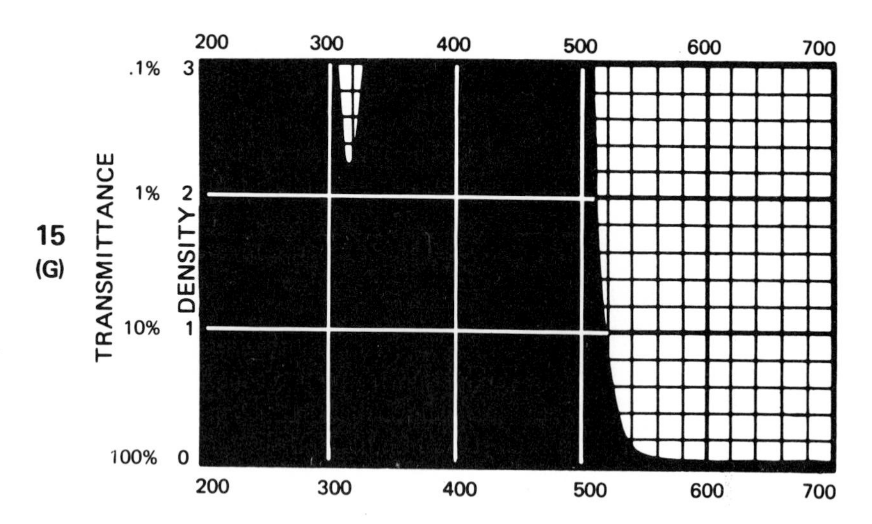

FIGURE 11 Wratten No. 15G filter (deep yellow). Filter factor is 2.0 with panchromatic and infrared emulsions to reduce heavy haze. (*Courtesy Eastman Kodak Co.*)

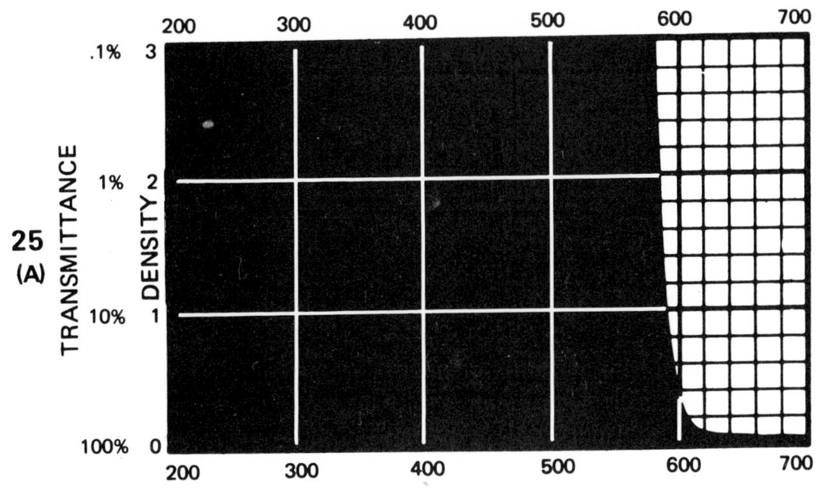

FIGURE 12 Wratten No. 25A filter (light red). Special effects to accentuate yellow-to-red objects on ground with panchromatic film. Strong haze-cutting ability; used also with infrared film. Filter factor is 4.0 with panchromatic film and 2.0 with infrared film. (*Courtesy Eastman Kodak Co.*)

Kodak packs emulsion-correction filters with Ektachrome Aero film (E-3 process) to restore color balance if the latter is lacking, and includes instructions about appropriate haze filters, depending on altitude and haze conditions encountered.

Sharp-cutting filters have been used to break up the visible and near-infrared portions of the spectrum into 9 to 16 nonoverlapping

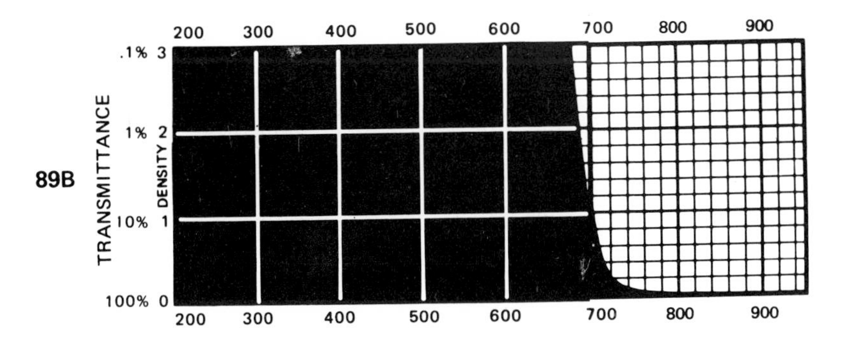

FIGURE 13 Wratten No. 89B filter (dark red). Cuts out all visible wavelengths. Used exclusively with infrared film. Filter factor is 3.0. (*Courtesy Eastman Kodak Co.*)

bands. The photography is by a special camera having 9 or 16 matched lenses, each covered by a different filter. A simultaneous exposure is made by means of a focal-plane shutter. The resulting print—made up of 9 to 16 separate photos—is called a multispectral photograph (see Figure 9, Chapter 4). Marked tonal differences usually occur in at least 3 or 4 of the photos, but less information appears available from the other bands. By computing density differences of these 3 or 4 significant bands and manipulating processing techniques in the darkroom, a single color print can be made on which all images are registered. This enhanced photo, while not representative of actual conditions, may show up subtle differences in images not detectable by conventional photography.

One disadvantage in the use of this technique is that sharp-cutting filters that mask out some wavelengths and transmit within narrow limits are usually quite dense. Therefore, long exposures are needed, and this introduces blurring and loss of resolution.

Aerial Cameras, Aerial Lenses, and Image Motion

Aerial Cameras and Lenses Aerial cameras have been developed for specific purposes such as mapping and reconnaissance. They may also be classified as to angular field of coverage, focal length, or type (frame, strip, panoramic, and infrared).

Mapping cameras are designed to produce maximum positional accuracy with reasonably good resolution capabilities, whereas reconnaissance cameras should provide maximum image resolution and moderately good positional accuracy.

The *Manual of Photogrammetry*[23] provides an excellent discussion of aerial cameras and lists 25 single-lens mapping cameras, 19 of them having a 229 × 229-mm (9 × 9-in.) format and the remaining ones a 178 × 178-mm (7 × 7-in.) format. It also lists 23 single-lens-frame reconnaissance cameras with formats varying in size from 57 × 57 mm (2¼ × 2¼ in.) on 70-mm film to 229 × 457 mm (9 × 18 in.) on 221-mm (9½-in.) film.

The C. Zeiss and Wild Heerbrugg precision cameras and lenses deserve special mention. Their lens designs accept color, infrared, and panchromatic films without the focal-length changes that are required by many other camera systems. The wide-angle 152-mm (6-in.) Pleogon A (Zeiss) and Universal Aviogon (Wild) lenses can be used both for reconnaissance and for mapping missions because of their low-distortion and high-resolution (40 line pairs/mm) qualities. The Zeiss camera has an electrically operated between-the-lens shutter with speeds to 1/1,000 sec; the RC-8 Wild Heerbrugg camera has shutter speeds to 1/700 sec. Two American-made lenses, the Geocon I and Geocon IV (designed by J. G. Baker), also can be used with all types of film and have been successfully mounted in two Fairchild cameras, KC-4 and KC-6A, respectively; resolution and distortion parameters are similar to those of the Zeiss and Wild cameras.

Forestry and agricultural users of aerial photography should specify precision cameras and lenses in aerial photographic work that requires high-quality imagery. The 1 : 20,000 photography available to most agriculturalists has been taken with cameras having relatively low-resolution capabilities. Indeed, some recent aerial photographs exhibit poorer image quality than some taken 20 years ago. Such factors as camera vibration, poor printing, and improper processing techniques may contribute to this decline in standards, but all future photographs should be taken with a high-resolution system. In flat-to-rolling terrain, a 152-mm (6-in.) focal length is adequate; a 210-mm (8¼-in.) focal-length precision lens is available and would be more desirable for photography in mountainous areas. Should scales of 1 : 12,000 or larger be used in rough terrain, a 305-mm (12-in.) focal-length lens is needed. C. Zeiss manufactures a 305-mm (12-in.) precision camera that has an average weighted area resolution (AWAR) of 40 line pairs/mm and is comparable with the above-mentioned lenses. Perkin-Elmer has just designed and produced an advanced 152-mm (6-in.) lens that resolves 100 line pairs/mm.

Aerial cameras used for reconnaissance photography should be equipped with spectral sensitometric wedges to permit quantification of reflected solar energy. No commercially available cameras are so equipped. By exposing an 18-step gray-scale wedge through the lens and shutter system and imprinting the scale along the edge of the film at the time of exposure, standards of absolute irradiance could be established. Tonal values of a particular crop, for example, could be related to a position on the 18-step wedge. If solar-energy inputs were different from one flight to the next, they would be reflected in the gray scale. Adjustments could be made also during the printing operation to establish uniform prints whose tonal values would be closely related.

Manufacturers of thermal sensors found a similar standardization procedure necessary to relate gray-scale values to temperature. Gray scales are incorporated, for example, in the Barnes Engineering Company Model 12-600 infrared camera. A calibrated heat source is used also in the thermal optical-mechanical scanners developed by a subdivision of The University of Michigan (see Chapter 3).

A second-step wedge illuminated by an artificial light of calibrated intensity within the camera would be useful to standardize film development. Thus, if film were overexposed but properly developed, the spectral sensitometric wedge would be darker than normal, but the processing wedge would match other standards previously established. Such spectral wedges would be extremely useful for both color and black-and-white films.

Panoramic aerial cameras attempt to combine wide angular coverage and high resolution in one sensor. Sometimes the coverage is horizon to horizon in a direction normal (perpendicular) to the aircraft flight line. The panoramic principle has been used by commercial photographers for many years to get pictures of large groups of people who are arranged in slightly semicircular (or straight) lines about the camera. The camera then rotates mechanically about its vertical axis. The resulting photograph is long and narrow but of good quality. In the air, the panoramic cameras transfer images to the focal plane by rotating a prism or lens. The forward motion of the aircraft requires either rapid advancement of film between exposures or continuous film movement to obtain complete coverage along the line of flight (Figure 14).

Optical transforming printers are available to relieve positional distortion from the original panoramic negative. Six panoramic cameras are listed in the *Manual of Photogrammetry*[23] and have formats from 57 × 254 mm (2¼ × 10 in.) to 108 × 1041 mm (4¼ × 41 in.). All these cameras can operate at high shutter speeds, some to 1/10,000 sec; and one has image-motion compensation capability. For agricultural or forest coverage, simpler solutions for determining area of coverage and identification of crops or trees are available with vertical photography. Too often, forest vegetation would be hidden in the oblique portion of the picture in rolling or mountainous terrain if taken at medium photographic scales (1 : 20,000). If panoramic cameras are used from spacecraft, terrain features would have little effect in blanking out areas because of the narrow angle of view.

Aerial strip cameras or "Sonne" cameras were perfected during World War II to provide maximum information along a narrow strip by low-flying, high-speed aircraft. Strip cameras move film past an adjustable shutterless opening at an image speed on the focal plane that is commensurate with aircraft ground speed. This kind of camera has been used mainly for military reconnaissance. Distortion along the line of flight can often be serious even though images are sharp and identifiable. Again, for vegetation analysis, single-frame photography taken from slow aircraft with cameras having fast shutters is preferable to "Sonne" strip photography.

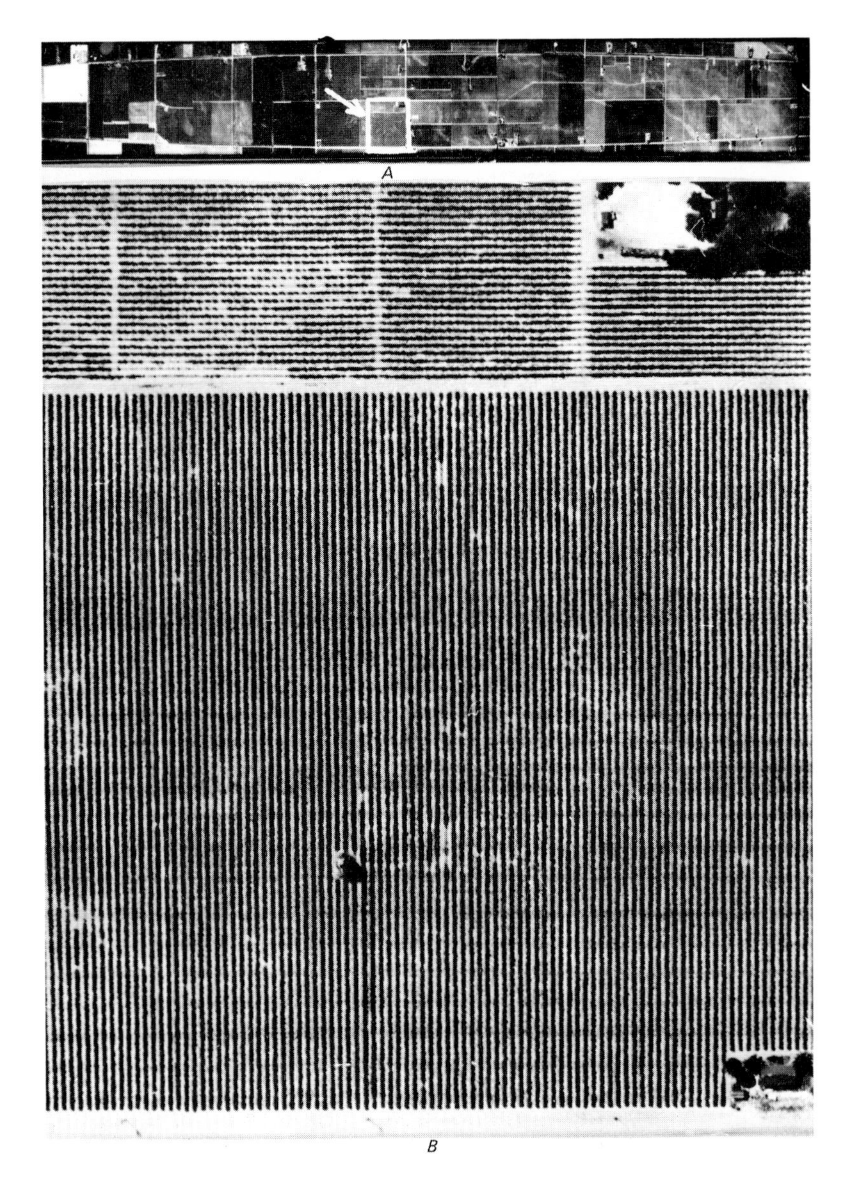

FIGURE 14 Panoramic photograph of vineyards in the San Joaquin Valley, California, taken at scale 1 : 17,000 lens focal length = 12 in. Note sharpness of enlarged portion (B); this high resolution is characteristic of panoramic photography. (*Courtesy Itek Corp. and California Crop and Livestock Reporting Service.*)

Image Motion The movement of the aerial platform produces some movement at the focal plane of the camera. This movement, in turn, induces blurring and degradation of the photographic image. The amount of image motion for vertical photography can be calculated as follows:

$$\text{Image motion (ips)} = 1.467Vf/H,$$

where 1.467 = conversion factor (1 mph = 1.467 fps)
V = velocity (mph)
f = focal length of camera (in.)
H = altitude above ground (ft)

The faster the platform moves, the longer the focal length; and the lower the altitude, the higher the rates for image motion. For aircraft flying more slowly than 200 mph and for scales of 1 : 5,000 or smaller, cameras equipped with fast shutters (1/500 sec or faster) reduce image motion to reasonable limits. Fast shutters also provide the cheapest and most dependable way to cut down on image movement. Conversely, at very high speeds or low altitudes, some means of moving the film or images a proper amount at the focal plane must be employed to overcome blurring. Methods to overcome image motion are discussed by Heller *et al.*[24] and are being used with modern cameras in jet aircraft; Figure 15 illustrates the effectiveness of a fast shutter (1/1,000 sec) as compared with shutter speeds (1/200 sec) commonly used with aerial cameras.

Aerial cameras with focal lengths up to 6,096 mm (240 in.) have been designed and successfully operated, but if such a camera were carried on a satellite circling 120 nmi above the earth at 18,000 mph, the pictures would scale about 1 : 36,000 and have 8.7 in. of image movement per second. A shutter operating at 1/1,000 sec would be needed to reduce image movement to barely tolerable limits—about 0.009 ips or about 4–5 line pairs/mm. This is scarcely sharp photography, and some form of image-motion compensation would be desirable. Cameras with shorter focal length operating from orbital altitudes would cut down image motion, but the scale would be appreciably smaller; for example, a camera with a 305-mm (12-in.) focal length would produce a scale of about 1 : 730,000 (1 in. = 12 miles) and on standard 229 × 229-mm (9 × 9-in.) photographs would cover 108 miles on a side, or about 11,000 square miles. The result would be truly synoptic photography, useful for broad-area coverage. If resolution were as good as that of World War II photography (on the average, 10 line pairs/mm), an interpreter could resolve objects no smaller than 250 ft unless the object

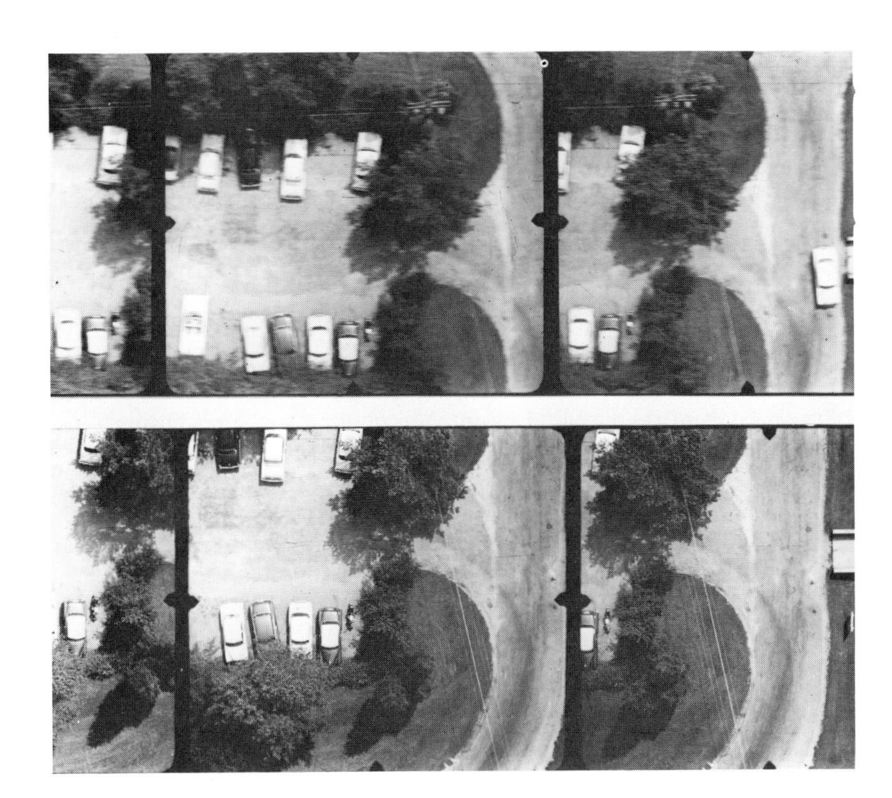

FIGURE 15 Stereo pair (70 mm) of a parking lot made from Super Anscochrome transparency. Upper pair was blurred by image motion which occurred while normal shutter speed (1/200 sec) was used. Image motion was eliminated on lower pair by a high-speed shutter (1/1,000 sec). Scale 1 : 500. Note lost detail of trees, automobiles, and wires on upper photo. (*Courtesy U.S. Forest Service.*)

were unusually long or had high luminance. With present-day equipment and materials, resolution should be at least 4 times better than that of World War II photography; this would permit objects of 60 ft to be resolved.

Resolution

The factors described in the preceding sections contribute to the final quality of aerial imagery. Under laboratory conditions, high resolution values can be attained. However, aerial photographs are not taken under such conditions. Aircraft or camera vibration, atmospheric haze, sudden illumination changes that vary target luminance, and other disturbances all contribute to a degradation of the image.

A standard U.S. Air Force bar target (Plate 4) is often used to measure resolution; its bars are known to decrease in size as the sixth root of 2.

Katz[25] reports that a reliable way to assess resolution is to assign threshold values to the factors affecting image quality and then to sum the reciprocals of these values in order to determine the resolution (R) of the final image:

$$\sum_{i=1}^{z} \frac{1}{R_{\text{image}}} = \frac{1}{R_{\text{film}}} + \frac{1}{R_{\text{image motion}}} + \frac{1}{R_{\text{lens}}} + \cdots \frac{1}{R_z},$$

where z is the total number of factors.

Kodak Plus X aerial film resolves 90 line pairs/mm at 6.3 : 1 target luminance. A camera with a focal length of 152 mm (6 in.) and a 1/1,000-sec shutter in an aircraft moving at 200 mph, 10,000 ft above the ground, would have image movement of about 0.0002 ips, which computes to about 50 line pairs/mm. A Zeiss Pleogon A lens with a 152-mm (6-in.) focal length resolves 40 line pairs/mm on high-contrast targets with panchromatic film. If these values are substituted into the equation above,

$$\frac{1}{R_{\text{image}}} = \frac{1}{90} + \frac{1}{50} + \frac{1}{40} = \frac{1}{18}, \text{ or} \sim 18 \text{ line pairs/mm.}$$

From this kind of analysis, one can calculate that each element has a measurable effect on reducing resolution of the system. Obviously, the interaction of all elements always produces a lower system resolution than the poorest element in the system.

The modulation-transfer function is another mathematical tool to study the performance of a photographic system. Instead of using bar targets to measure resolution, this function is obtained by making exposures of test objects whose luminance varies sinusoidally with distance. A microdensitometer is then used to measure the modulation of the frequencies from the densities on the film. From these traces, ratios are calculated between the amplitude of the modulations of the test objects and those of the traces. The ratios represent the modulation transfer at each frequency. A curve is then drawn comparing modulation with frequency and is called the modulation-transfer function. The advantages of this function over the use of threshold contrasts as determined from bar targets is that the modulation-transfer function shows the modulation at all frequencies. It is not as commonly used as the bar target to determine resolution because of unavailability of sinusoidal targets and of the need for a microdensitometer to carry out the tests.

Despite all the possible degrading factors that enter into the production of the photographic image on the emulsion, the aerial photograph provides the greatest amount of information of any sensing system now available. Katz[25] has calculated that a 229 × 229-mm (9 × 9-in.) aerial photograph has about 81 million bits of information available if the photograph has resolution as low as 10 line pairs/mm based on 15 definable gray-scale values. Swanson[26] states that positive color-transparency film available today can, with proper handling, achieve as high resolution as panchromatic emulsions. Thus, color film should provide more information than black-and-white film because of the addition of hues and chromas. Color films have shown themselves to be surprisingly useful in the 70-mm Hasselblad photography taken by the Gemini astronauts. According to White,[27] water depth was determined roughly by the U.S. Navy Hydrographic Office in the upper Gulf of California from one color photograph (Plate 5). Other space color photography near the Salton Sea area of California (see Figure 17, Chapter 4, for black-and-white reproduction) clearly defines agricultural fields in the Imperial Valley and with simple rectification could provide a small-scale map covering 8,000 square miles.

Scale

The scale of photography affects the resolution of the image. Photographic scale is expressed as a fraction or ratio, the numerator representing the photographic image size and the denominator, the object size on the ground—always in the same units (feet, meters, inches, etc.). The following adjectives are the author's, but they may help the reader to put scale in perspective:

1. Very large scale (obtained from helicopters, stationary towers, cherry pickers, etc.): 1 : 100 to 1 : 500
2. Large scale: 1 : 600 to 1 : 2,000
3. Medium scale: 1 : 5,000
4. Normal scale: 1 : 12,000 to 1 : 20,000 (most common scales available to foresters and agriculturalists)
5. Small scale: 1 : 30,000 or less
6. Very small scale (taken from orbital altitudes): 1 : 100,000 to 1 : 2,500,000

In tests at Wright-Patterson Air Force Base, Katz[25] found that resolution was always best when scales (or images) were large. By enlarging small-scale photography of test targets to the same size as large-scale originals, he found that comparable resolutions were not obtained. The large-scale originals were always vastly superior; perfect reciprocity does not exist, and thus scale and resolution are not tradeable. If in doubt, "Always take the larger scale," he advises. Image enlargements should be made by optical means—microscopes, magnifying stereoscopes, hand lenses, etc.—not photographically.

Scale is important to vegetation analysts, and the scale to be selected depends upon the extent to which they must make accurate judgments about the photographic images. For example, foresters can usually distinguish forest from nonforest land and tell something about tree height and tree density on 1 : 20,000 photographs, but they can seldom identify individual trees accurately until color film is used and scales approach 1 : 1,584.[11] Boundaries of agricultural fields can be well delineated on normal-scale photographs, but the exact crop being grown is almost impossible to differentiate among a variety. Colwell[28] found, by using various combinations of filters and films, that a number of crop species responded variously in different regions of the electromagnetic spectrum. By comparing photographs of the same scene on color, infrared color, infrared, and panchromatic film, he could discriminate several crop species. This point is illustrated for rangelands in Chapter 4 (Figure 34) and in Plate 15. When this technique is extended by images resolved by optical-mechanical scanners in other parts of the electromagnetic spectrum, still more discrimination appears possible.

INTERPRETATION OF PHOTOGRAPHIC IMAGERY

Subject-matter specialists have been responsible for gathering data from aerial photographs, relating them to ground information, and extrapolating their findings to broader areas. For example, from aerial photographs, oceanographers study sea-ice patterns and ocean currents; geologists pick out morphological structure of the terrain, rock types, and stream patterns; soil scientists define zones of similar soil types and parent material; agriculturalists map areas of farmland and describe crop acreages and expected yields; foresters delineate forest types and make mensurational estimates of timber volume. These specialists are trained in their own scientific disciplines. They have learned to use aerial photographs as a tool to make their own work more efficient.

Except in cases where narrowly defined objectives are specified—such as having less qualified persons make dot counts of agricultural land—most photo-interpretation work requires a high level of deductive reasoning and association with past experience and earlier training. The photo-interpretation literature[1] is documented with examples from many disciplines of the scientist's ability to cover more area and do his job more effectively by using aerial photos than by using ground methods alone. But whatever its efficiency, photo interpretation is time-consuming, and the highly trained personnel needed for it are in short supply. Therefore, investigations are under way to determine those parts of the job that can be automated with scanners, computers, and pattern-recognition techniques, those that require a man–machine interface, and those that man must do alone (see Chapter 9). The degree of success of this effort should be evident in a few years. Some jobs defy automation and computer analysis. One task that automated photo interpretation may be able to do is to reject large areas needing no inspection by the human interpreter and to select for sampling the areas that need deductive reasoning.

Human Photo Interpretation

The photo interpreter should have normal binocular vision, with little or no color-vision defect—particularly when color photographs or transparencies are part of the multispectral analysis. Also, he should be trained in orientation and use of aerial photographs, stereoscopes, and measuring equipment. He should be motivated to get maximum information from the aerial photographs.

Production output by photo interpreters depends entirely on the objectives of a particular job. Suppose the mission is to search for and count vehicles in the open or count discolored dying trees in a green forest. The interpreter can accomplish this task at a faster rate than if at the same time he has to identify and measure the length of the vehicles or measure the heights and densities of the dying trees. Scale of photography, resolution of the images, and photographic tone or color contrast of the object against its background also affect the rate at which interpretation can proceed. An example of the time required for an interpreter to examine a 3,600-acre forested area for insect-damaged timber on color transparencies is shown for three scales with their accompanying accuracies:[15]

Photo Scale	Total Photos	Time (hr)	Accuracy (%)
1:3,960	208	13.0	82
1:7,920	47	4.9	65
1:15,840	19	2.8	56

Another example can be cited from Aldrich's unpublished study[29] on estimating timber volume from stereo and nonstereo viewing of Agricultural Stabilization and Conservation Service 1 : 20,000 panchromatic photography in 10 counties in North Carolina. In the nonstereo phase of the study, interpreters used a crown diameter–crown density viewing aid; in the stereo phase, they used the viewing aid but also measured tree height with a Zeiss parallax bar under an old Delft scanning stereoscope. The interpretation time comparison:

Method	Time (hr)	Ratio
Nonstereo	24	—
Stereo	200	8.3:1

Thus, the addition of the tree-height measurement and stereo examination must provide an eightfold increase in accuracy of estimate to be worth the added time differential. By way of comparison: if the work were done by ground examination, an estimate of 2,600 hours would not be unrealistic. The time needed to make the stereo estimate is greater than that for the nonstereo, but much shorter than the time required to gather the same data on the ground.

The reader should refer to Chapter 3 in the *Manual of Photographic*

Interpretation[1] for a complete review of techniques, visual requirements, human factors, training, measuring techniques, equipment, and psychological factors involved.

Automatic Photo Interpretation

The idea of having a machine that can work day and night scanning and evaluating aerial photographs is very appealing, particularly when one considers the volume of photographs or video tape that could be turned out by an earth-circling satellite in good weather. Combining the matrix of multispectral photographic data with advances in photogrammetry, optical-electronic scanners, and computers would make it possible to collect, interpret, and analyze detailed data from aerial photographs on a mass scale.

Various instruments can scan images for differences in optical density. Moore *et al.*[30] at the National Bureau of Standards adapted a precision lathe with a photomultiplier tube to send data to an analog-to-digital converter and thence to a magnetic-tape unit (Figure 16). Rosenfeld[31] compared six land-use classes by projecting a video trace across photos and onto an oscilloscope, where they were photographed. Doverspike *et al.*[32] used a GAF-Ansco Model 4 microdensitometer with

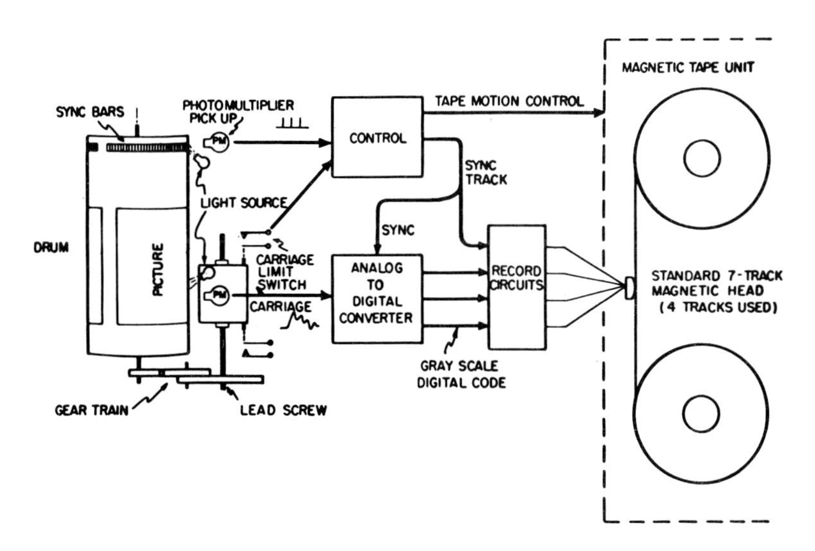

FIGURE 16 Scanning system developed by National Bureau of Standards to read out density values from aerial photographs and record these values digitally on magnetic tape. (*Courtesy National Bureau of Standards.*)

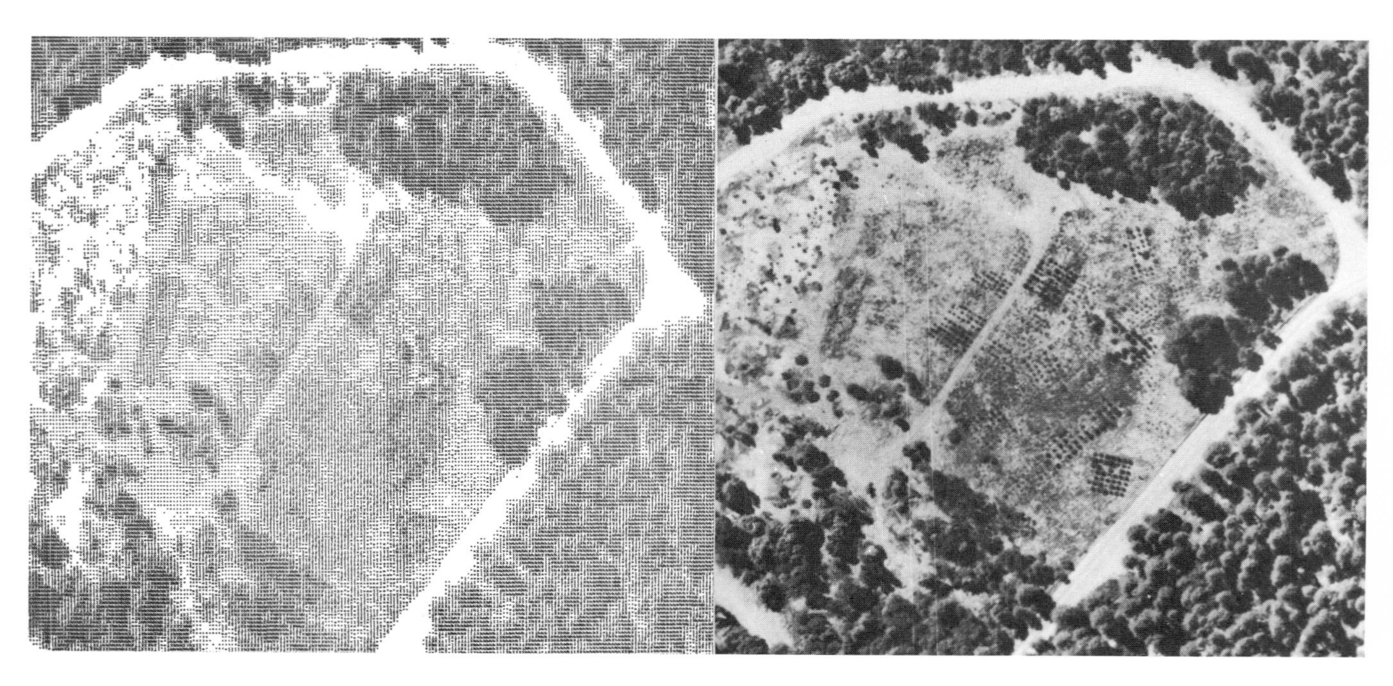

FIGURE 17 Computer printout (*left*) of an aerial photograph (*right*). Alphanumeric characters are used to approximate the original density range. (*Courtesy U.S. Forest Service.*)

appropriate filters to identify types of land use by reading densities of the three dyes (red, green, and blue) incorporated in positive color film. At IBM,[33] reflected brightness values were read from the photo and averaged into larger groupings to permit correlation of the digital groupings from one photo to the next. Langley[34] demonstrated that an aerial photograph recorded from the National Bureau of Standards scanner onto magnetic tape could be printed so that the digital image was visible to the eye (Figure 17). This is the first phase of the system that also includes: (1) rectification of the digital image from tilt displacement, (2) rectification of the digital image from relief displacement, and (3) analytical interpretation of the digital image map. Colwell[35] has summarized the advantages and disadvantages of using human and mechanical data extraction from photos.

Evaluation of Photographic Sensing Systems

The following discussion is not intended to favor photographic systems over other sensing systems. In fact, a combination of sensors would undoubtedly yield the most information to the resource specialist. When discrimination-analysis techniques and accuracy parameters are better established (see Chapters 3 and 9), the system that provides the desired information fastest will be used. The advantages and disadvantages of photographic sensing relate to the present state of the art; however, improvements in all systems can be expected.

Advantages

1. Photographic systems have superior resolution capabilities over other systems such as television or optical-mechanical scanners. A comparison of resolution in terms of object size is shown below for a scale of 1 : 5,000 to 1 : 100,000:

	Scale	
Sensor	1 : 5,000	1 : 100,000
Photographic at 50 line pairs/mm	2 in. (50 mm)	40 in. (1,000 mm)
Optical-mechanical scanner		
1 mrad	5 ft	100 ft
3 mrads	15 ft	300 ft
Television (19-in. receiver)		
525 TV lines	15 ft	300 ft
5,000 TV lines	1.6 ft	32 ft

To provide some perspective: Tiros (television infrared observation satellite) has resolution capabilities of only 1–3 km. This resolution is good enough for cloud studies but not for detecting life on earth; no cultural features (roads, cities, etc.) show up. Part of the poor resolution can be attributed to the altitude above the earth; for example, Tiros 10 is in an orbit 530 miles above the earth. The Apollo satellites are designed to operate about 120 miles above the earth.

2. Photographs from precision cameras provide good spatial orientation and require little rectification.

3. Human interpreters can identify vegetation from normal scales, make measurements on photographs, and may be able to make gross third-order measurements on satellite photography.

4. Aerial photography and film processing are relatively simple operations. They do not rely on complicated electronic circuitry and test equipment.

5. Photographic systems require minimal electric power and space.

6. Data can be processed on a space vehicle by rapid-processing techniques, or they can be rapidly processed on earth. Better resolution is obtained if original film is returned to earth than if films are scanned by television and retransmitted to earth.

7. Equipment is low in cost compared with that for other systems and lends itself to automated interpretation.

Disadvantages

1. Clear weather—no clouds—and sunlight are required.

2. There may be a delay before film or photographs are processed; this is not true of real-time systems, such as data transmission by telemetering and by television systems.

3. Systems used in space vehicles must be shielded from radiation to prevent fogging of film.

4. Procedures must be worked out for human interpretation and also automatic scanning, as well as storing and computing—all for large collections of data.

5. Since photography senses only in limited portions of the electromagnetic spectrum, use of instruments sensitive to other portions provides additional data sources.

References

1. American Society of Photogrammetry. 1960. Manual of photographic interpretation. George Banta Co., Menasha, Wis. 868 p.

2. Carman, P. D., and R. A. F. Carruthers. 1951. Brightness of fine detail in air photography. J. Opt. Soc. Amer. 41:305–310.
3. Brown, D. R., Comdr. 1952. Natural illumination charts. U.S. Navy Rep. No. 374-1.
4. Specht, M. R., N. L. Fritz, and A. L. Sorem. 1966. The change of aerial camera exposure with solar altitude. Photogr. Sci. Eng. 10(3):150–155.
5. Condit, H. R., and F. Gumm. 1964. Spectral energy distribution of daylight. J. Opt. Soc. Amer. 54:937–944.
6. Smith, J. T., Jr. 1962. Color—A new dimension in photogrammetry. Photogram. Eng. 29:999–1013.
7. Brock, G. C., D. I. Harvey, R. J. Kohler, and E. P. Myskowski. 1965. Photographic considerations for aerospace. Itek Corp., Lexington, Mass. 122 p.
8. American Society of Photogrammetry. 1968. Manual of color aerial photography. George Banta Co., Menasha, Wis.
9. Fritz, Norman L. 1965. Film sees a new world of color. Citrus World. 2(2).
10. Munsell Color Company. Munsell book of color. Ed. 1920–60. Munsell Color Co., Inc., Baltimore, Md.
11. Heller, R. C., E. G. Doverspike, and R. C. Aldrich. 1964. Identification of tree species on large scale color and panchromatic photographs. U.S. Dep. Agr. Handbook 261, 17 p., illus.
12. Evans, Ralph M. 1948. An introduction to color. John Wiley, Inc. New York. 340 p.
13. Tarkington, R. G., and A. L. Sorem. 1963. Color and false-color films for aerial photography. Photogram. Eng. 29:88–95.
14. Ray, R. G., and W. A. Fischer. 1957. Geology from the air. Science 127(3277): 725–735.
15. Wear, J. F., R. B. Pope, and P. G. Lauterbach. 1964. Estimating beetle-killed Douglas-fir by aerial photo and field plots. J. For. 62:309–315.
16. Heller, R. C., R. C. Aldrich, and W. F. Bailey. 1959. An evaluation of aerial photography for detecting southern pine beetle damage. Photogram. Eng. 25:595–606.
17. Mikailhov, V. Y. 1960. The use of colour sensitive films in aerial photography in U.S.S.R. Trans. Can. Nat. Res. Council.
18. Cooper, C. F., and F. M. Smith. 1966. Color aerial photography: toy or tool. J. For. 64:373–378.
19. Harvey, D. I., and E. P. Myskowski. 1965. Physics of high altitude aerial photography. Photography. Considerations for Aerospace. Itek Corp., Lexington, Mass. 122 p.
20. Eastman Kodak Co. 1961. Manual of physical properties of Kodak aerial and special sensitized materials. Eastman Kodak Co., Rochester, N. Y.
21. Eastman Kodak Co. 1960. Physical properties of Kodak Estar-base films for the graphic arts. Eastman Kodak Co. Pamphlet Q-34.
22. Eastman Kodak Co., Wratten filters, Eastman Kodak Co. Pub. No. B-3.
23. American Society of Photogrammetry. 1966. Manual of photogrammetry. George Banta Co., Menasha, Wis. 1199 p.
24. Heller, R. C., R. C. Aldrich, and W. F. Bailey. 1959. Evaluation of several camera systems for sampling forest insect damage. Photogram. Eng. 25: 137–144.
25. Katz, A. 1960. Observation satellites: problems and prospects. Astronautics, in 6 parts, April, June, July, August, September, October.

26. Swanson, L. W. 1964. Aerial photography and photogrammetry in the Coast and Geodetic Survey. Congr. Int. Soc. Photogram., Lisbon, Portugal.
27. White, R. 1966. Atlas of aerial photography of the earth. Published by U.S. Army ERDL GIMRADA, Ft. Belvoir, Va., under contract from OSSA, NASA. 2 vols.
28. Colwell, R. N. 1964. Aerial photography—a valuable sensor for the scientist. Amer. Sci. 52.
29. Aldrich, R. C. 1967. Cubic foot volume stratification on single 1 : 20,000 scale panchromatic aerial photographs. Preliminary report. Pac. SW For. Range Exp. Sta., U.S. Forest Service, Berkeley, Calif.
30. Moore, R. T., M. C. Stark, and L. Cahn. 1964. Digitizing pictorial information with a precision optical scanner. Photogram. Eng. 30:923–931.
31. Rosenfeld, A. 1962. Automatic recognition of basic terrain types from aerial photographs. Photogram. Eng. 38:115–132.
32. Doverspike, G. E., F. M. Flynn, and R. C. Heller. 1965. Microdensitometer applied to land use classification. Photogram. Eng. 41:294–306.
33. IBM Corporation. 1964. Optimized digital automatic map compilation system. Final Tech. Rep. U.S. Army ERDL GIMRADA, Ft. Belvoir, Va. 136 p.
34. Langley, P. G. 1965. Automating aerial photo-interpretation in forestry—how it works and what it will do for you. Proc. Ann. Meeting Soc. Amer. For., Detroit, Michigan.
35. Colwell, R. N. 1965. The extraction of data from aerial photographs by human and mechanical means. Photogrammetria 20:211–228.

3

Imaging with Nonphotographic Sensors

MARVIN R. HOLTER University of Michigan, Ann Arbor, Michigan
Contributing Authors: M. BAIR, J. L. BEARD, T. LIMPERIS, R. K. MOORE

INTRODUCTION

Since about 1945, a variety of nonphotographic image-forming systems has been developed to detect radiation reflected or emitted (or both) from a remote scene. These filled a void left by the problems inherent in the use of that most common of optical imaging devices, the camera.

Cameras, although small, light in weight, and relatively simple to use, have two serious limitations in remote-sensing applications. First, the output is a photograph, which is awkward to telemeter or to process for discrimination analysis. Second, photographic film is limited in spectral response to the region from the near ultraviolet to the near infrared. Therefore nighttime operation is almost impossible unless artificial light sources are used for illumination. Moreover, clouds, fog, and smoke are opaque in this spectral region, so that "seeing" the ground from very high altitudes is sometimes difficult and often impossible.

Nonphotographic sensors operate in portions of the electromagnetic spectrum from the microwave to the ultraviolet region. Infrared, passive-microwave, and radar sensors operate under both day and night conditions, and radar sensors are not seriously hindered by clouds and bad weather. Since the data are collected in electrical form, they are easily transmitted to a remote location. In addition, signals for discrimination analysis can be processed with electronic circuits. These sensors are generally more complicated, are larger, and have lower spatial resolution than comparable photographic equipment.

Spectral Regions

The electromagnetic spectrum extends from radio frequencies (fractions of a cycle per second) to cosmic rays (frequencies exceeding 10^{26} cps). The spectrum is commonly divided approximately as shown in Table 1.

In the passive-microwave and radar* regions, centimeters and millimeters are commonly used for wavelength units, and megacycles (10^6 cps) and gigacycles (10^9 cps) are used for frequency units. Recently the name Hertz has become widely accepted as a substitute for cycles per second (1 Hertz = 1 cps). In the optical portion of the spectrum (far infrared through ultraviolet), micron (μ) (10^{-6} m) and angstrom (Å) (10^{-10} m) are most commonly used for wavelength. The dividing line between the far infrared and millimeter regions depends largely upon the techniques and instrumentation employed. For example, some modern, solid-state, infrared-radiation detectors have spectral responses that extend from the far infrared into the millimeter region. Conversely, the "rf plumbing" (waveguide) technique used in systems sensitive in the millimeter region has been employed in the far-infrared portion of the spectrum.

Imagery from radar, which is in the microwave region, was not produced in quantity until after World War II, though early bombing radars that mapped land masses might have been called imaging radars. Although great strides were made during this war in pushing forward the frontiers of the use of the electromagnetic spectrum, at the war's end the regions between the intermediate infrared and microwaves remained virtually unexplored. Passive-microwave and radar imagery, because of the longer wavelengths, cannot have as high a resolution as infrared or photographic images. The exception is the special technique of coherent synthetic-aperture radars. The theoretical limit in resolving power of an optical system or radar antenna with a circular aperture can be stated as being $1.22\ \lambda/D$, where λ is the wavelength and D is the diameter of the aperture. To avoid diffraction or sidelobe effects, D should be many times larger than λ. At optical fre-

*Passive systems utilize natural radiation from observed objects; active (radar) systems use an electromagnetic source to illuminate an object or scene artificially.

TABLE 1 Wavelength and Frequency Ranges of Operation for Remote Sensors

Spectral Region	Wavelength	Frequency	Common Applicable Imaging Sensors
Microwave			
Active (radar) Passive: Decimeter (UHF)	10–100 cm	3×10^9 to 3×10^8	Scanning antennas with radio-frequency receivers
Active (radar) Passive: Centimeter	1–10 cm	3×10^{10} to 3×10^0	
Active (radar) Passive: Millimeter	0.1–1 cm	3×10^{11} to 3×10^{10}	
Optical			
Infrared: Far Infrared	8–1,000 μ	3.75×10^{13} to 3×10^{11}	Scanners with infrared detectors
Infrared: Intermediate Infrared	3–8 μ	1×10^{14} to 3.75×10^{11}	Various image tubes (not very satisfactory)
Infrared: Near Infrared	0.780–3 μ	3.85×10^{14} to 1×10^{14}	Photographic film to approximately 1 μ Scanners with infrared detectors Various image tubes
Visible	0.380–0.780 μ	7.89×10^{14} to 3.85×10^{14}	Photographic film Scanners with photomultiplier detectors Television
Usable: Near Ultraviolet	0.315–0.380 μ	0.952×10^{15} to 7.89×10^{14}	Photographic film (quartz lenses) Scanner with photomultiplier detectors Image-converter tubes
Usable: Middle Ultraviolet	0.280–0.315 μ	1.08×10^{15} to 0.952×10^{15}	
Far Ultraviolet	0.010–0.280 μ	3.0×10^{15} to 1.08×10^{15}	These wavelengths do not penetrate the earth's atmosphere significantly so are not useful for agricultural remote sensing
Vacuum Ultraviolet	0.004–0.010 μ	7.5×10^{16} to 3.0×10^{15}	

quencies this is achieved with apertures of reasonable physical size. Unfortunately, to obtain equivalent ratios at microwave frequencies, the antenna must be extremely large. Until discovery of the synthetic-aperture technique, obtaining high-resolution imagery at microwave frequencies did not appear promising. The synthetic-aperture radar (see page 93) in effect, allows effective λ/D ratios to approach those of optical systems while physical antennas are kept down to practical size. As a result, active microwave imagery resolution approaches that obtained by optical systems. In addition, radar systems operating at wavelengths longer than 3 cm have the great advantage of all-weather, day-or-night capability and are not limited by clouds. Perhaps most important of all, their resolution is not limited by range, an extremely important factor when the distance between the imaging sensor and the scene is great. In general, however, the sizes, power requirements, and complexity of synthetic-aperture radar systems are far greater than those required for optical-frequency equipment.

In the millimeter wavelength region (0.1–1 cm) and in the centimeter wavelength region (1–10 cm), there is sufficient self-emitted thermal radiation that passive sensing is possible, although not with the same temperature sensitivities as in the infrared region. However, passive systems are available in these regions.

Between 100 μ and about 1 mm lies the unexploited and virtually unexplored millimeter and submillimeter wavelength region. Although progress is being made to develop this part of the spectrum, it is not expected to produce any significant advances for imaging purposes, at least until improved detectors and microwave sources are developed. Lasers hold promise as sources of submillimeter wave energy, and quasioptical methods may ease component requirements, but the development of practical submillimeter devices is in the future.

Since World War II, with the wealth of technology and experience in infrared molecular spectroscopy as a foundation, there has been an extensive development of imagery systems in the near infrared (0.78–3.0 μ), in the intermediate infrared (3–8 μ), in the short-wavelength end of the far infrared (between 8 and 14 μ), and to a lesser extent, in the far infrared between 14 and 100 μ. The latter part of the far-infrared spectrum is of less interest because of quite uniformly poor atmospheric transmission and the small number of satisfactory detectors.

Photographic sensors in the visible region (0.380–0.780 μ) are discussed in Chapter 2. Since about 1958, it has been found useful to supplement these with scanned photomultiplier imaging systems, which are discussed in this chapter (page 148) and in Chapter 9, where the full reason for interest in them for use in multispectral systems is brought out.

Television systems are obviously visible-region sensors also, but they are not discussed in this book. They have significantly less resolution capability than photographic sensors and do not appear to be as compatible as scanned photomultiplier systems for use in multispectral systems also containing nonvisible detectors. Therefore, they appear to be somewhat outside the main trend of interest in the present discussion.

Until about 1958, few nonphotographic imaging devices were used in the ultraviolet region, and only since then have any of these devices been used to gather imagery in the ultraviolet. The principal reason was the lack of interest on the part of the military, which has supported nearly all the research in nonphotographic imaging systems. Further, it has been experimentally determined that the atmosphere strongly absorbs and scatters radiation of wavelengths shorter than approximately 3,000 Å, thus limiting remote sensing to the near-ultraviolet region. In the decade after 1958 interest in this region increased; scanning systems with ultraviolet photomultiplier detectors are discussed in this chapter (page 157) and in Chapter 9 in connection with multispectral systems.

Types of Systems

Intensive development of radar sensors has been going on since World War II, and a wide variety of types exists. Many of these types do not have any obvious applicability to agriculture. The side-looking airborne radars, especially those employing synthetic-aperture techniques (see page 93) to produce finer resolution than the physical dimensions of the antenna would permit by normal means, promise to be a very useful tool for agricultural applications. At radar wavelengths these systems are less troubled by atmospheric and poor-weather attenuation of the signals than sensors in any other region, and being self-illuminating, they are not dependent on solar illumination or self-emitted radiation from the objects being viewed. These systems view at right angles to the aircraft or spacecraft flight line. In that direction, i.e., the range direction, scanning is accomplished by the time the emitted pulse travels from the antenna to the ground and back to the antenna. Thus, time will differ for objects at different distances. Scanning along the direction of the vehicle flight line is accomplished by the vehicle motion, which causes the position of each transmitted and received pulse to be displaced slightly from the previous one. The usual system has a transmitter, an antenna, a detecting element followed by a superheterodyne receiving section, and a display. When the synthetic-aperture resolution-enhancing technique is employed, the display is replaced by a recording mechanism, usually photographic film or magnetic tape. The recorded signals are subjected to a sophisticated processing, and the resulting enhanced image is displayed on a cathode-ray tube (CRT), which is photographed. Present-day transmitted powers and receiver sensitivities that can be achieved from aircraft or spacecraft are very high and quite capable of producing imagery of near-photographic quality. The available experimental results indicate the strong possibility of a high degree of utility in agricultural applications.

State-of-the-art passive-microwave sensors are not as well developed as sensors for the other spectral ranges (see page 116). At passive micro-wavelengths, self-emitted radiation is much less than in the infrared, and as a consequence the highest sensitivities and lowest internal noise powers are required of the receivers. The systems presently available resemble radar receiving systems but of course have no transmitters. They have a receiving antenna that is scanned; a detector, which may be one of several different types; a more-or-less conventional superheterodyne receiver section; and finally, a display usually consisting of a CRT, which is photographed. Very often the detector is caused to view alternately an internal calibrated source and the scene to be examined, and, as a means of overcoming instabilities and providing higher performance, only the difference signal is recorded. This is called the Dicke radiometer technique. Detector sensitivities and internal noise are the present limiting factors, and improvements may be expected. Present technique makes possible temperature sensitivities of 0.5° C and resolutions of 10 mrads, i.e., a 10-ft-diameter spot from an altitude of 1,000 ft, and correspondingly larger at higher altitudes. A slight extension of present techniques would provide a temperature sensitivity of 2° C and a ground resolution of 0.4 mile from an altitude of 200 miles. Hence, because of the sensitivity limitations, the passive-microwave sensors have the coarsest resolution limitation of any of the sensors discussed (assuming that synthetic-aperture techniques are used for the radar sensors). The utility of these sensors for agricultural purposes has not yet been well explored; pending further work, no definite conclusions can be drawn.

Nonphotographic imaging in the ultraviolet, visible, and infrared is accomplished by scanning and nonscanning systems. The scanning system (see page 133) has a rotating mirror that causes the sensor to view portions of the scene sequentially in some regular manner. Nonscanning devices generally view a larger field and convert the radiation from it into a visible image without the use of moving mechanical parts. During World War II, interest in image systems increased because of the military need for an improved night-vision capability. One of the resulting developments was a near-infrared image tube that was used in devices such as the sniperscope—a night-viewing instrument mounted on a rifle. The success of the near-infrared tube, which extended vision to wavelengths longer than 1 μ, stimulated the development of other such devices. Experiments were conducted to produce imagery at wavelengths as long as 10 μ. Superconducting niobium bolometers were used in Nipkow scanners to produce crude pictures in the far infrared. Because of the extreme difficulties and danger in handling the superconducting detector coolant, liquid hydrogen, and because of the lack of a better detector in this region, research was dropped until after the war.

During the early 1950's the Haller-Raymond and Brown Company, under Air Force sponsorship, developed the "Reconofax," a scanning system operating in the visible region. It led the way to many other visible and infrared systems. The Reconofax was an airborne scanning system that used a photomultiplier detector and produced maps of the ground scene at night using either moonlight or its own internal arc source to illuminate the instantaneous field of view. Its imagery lacked resolution when compared with that produced by direct photographic methods and flash bombs. Later modified to use a lead sulfide detector, the Reconofax was able to detect the radiation emitted by hot objects such as factory furnaces and stacks as well as reflected solar energy.

During the same period, the Servo Corporation of America, also under an Air Force contract, developed the TRD (Thermal Reconnaissance Device). Its main purpose was to detect the presence of hot targets, which in turn triggered airborne, night-photographic flash equipment. However, the hot-target detector (a thermistor bolometer) had sufficient sensitivity to make infrared maps (images) by direct thermal mapping. With an angular resolution of approximately 1/4° and a capability of detecting temperature differences of about 1° K, it produced maps that clearly indicated the potential of infrared reconnaissance devices.

Other scanning systems using thermal detectors were also developed; however, their performance, like that of the earlier systems, was limited by the lack of sensitive detectors that could respond to rapidly fluctuating radiation intensities. The subsequent development of highly sensitive detectors with time constants shorter than 1 μsec (i.e., detectors that could respond to time fluctuations in radiation intensity as short as 10^{-6} sec without a degradation in performance) led to dramatic improvements in remote sensors. With multielement detectors, nonphotographic systems could be built to operate in the infrared with resolution approaching the Rayleigh limit—the theoretical limit of performance. Since all the optical systems are commonly reflective, these scanners can be used in the visible and ultraviolet regions by the installation of a suitable photomultiplier detector.

Today the appearance of imagery obtained with scanners in the infrared, visible, and ultraviolet approaches photographic quality, and many components of nonphotographic devices have been improved. Compact cryogenic systems, miniaturized electronics, and special displays have led to reliable, practical, and reasonably simple systems for gathering such imagery.

Radar Imaging

Radar is unique among the commonly used sensors in that it provides its own illumination—radio waves. Because the properties of the source are accurately known, the time for a wave to travel to a remote object can be measured by comparing the time the wave left the radar with the time it returns. Using the known speed of electromagnetic waves in space (3×10^8 m/sec), this time measurement can be converted into a distance measurement. Furthermore, the wavelength of the returning wave can be compared with that transmitted to determine the Doppler shift, and from it the relative velocity of radar and object can be calculated. Both the distance-measurement and velocity-measurement properties of radar are used in imaging systems to improve the resolution over that possible with a passive system.

Radar history apparently goes back to the 1920's, when observations were made of fluctuations in radio signals associated with passage of airplanes nearby. By the outbreak of World War II, German, British, French, American, and Japanese military groups had developed radars for detection and location of aircraft and, in some cases, ships. From these uses the acronym radar (*ra*dio *d*etecting *a*nd *r*anging) was devised. Imaging radar became a reality early in that conflict, but the quality of the images obtained by shipboard and airborne radars was unsuitable for observation of terrain properties other than land–water boundaries, major mountains, and cities. In the 1950's, imaging radars with much better pictorial abilities were developed for military reconnaissance, and the radars used today for studying earth resources are an outgrowth of these systems.

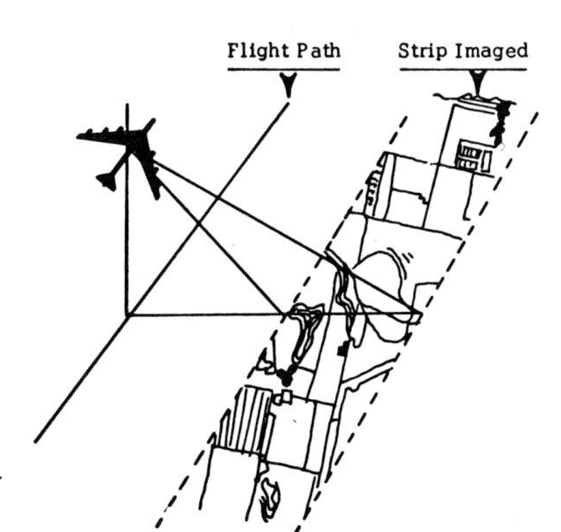

FIGURE 1 Diagram of application of side-looking airborne radar.

Radar's greatest advantage is that aerial surveys may be conducted on a preset time schedule without waiting for clear weather, without concern for time of day, and without concern for problems of interpretation associated with differences between day and night thermal patterns. This advantage is due to the wavelength region of radar (between about 0.5 cm and 1 m), since waves of these lengths can penetrate clouds and smoke, and waves longer than 2 cm or so can penetrate fog and precipitation.

Imaging radar systems take many forms, but the one most likely to be useful for agricultural surveys is the side-looking airborne radar (SLAR). In this system, a continuous strip image is recorded on photographic film. The terrain imaged is to the side of the vehicle carrying the radar, and a strip directly beneath the vehicle cannot be imaged readily. Such a system is illustrated in Figure 1. Strips may be imaged on both sides of the vehicle or only on one side. Such a strip may extend to the side from a distance about half the altitude out as far as 100 km or more. The vehicle carrying the radar may be an aircraft or a spacecraft.

Fundamental Principles of Radar Echoes

Radar returns depend upon reradiation (reflection) of energy supplied by the radar system. The strength of the reradiation depends both on the properties of the transmitted electromagnetic field and the properties of the irradiated object. These properties are summarized in the list below. They apply also to self-emission, except that temperature and other causes of self-emission must be added.

Parameters of source (or receiver)
- Wavelength
- Polarization
- Direction

Parameters of surface
- Dielectric and conducting properties, including quantum resonances
- Surface roughness in wavelength units
- Physical resonances
- Surface slopes
- Subsurface effects
- Scattering area

The electromagnetic field generated by the radar transmitter is determined solely by the instrumentation. The wavelengths most commonly used for present-day operating imaging radar systems are between 0.86 cm and 3.3 cm, although experimental systems operating at wavelengths of up to 70 cm are producing synthetic-aperture images. Present radars are monochromatic; that is, they use only a single wavelength of great spectral purity compared with the best obtainable with passive systems using filters in the optical and infrared regions. Spectral purity of radar systems is comparable with that of lasers.

The direction at which the transmitted energy strikes depends on the position of the object being viewed with respect to the radar. This configuration is a matter of choice. The *angle of incidence* (measured with respect to the vertical) may range from zero (straight down) to 90° (horizontal). In practice, the vertical is avoided because geometric considerations make it very difficult to discriminate ground distance by time measurement near the vertical; near-horizontal incident angles are avoided because of shadowing. Normally, airborne imaging radars operate between incidence angles of 20° or 30° and 75° or 80°. The *azimuth angle* may be chosen anywhere from straight ahead to directly astern of the aircraft, but most high-resolution systems look directly to the side. The *viewing angle* is determined by the flight direction once the azimuth angle with respect to the aircraft has been fixed. For some ground objects, this angle can be important; i.e., there is a difference at sea between observation upwind, downwind, and crosswind.

The reradiation for any particular surface depends on its dielectric and conducting properties. Since the dielectric constant of water is very high, and plants often have high moisture content, many plants are good reflectors.

The degree of roughness of the surface is an even more vital factor in setting radar-return strength. Numerous theoretical studies of radar scatter from statistically described surfaces are largely irrelevant for agriculture because of the complexities of the shape of growing plants. Only one mathematical description for growing crops has been attempted.[1]

The presence of resonant structures within surfaces can give rise to strong radar signals from apparently small objects if they are properly oriented, and it can also result in weaker signals with improper orientation. Surface slopes are, in a sense, measures of roughness, but they deserve separate listing because of their importance.

If the radar signal can penetrate a significant distance into the surface material, the signal observed will then be determined by the combination of the surface and the subsurface parameters. This is frequently true with crops, where the "surface" is the top of the vegetation and penetration to the soil is likely.

The scattering properties of the ground as a radar target are usually expressed in terms of the average differential scattering cross section σ^0. The signal received by the radar is proportional to σ^0 times the area contributing. All other factors in the radar power equation are either parameters of the radar system or of the geometric relations between radar-carrying vehicle and target area. The differential scattering cross section is a function both of the ground properties and of radar parameters such as wavelength and look angle.

As radar moves past a particular terrain element, its instantaneous cross section fluctuates over a wide range because of changes in the relative phase of signals from different parts of the element. This results in an "antenna pattern" for each surface element, as indicated in Figure 2. The signal fluctuates rapidly (fades) as the radar passes through this pattern. Thus, individual looks at targets can lead to erroneous conclusions, and averaging must be used to overcome the effects of fading on measurement precision. In optics, the same effect causes a laser-illuminated picture to be "speckled."

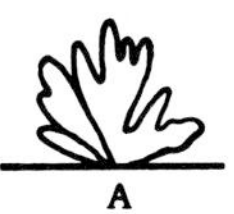

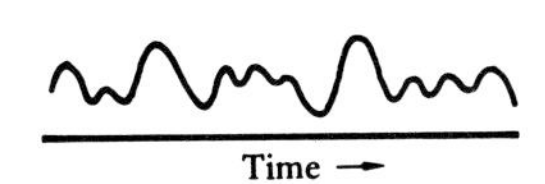

FIGURE 2 Antenna pattern and resulting fading. Left: strength of signal from a target at *A* is proportional for each direction from *A* in the diagram. Right: signal received in an aircraft going past a target at *A* but always pointing toward it.

Rather than treating each patch of ground as having a single pattern, an approach based on superposition of the patterns of many flat facets is sometimes employed. The two methods give equivalent results. A rough surface is approximated by a collection of essentially flat (within a small fraction of a wavelength) facets of different sizes. "Rougher" parts of the ground require small facets, and smoother parts can permit larger facets. Backscatter patterns for facets of different sizes are indicated pictorially in Figure 3. The narrow pattern goes with a facet many wavelengths across, and the wide pattern goes with a facet that is a fraction of a wavelength across.

The direction of largest signal return is that having the most facets perpendicular to it. Returns come from the sea and from plowed fields at angles near grazing (the angle as it approaches 90°; see Figure 38), for which no facets are perpendicular. Presumably, these are from the edges of the broad patterns for small facets.

Plants present a more complex situation than either plowed fields or oceans. Cylindrical scatterers, in most cases, behave to some extent like flat facets; that is, the maximum signal is returned if the axis of the cylinder is perpendicular to the direction of the incident wave (at least if the wave is polarized properly). Thus, radar return from plants with simple vertical stems should be highest near grazing incidence. Plants with complex structure, like corn stalks or apple trees, provide strong returns over a wider range of angles.

Return from plants is also wavelength-sensitive. At millimeter wavelengths, nearly all plant surfaces are "rough," except, perhaps, leaves. Leaves that are relatively flat can act as facets at these wavelengths, and a plant having most leaves almost horizontal should give strong

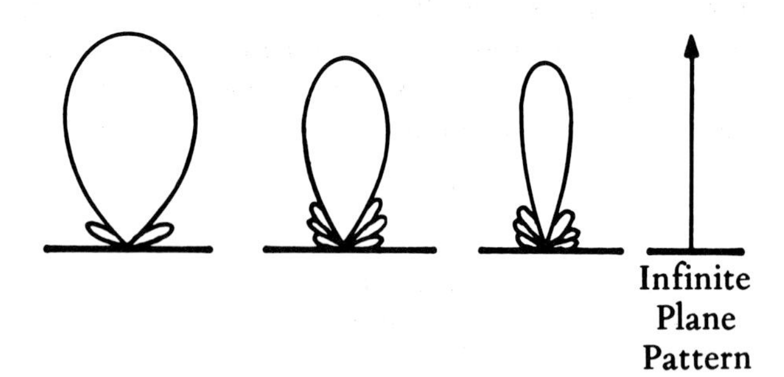

FIGURE 3 Facet patterns. Uniform normally incident illumination. Note that small facets are almost nondirective. Large facets give strong signals in a single direction only.

returns for vertically incident radar signals. The leaves are only a fraction of a longer wavelength across, and the signals from them are almost nondirective. At still longer wavelengths (meter range), leaves are so small that they contribute very little to the signal return at all.

Moisture content of the plant can influence its radar return significantly, although a saturation level apparently exists beyond which additional moisture makes no difference. A completely dry (dead) plant, however, cannot return signals as well as a living, wet plant. Preliminary studies of this relationship have been reported,[2] and returns from wet and dry fields indicate this effect.[3]

The polarization of the transmitted energy is defined as the direction of its electric field vector, and it is determined by the antenna design. The receiving antenna may be designed to respond only to the same polarization transmitted, but it may also be designed to respond to electric field vectors in other directions. When separate receivers are used for different polarization directions, the shift in polarization by an echoing object may be observed and used as a discrimination feature.

Polarization is a significant factor in radar-return strength. For straight-wire antennas, this direction is either parallel to the wire or at least in a plane passing through the wire. For other types of antennas, the direction of polarization is set by less obvious factors.

Study of polarization influence on radar signals returned from vegetation has been reported recently.[4] Preliminary observations of centimeter wavelength images show clearly that for some crops, returns are significantly different when signals received have the same polarization as those transmitted and when signals received have a polarization at right angles to that of the signals transmitted; these difference do not exist for other crops. Differences between signals with horizontal polarization, both transmitted and received, and signals with vertical polarization, both transmitted and received, have not been so clearly identified, but they are believed to exist.

A signal polarized in the direction of stalks and striking them at normal incidence should give a stronger return than a signal polarized at right angles to the stalks.

Cross-polarized returns (received polarization different from transmitted polarization) are usually stronger on surfaces having elements with small radii of curvature and on surfaces so rough that multiple scatter is easy. This should carry over to returns from plants as well.

Radar return from vegetation is more complicated than radar return from surfaces, and return from surfaces is not yet fully understood. Nevertheless, empirical determination of differences in return from vegetation as a function of wavelength, polarization, incident angle, and moisture indicates the sensitivity of radar to many significant plant differences. More research is needed to catalog this sensitivity adequately.

Atmospheric Effects on Radar

Serious atmospheric effects on radar are confined to the shorter wavelengths used for radar (usually wavelengths longer than 3 cm are relatively unaffected). The atmosphere only slightly absorbs radar signals at most of the wavelengths used. Radar echoes from precipitation and clouds present more of a problem than absorption, as they might obscure a desired echo. Cross-polarized returns from clouds and rain are so small that they never exceed ground echoes, so cross-polarization may be used if precipitation echoes are likely to be a problem.

Molecular absorption of radar signals is seldom a problem. Most radars operate at wavelengths longer than the 5-mm wavelength of the first oxygen-absorption line. A water-vapor-absorption line that occurs at about 1.35 cm must be avoided, but no absorption lines occur at longer wavelengths.

Echoes from precipitation are proportional, for a single drop, to

$$\text{Diameter}^6/\text{Wavelength}^4.$$

Thus, for most wavelengths, clouds give little echo, but precipitation gives somewhat stronger echoes because of the larger particle diameter of raindrops. This property of moisture-particle echoes is used in weather radar to distinguish regions of precipitation from regions of cloud.

Absorption due to liquid water droplets is also inversely proportional to wavelength^4, but is only proportional to the third power of diameter, and hence is directly proportional to mass. Thus, total absorption is set by the mass density of water droplets, regardless of their size distribution. Some examples are given in Figure 4. In heavy rain this can have a significant effect on radars in the 1-cm-wavelength region, but the effect is negligible for wavelengths beyond about 8 cm. A more extensive treatment is given by Bean and Dutton.[6]

Fundamentals of Radar Operation

A radar system consists of the five basic elements shown in Figure 5. Since most imaging radars use pulse transmissions, a brief description of operation of a pulse system follows. The cycle starts with generation

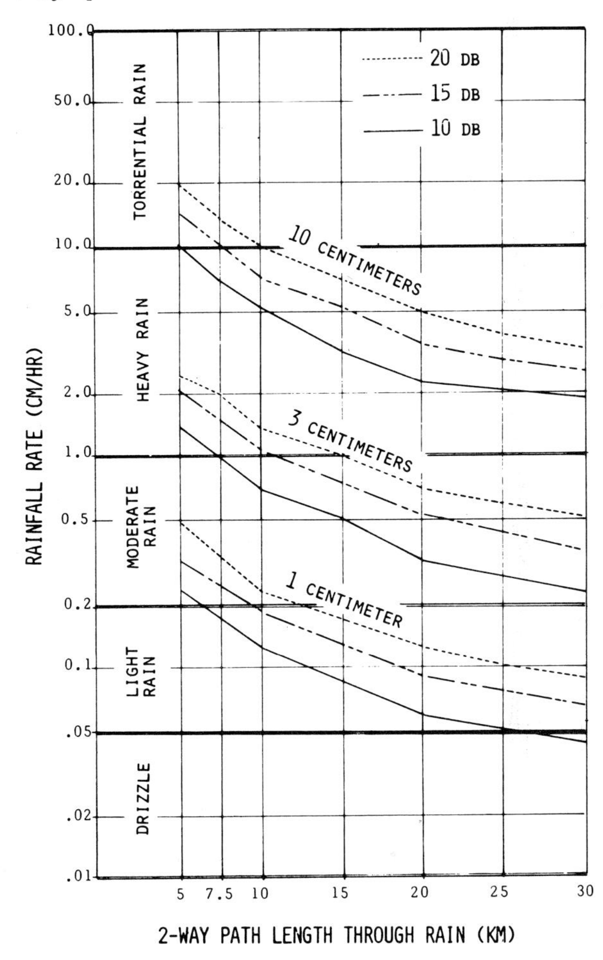

FIGURE 4 Attenuation of 1-, 3-, and 10-cm radar by rainfall. The attenuations given are very conservative, being the most extreme determined by Medhurst.[5]

in the transmitter of a short burst of sinusoidally oscillating voltage at the carrier frequency (3×10^8/wavelength in meters). This burst, or pulse, varies in length from about 0.01 to 0.1 μsec. For most radars the power during the pulse is many kilowatts. The pulse is radiated from the transmitting antenna, and the waves travel through space to the ground. Waves reflected from different parts of the ground arrive at the receiving antenna at different times, depending on the distance from the radar. The synchronizing system that had initiated the trans-

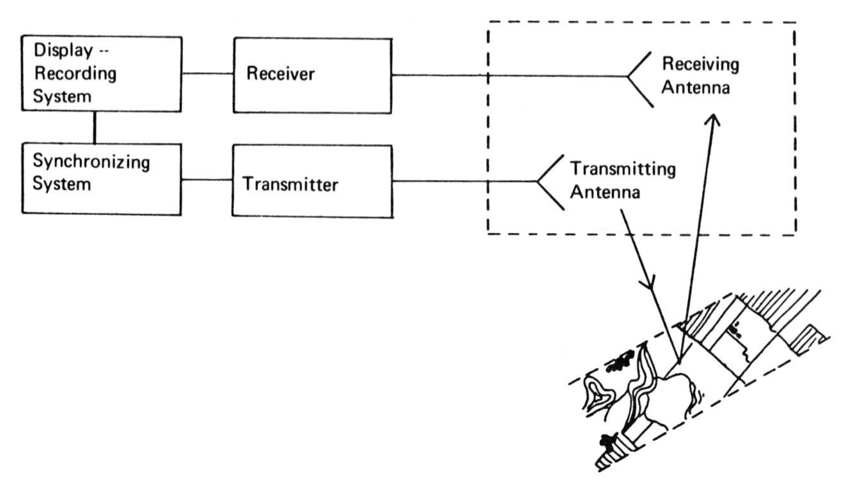

FIGURE 5 Diagram of the five basic elements of a radar system. NOTE: the antenna system may use separate or common transmitting and receiving antennas. Common-antenna system contains switch to connect antenna to transmitter and receiver alternately.

mitter pulse initiates action of the display or recording system either simultaneously with transmission or after a long enough delay that the first echoes are just arriving at the receiver. The receiver amplifies the returning signals (now at a power level often as small as 10^{-12} W) to a level suitable for operation of the display or recording systems.

Because of the many possible radar transmitting-signal designs, and the many ways one can relate transmitted and received signal, the design of a radar system is quite complex, and the problems are only touched on here. Because of these many possibilities, design compromises of many types must be made, and systems can be tailored to specific needs.

Limitations are placed on radar systems by available antenna space, by available power, by stability of the vehicle carrying the radar, by available data storage or telemetry, and by other factors. Transmitting and receiving systems themselves seldom limit performance within the wavelength regions commonly used. Thus, limitations on system performance are specified by the environment in which the radar operates, so design compromises are based on the environment. Probably the most severe restriction on radar systems is the limited space available for antennas. Accordingly, much of this discussion is devoted to techniques used to overcome this environmental limitation.

Information sensed by a radar is listed below. Numerous properties of the remote object may, of course, be inferred, but the information listed is, in fact, all that is sensed directly for a given wavelength and polarization. Because range and velocity measures are made independently of angle measurement, the radar system can have better resolution than a passive system that must determine position exclusively by angle measurements.

Signal strength Angle to object	same as passive sensors
Distance to object, by time measurement Relative velocity of object of Doppler measurement	unique to radar

Angles are measured by electromagnetic sensors by techniques depending on differences in phase shift along different paths. Lenses and antennas both depend on this principle. The precision of an angular measurement depends on the size of the aperture relative to the wavelength being used. Thus, an aperture of a given size permits better angular measurement at short wavelengths than at long ones.

Radar angular-measurement ability is set by antenna patterns. The antenna pattern shows the strength in different directions of a signal transmitted from an antenna. It also shows the ability of the antenna to receive from those directions. An ideal antenna pattern might transmit (or receive) equal signals over an angular interval about its pointing direction and zero outside this interval. Figure 6*A* shows such an ideal pattern (in one angular dimension). In practice, antenna patterns look more like that of Figure 6*B*. The angular interval over which the pattern exceeds half its maximum value is defined as the "beamwidth." The beamwidth (in radians) of an antenna (or a lens) is approximately

$$\text{beamwidth} = \frac{\text{wavelength}}{\text{antenna length}}.$$

Thus, the length of an antenna must be many wavelengths if the beam is to be narrow. In the centimeter, decimeter, and meter wavelength regions, this can be a very severe requirement; in the micron range it may not be as severe a limitation as equipment problems. For example, a 1-mrad beam can be achieved at 10 μ (infrared) with a 1-cm lens; for a 10-cm radar, a 10-m antenna is required.

The radar measures range by comparing the received signal with a

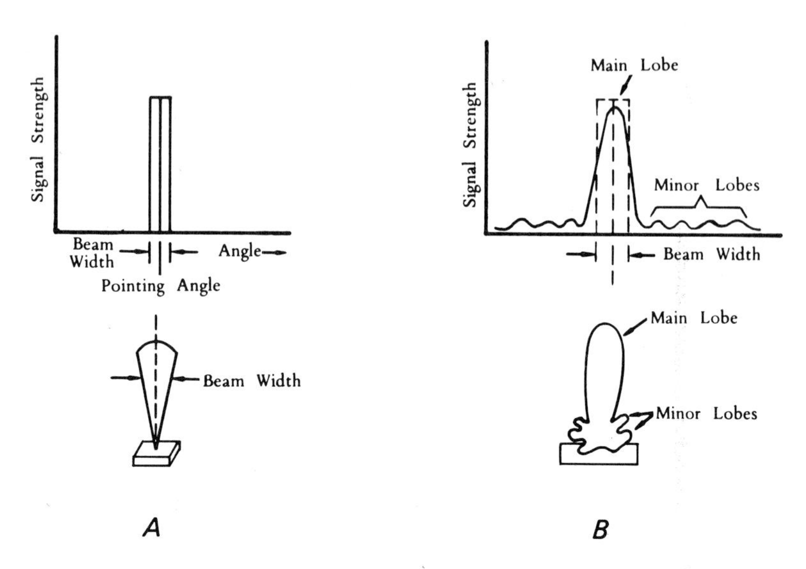

FIGURE 6 Diagrams of antenna patterns. *A:* idealized; *B:* realistic.

delayed sample of the signal transmitted earlier. The amount of delay required to obtain correlation is a measure of the range. Most radars used for imaging transmit short pulses of the microwave signal and achieve range resolution by separating the pulses returning from different parts of the ground by their differing times of arrival. Figure 7 illustrates the way a transmitted and received pulse would be displayed on an oscilloscope if the reflector were a single point. Since the velocity of the wave is known, the time t_d may be related directly to distance. Figure 8 shows the comparable oscilloscope picture for a signal returned from the ground. Here t_d is shown as the time delay for the signal from the closest point (directly under an aircraft radar); the

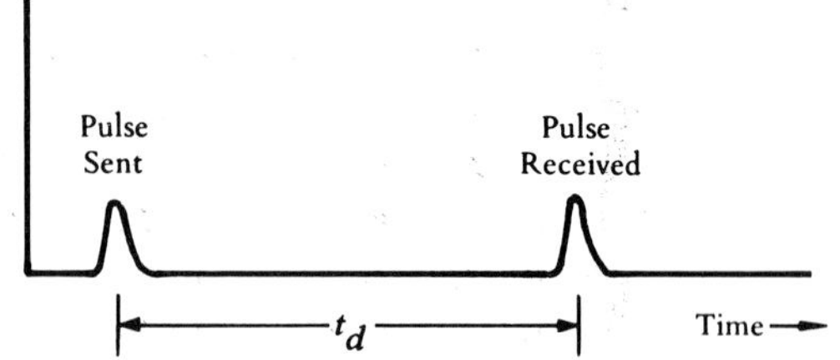

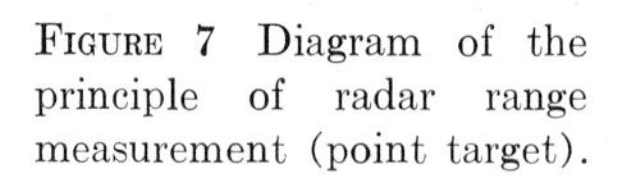
FIGURE 7 Diagram of the principle of radar range measurement (point target).

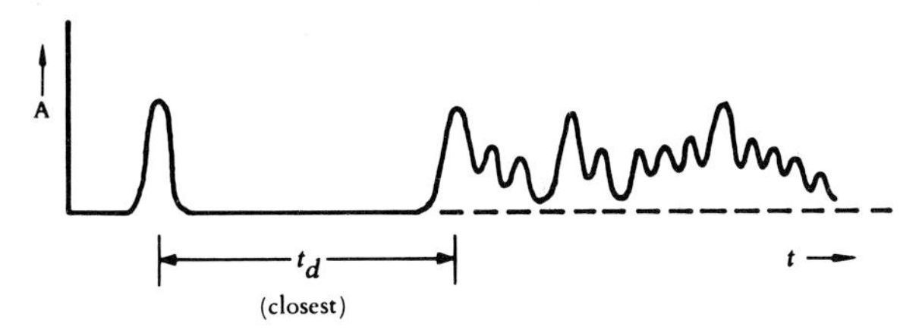

FIGURE 8 Diagram showing radar range measurement (extended target).

signals extending to larger delays come from ground areas farther away (slant range). Hence, the total return shown is for an area from beneath the aircraft to a point some distance away, and the signals received from different distant points may be identified simply by observing at the proper delay time. An image is produced by *simultaneously* looking at signals returning from many ranges and either moving the look direction by rotating an antenna or looking at new regions by moving forward while keeping an antenna fixed in a side-looking direction.

A radar measures relative speed, a quantity dependent on magnitude and direction of radar and target velocities. Multiple radars would be required to determine the velocity vector. Figure 9*A* illustrates what happens *for* a target coming toward the radar. The single frequency transmitted is received with a higher frequency. If the target were receding, the shift would be toward a lower frequency. Although only a single transmitted frequency is shown, the same shift occurs for all components of the transmitted signal.

For an airborne radar used against an area-extensive target, many Doppler frequencies are present. The reason is that the relative velocity between the radar and different points on the ground differs, depending upon the angles between the radar velocity vector and a line joining the point observed with the radar. Thus, the line spectrum indicated in Figure 9*A* for a single target becomes a distributed spectrum as in Figure 9*B* for the ground. The maximum relative velocity occurs along the flight track at the horizon. The relative velocity is zero on both sides and is negative behind the vehicle. Thus, the spectrum shown in Figure 9*B* corresponds to an antenna illuminating an area ahead of the radar since neither the zero frequency corresponding to the side nor the negative frequencies corresponding to the rear are present. This effect is used in the synthetic-aperture system for high resolution.

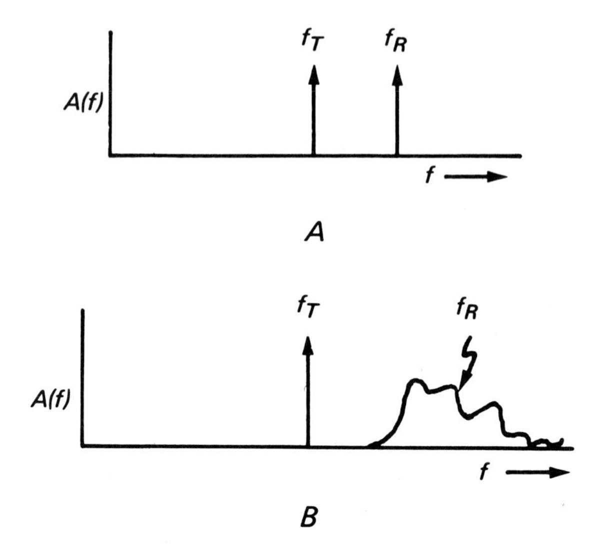

FIGURE 9 Diagrams of radar velocity measurement. *A:* single target; f = velocity; T = transmitted; R = received; $f_R - f_T$ = Doppler shift. Relative velocity toward radar makes $f_R > f_T$. *B:* Extended target. Signals simultaneously received from object with differing relative velocities toward radar.

Resolution

Figure 10 illustrates one definition of resolution. If two objects giving signals of equal intensity are sensed by a radar or other remote sensor as a single object, their spacing is less than the resolution distance. If the same two objects are sensed separately, their spacing is equal to or greater than the resolution distance. Usually, when two adjacent objects present signals of different intensity, they must be farther apart if the small one is to show up as distinct from the large one. Thus, the concept of resolution distance is not an easy one to

O O
|----D----|
Appear as One Object

O O
|-----D-----|
Appear as Two Objects

FIGURE 10 Diagram of one definition of resolution.

apply to natural surfaces for which some pairs of objects may be of equal intensity but others may be of greatly different intensity.

In imaging radar, the term "resolution" is often loosely used to mean distinguishable spot size as calculated on some arbitrary basis. The size of the spot may be determined by the half-power angular width of the antenna pattern and the time width of the pulse. This quantity is certainly related to the resolution distance and is of the same order of magnitude, but it is not necessarily equal to the resolution distance defined in the preceding paragraph.

A distinction should be made between *resolution, detectability,* and *precision.* It may not be possible to resolve two objects a hundred feet apart, yet one of the objects may be detectable even though it is only a foot across. Thus, the fact that an object is smaller than the resolution distance does not mean it goes undetected. For example, a metal fencepost may be resonant to the wavelength of the radar. If so, and if the illumination is at the correct angle, the fencepost will show up clearly, but one will not be able to distinguish between fenceposts spaced more closely than the resolution distance.

With an infinite signal-to-noise ratio, range can, in theory, be measured to any desired degree of precision, regardless of bandwidth and regardless of resolution. Thus, a radar with a resolution distance of 100 ft could, in theory at least, measure the range from the radar to the target with a precision of 1 ft.

Passive sensors must achieve all their resolution by angular measurements. Radar can improve on this resolution by using its range- and velocity-measurement capabilities. Figure 11 illustrates this. A passive system, or a radar system using continuous transmission with no modulation and no relative velocity, depends on angular resolution set by the antenna beam. If only angular resolution were used, the radar beam would have to be scanned as with the beam of a microwave radiometer or optical scanner. Since the radar can discriminate in range by other means, scanning is unnecessary. Figure 11*A* illustrates this point for a pulsed SLAR. Instead of a beam narrow in both dimensions, as for a radiometer, the beam is narrow in the direction along the flight path, but wide perpendicular to the flight path. Such a "fan beam" is achieved by an antenna long in the flight direction but short in the vertical direction. The antenna is fixed in position on the aircraft, and the effect of scanning in range is achieved by the different delay times for the signals returning from different ranges.

Where the required beamwidth calls for an antenna that is too large to be carried conveniently, its effective length is increased by the

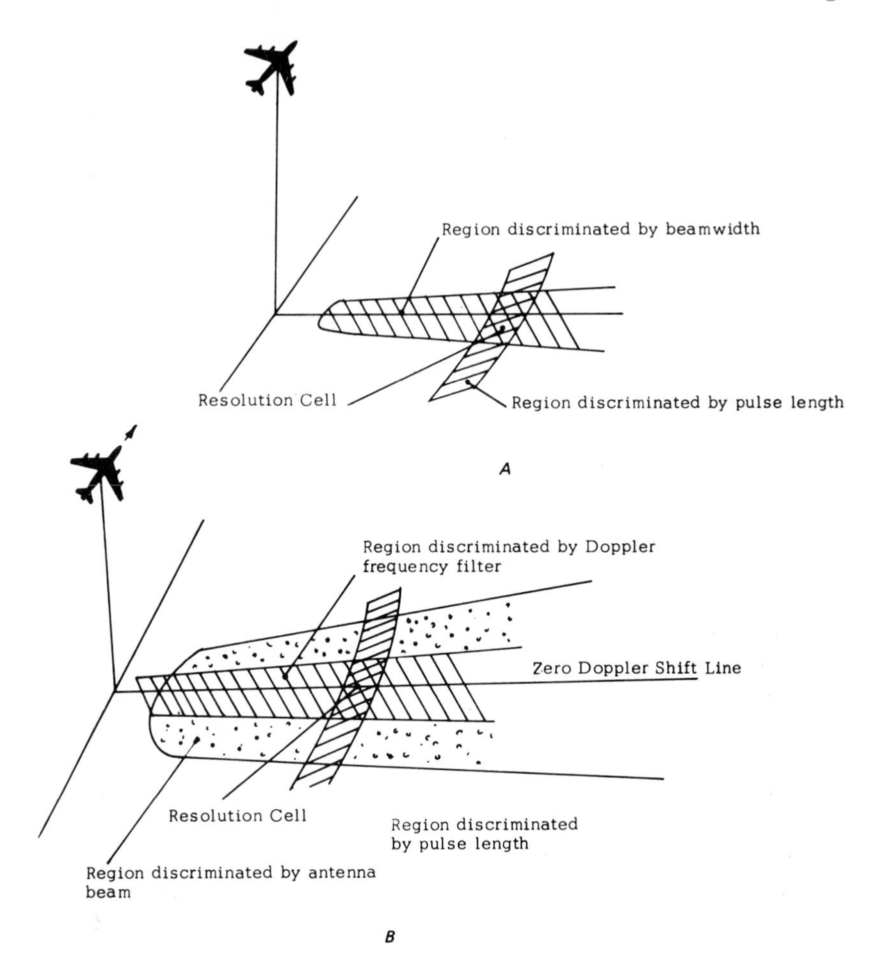

FIGURE 11 Diagrams of radar resolution. *A:* Real-aperture system; *B:* Synthetic-aperture system.

"synthetic-aperture" technique illustrated in Figure 11*B*. Here the resolution cell is not established by the antenna at all. Instead, range resolution is determined by pulse length, and azimuth resolution (along the flight path) is determined by a measurement of Doppler shift. If the radar looks directly to the side, the center of the resolved area is on the zero relative-velocity line having zero Doppler shift. By filtering so that only those signals experiencing a small positive and a small negative Doppler shift are detected, the resolution indicated can be achieved.

The term "synthetic aperture" arises from a different way of looking at the technique.[7] The operation involved in Doppler filtering may be thought of in different terms, and, indeed, it can be performed by an entirely different approach. Thus, if one stores the signals returned from a large number of pulses in some sort of memory, being careful to retain phase as well as amplitude information, he can process the signals in the memory in the same way that a large antenna processes them. For example, all the pulses transmitted during 300 m of flight may be combined to give a resolution cell of the same size as that achievable with an antenna 600 m long. Thus, although the real antenna might be only 1 or 2 m long, the synthetic antenna is 600 m long.

For remote-sensing uses, antennas and lenses with small apertures are ordinarily focused at infinity. If, however, the resolution desired is as small as or smaller than the aperture, the antenna or lens must be focused for the distance to the region imaged. Many times synthetic apertures are much longer than the desired resolution, so they must be focused. This can be achieved in the processing, although the simple Doppler filter associated with Figure 11*B* is insufficient to do the focusing. Focused synthetic-aperture radars can, in theory, achieve resolutions equal to half the length of the physical antenna used; unfocused systems are simpler but achieve resolutions only down to half the square root of wavelength times range.

Radar Presentations

Many types of presentations have been developed for radar systems. Some of these are particularly appropriate to special purposes such as airport landing systems or mapping radars. The highest resolution airborne systems use a fixed side-looking antenna or a synthetic aperture (equivalent to a longer side-looking physical antenna).

The presentation commonly used for these, at least for recording and later viewing, involves intensity modulation of a swept CRT beam by the signal. This is illustrated in Figure 12. The beam is swept in one direction only, so it always appears in the same line on the tube; thus, a single line is scanned rather than an area, as with television. The other dimension in the map is achieved by moving a film past this line in synchronism with the motion of the vehicle. Thus, the first sweep appears as a line on the film. When the second sweep comes along, the film has advanced so the second sweep appears as a different parallel line on the film. In this way a maplike image is produced. Similar techniques are used to produce images with infrared scanners.

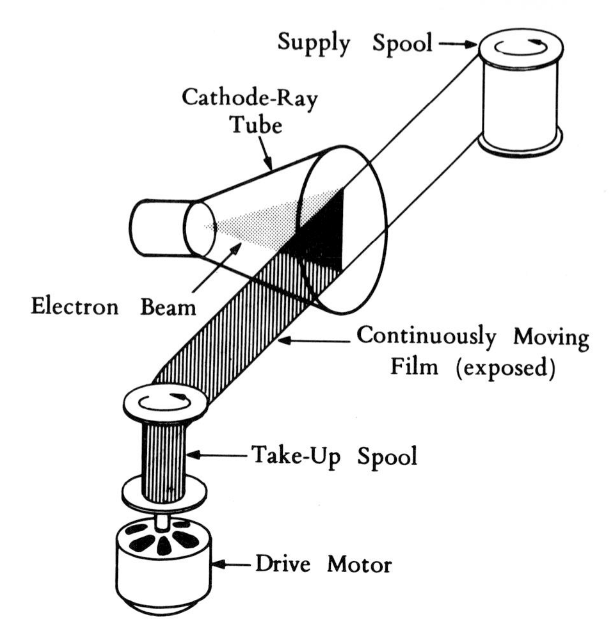

FIGURE 12 Diagram of recording technique for side-looking airborne radar. Electron beam scans vertically with position proportional to distance from flight track.

Radar System Choices

Most of the radar system-design choices of interest to the user relate to resolution. This in turn relates to the space available on the vehicle for an antenna, to the range to the farthest target, to the choice of wavelength, and to the complexity permitted by cost and reliability constraints. A side-looking radar with a "physical aperture" (no synthetic-aperture techniques used) is intrinsically much simpler than a synthetic-aperture system. An unfocused synthetic-aperture system is simpler than a focused system.

Wavelength choices depend on the application planned, although it is clear that multispectral systems are preferable to single-wavelength systems. Physical-aperture systems must operate at the shortest possible wavelengths to achieve good resolution, but as a result they pay the penalty of susceptibility to attenuation by and echoes from atmospheric liquid moisture. The optimum wavelengths for different agricultural applications have not been determined, but it is fairly clear that wavelengths should be in the centimeter region for crop distinction, whereas it appears the meter wavelengths may be superior for soil studies. Longer wavelengths (decimeter to meter) appear desirable for study of stream patterns and similar patterns under forests because they penetrate the tree cover better. Presumably, some intermediate wavelength may be shown to be best for timber studies.

Figures 13 and 14 have been prepared to relate these choices to the above considerations. Resolutions with two different antenna lengths (3 and 10 m) have been plotted for each of three types as systems versus wavelength and versus maximum slant range. Theoretical values have been used for these diagrams; one should not necessarily infer from this that these resolutions are actually attainable in the present state of the art. Furthermore, although the curves for focused synthetic aperture indicate, like the theory, that resolution is independent of range, the length of the synthetic aperture required is proportional to range, so vehicle-stability problems and equipment-stability problems are obviously more severe at the longer ranges. Thus, realistic focused synthetic-aperture curves would show degradation in resolution, especially at longer ranges.

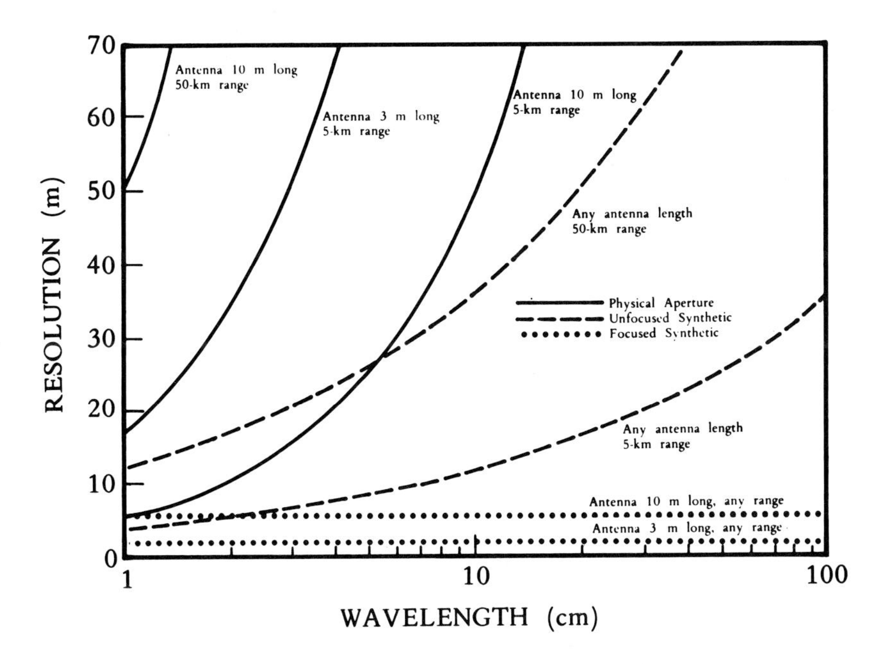

FIGURE 13 Design trade-off information for side-looking radars.

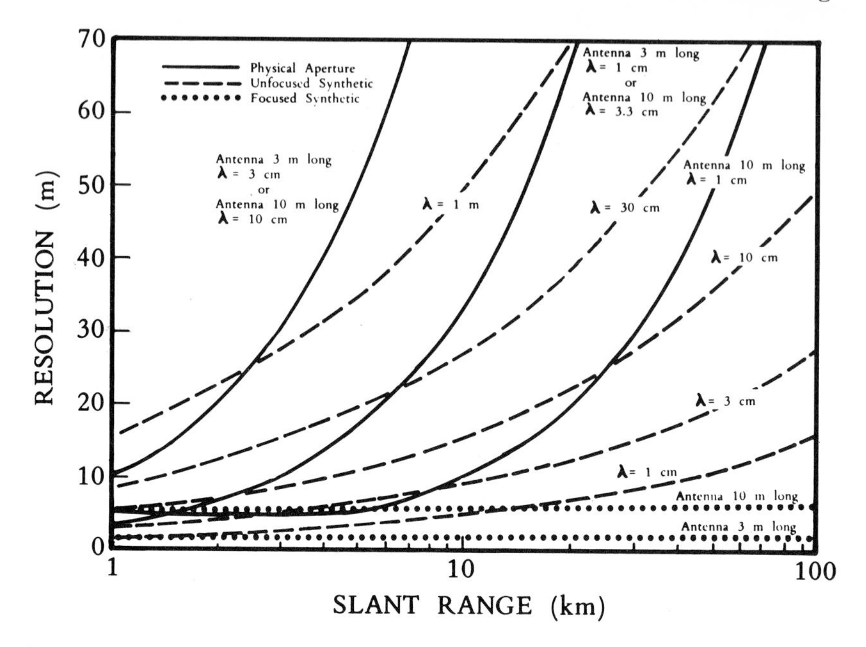

FIGURE 14 Design trade-off information for side-looking radars.

For many agricultural purposes, resolutions of 15–30 m should be adequate. Clearly, these cannot be achieved with any physical-aperture system with a wavelength of more than 1 cm at a 50-km range, and the aperture required at that range would be so large that it would be difficult to carry the antenna on existing aircraft. Hence, if a swath this wide is desired, some sort of synthetic-aperture system is called for unless a considerable relaxation in resolution is possible. A 3-cm wavelength unfocused system should be adequate to meet this resolution-swath requirement, but at 10 cm, a focused system is required. On the other hand, if a 5-km total slant range is sufficient (probably a 4-km swath at about 1-km altitude), a physical aperture 3 m long would permit a 1-cm system, and an aperture 10 m long would permit a 3.3-cm system. An unfocused synthetic system could meet this requirement at 30-cm wavelength. For longer wavelengths, a focused synthetic-aperture system is required.

Many other examples could be developed from these curves, and the curves themselves could be expanded. Relations used in plotting the curves are[7]:

$$\text{Physical-aperture system: } r = \lambda R/l$$
$$\text{Unfocused synthetic-aperture system: } r = \sqrt{\lambda R/2}$$
$$\text{Focused synthetic-aperture system: } r = l/2$$

where r = resolution distance
R = slant range
l = antenna physical length
λ = wavelength

To achieve reasonable resolution from space, a synthetic-aperture system is obviously required. Since, in theory at least, synthetic-aperture focused systems have resolution independent of range, the resolutions indicated might be achieved. A system proposed to NASA in 1965 would have 15-m resolution at wavelengths of 60, 15, and 3.75 cm from orbital altitudes. For various complex reasons, the swath width of such systems is limited. The system proposed would have a swath width of 40 km, based on an antenna 8 m long. A shorter antenna could theoretically permit better resolution, but it would also reduce the permissible swath width. Techniques have been devised for overcoming this swath-width limitation, but they are costly in power and in complexity. The vast quantity of information that can be gleaned from a spaceborne system sweeping out an area 8 $\times$ 40 km/sec makes it appear that the additional complexity is not warranted for some time, as analysis of images obtained with the 40-km width will be a staggering task.

The Unique Role of Radar in Agriculture

Radar has certain characteristics important for agricultural surveys. Some of these are due to the nature of the radar system (usually an active device that operates at single frequency employing a coherent radiation technique), and others are due to the spectral region commonly used.

VALUES INHERENT IN RADAR SYSTEMS

1. Images are produced under the same conditions day and night, regardless of temperature history. This is true for soils and variations in illumination conditions. However, plant leaves do exhibit changes in configuration with variations in temperature, illumination, and wind.

The influences of these adaptive plant changes on radar returns have not yet been explored.

2. Wide areas may be imaged simultaneously with essentially uniform resolution.

3. Polarization of illumination may be controlled.

VALUES DUE TO USE OF RADAR SYSTEM IN CENTIMETER–METER WAVELENGTHS

4. Images may be produced in any weather.

5. Crop cover may be penetrated to give soil response (longer wavelengths).

6. Tree cover (orchard and forest) may be penetrated to show underlying cultivation, drainage, and geologic patterns (longer wavelengths).

7. Additional multispectral discriminations between crops, crop states, soils, etc., may be added to those in other wavelengths.

For quick surveys with aircraft, item 2 of the list above is significant, since the useful swath width from aircraft using radar is greater than for other sensors. If the sensors are carried in spacecraft, however, the swath-width advantage of high-resolution radar is lost. The other advantages remain.

Radars operating in the decimeter (and perhaps meter) range give a composite image in which the soil, or other surface, plays a large role, for the waves are attenuated only slightly in passing nearly vertically through vegetation. With the shorter radar wavelengths, the plants themselves provide more attenuation, so the signal is primarily due to vegetative cover where there is any. Thus, combinations of radars operating at short and long wavelengths should provide information on both soil and crop characteristics.

Item 7 in the list above is, of course, quite significant in multispectral analysis to permit identification of terrain properties strictly on the basis of remote sensing, for each portion of the wavelength spectrum (from ultraviolet to long waves) is sensitive to different properties of the surface.

Examples of Radar Imagery

Radar systems presently available for high-resolution imaging have all been developed for military applications. Accordingly, examples are unavailable for many of the more modern systems, especially the synthetic-aperture systems. The images presented here are indicative of the type available in the spring of 1968 for general use by the scientific community.

Using synthetic-aperture techniques, images of the types shown should be feasible with systems carried in orbit. An orbital test has not been conducted, but all calculations indicate that an orbital system with these capabilities is quite feasible.

Several examples of radar imagery obtained in agricultural areas are given here, and others may be found in Chapters 4 and 6. The examples here are taken at wavelengths of 3 cm and shorter, using both real-aperture and synthetic-aperture radars.* Details of the radar system characteristics are given (where available) in the discussion appropriate to each of the accompanying illustrations.

Figure 15 is an image of the outskirts of Indio, California, taken on November 4, 1965, with the AN/APQ-97, a *K*-band, real-aperture radar. The area is at the upper end of the Imperial Valley, north of the Salton Sea. A wide variety of irrigated vegetable crops is grown throughout the year. A narrow line of radar shadow along one edge of some bright fields suggests orchards. Sugar beets are also grown here. For an example of *K*-band imagery with two different polarizations coded with different colors, see Plate 11.

Figure 16 is an image of farmland and urban settlements south of Detroit, Michigan, taken with the AN/UPD-1 (XH-1) in the spring of 1960. This is an unfocused, synthetic-aperture, *X*-band system. Note the variations in gray scale of the different agricultural fields and the way in which trees and hedgerows sharply delineate the field boundaries. Some lake ice is still visible north of Stony Point.

*Because of technical and economic considerations, radar development has tended to cluster at and around a few frequencies. These bands are designated by letters:

Band	Nominal Ranges (mc)	Wavelength (cm)
P	220–390	133–77
L	390–1,550	77–19
S	1,550–5,200	19–5.8
C	3,900–6,200	7.7–4.8
X	5,200–10,900	5.8–2.7
K	10,900–36,000	2.7–0.83

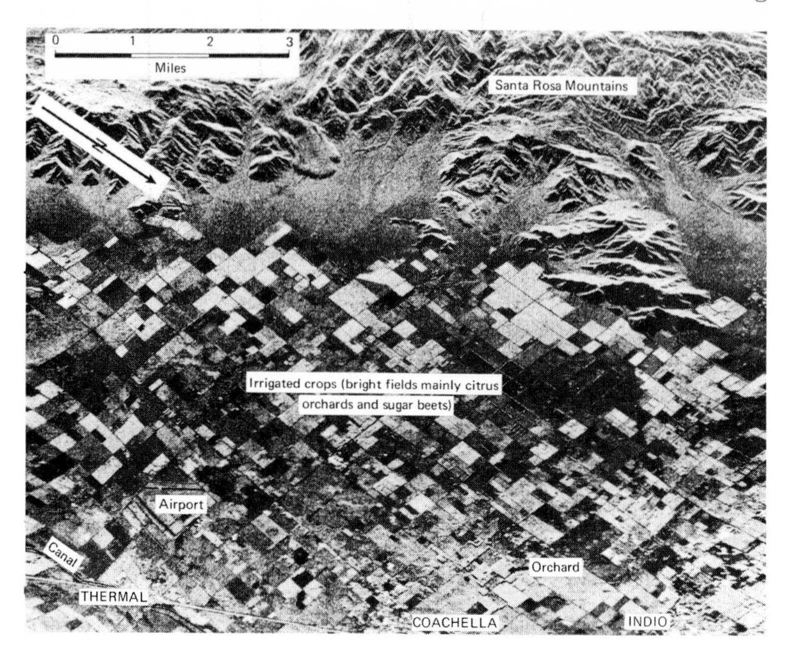

FIGURE 15 Radar imagery taken on the outskirts of Indio, California.

Figure 17 is included to give an indication of the broad coverage possible with a radar set. It is an image of the Grand Lake–White Lake region of Louisiana, taken with the AN/APQ-69 radar. This is a real-aperture X-band radar system capable of covering up to 50 nmi on each side of the aircraft from altitudes up to 70,000 ft. The presentation is a true ground-range sweep. Resolution in range is about 50 ft everywhere on the image, but the resolution and azimuth are 688 ft at 50 nmi, 344 ft at 25 nmi, and 140 ft at 10 nmi. A synthetic-aperture system with the same swath width would retain more-or-less constant azimuth resolution at all distances. The area north and east of White Lake in Vermilion Parish, Louisiana, is dominated by cultivated land mainly in rice, corn, sugarcane, and cotton. Variations in gray scales of the cultivated fields reflect the difference in crop type and crop growth stage and indicate that this X-band system has a sensitivity to difference in crop type and stage similar to that for the K-band systems.

Figure 18 is an AN/APQ-97 K-band image of the Baldwin area in eastern Kansas, obtained September 15, 1965. This is an area of mixed agriculture. Pasture and trees dominate the uplands. Hedgerows and differences in gray scale help to delineate fields. Trees, giving a raised, coarse-textured appearance, line most of the watercourses. Corn, wheat, alfalfa, soybeans, and sorghum are common crops. The lightest toned fields are usually corn; the very dark fields are bare or in stubble. Intermediate tones are usually soybeans, sorghum, or alfalfa. Many small water bodies, such as the common farm ponds, can be seen on the image.

Passive-Microwave Imaging

The preceding section is a discussion of radar sensors that operate at wavelengths longer than 0.5 mm, use an artificial electromagnetic source to illuminate the ground scene, and then make use of echoes for detection. Passive-microwave sensors operate in the same spectral region as very-short-wavelength radar, 0.1 mm–3 cm, but they do not

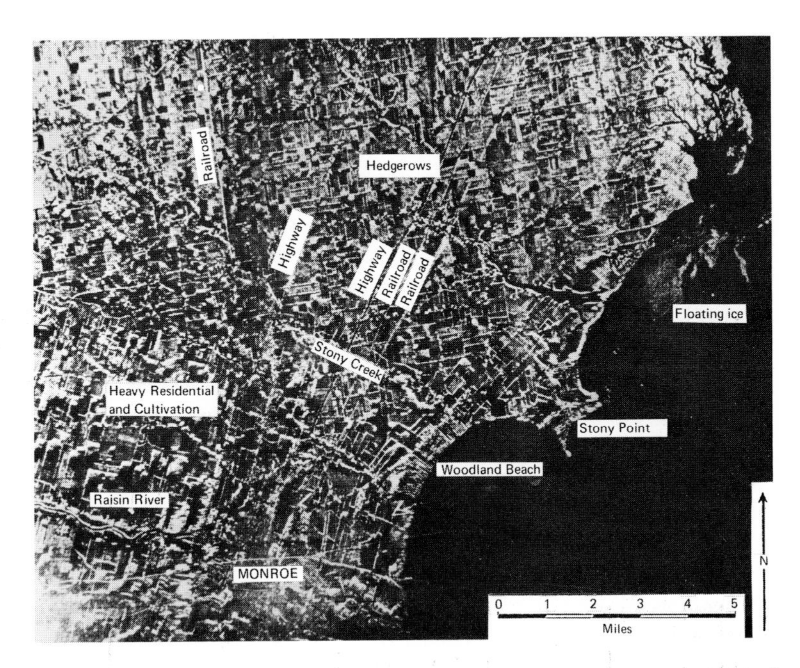

FIGURE 16 Radar imagery of farmland and urban settlements south of Detroit, Michigan.

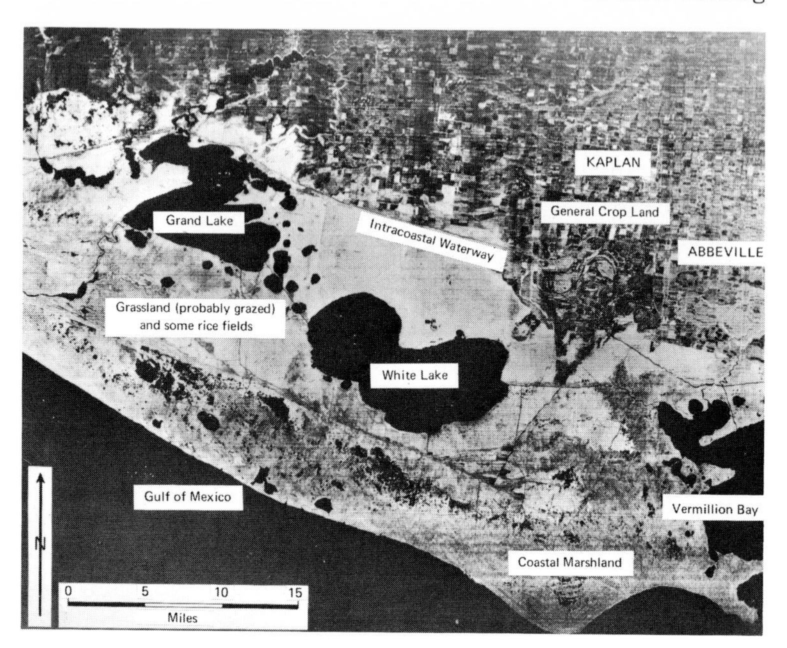

FIGURE 17 Example of broad coverage possible with radar imagery. Grand Lake–White Lake region, Louisiana.

illuminate the scene artificially; instead, they make use of the natural radiation emanating from observed objects. The important properties of objects that determine the character of this radiation are emittance, transmittance, reflectance, and object temperature. The character of external natural radiation sources that contribute to the reflected and transmitted radiation are also important. These components are shown schematically in Figure 19. Treated in detail in the subsequent sections are the basic radiation principles and the instrumentation used in these sensors. Also discussed is the general applicability of passive microwaves to remote sensing for agriculture.

Basic Radiation Principles

Planck's fundamental radiation equation[8] gives the spectral distribution of radiation flux from a perfect radiator (called a blackbody) with a uniform temperature T_B. This theoretical equation of Planck agrees with experimental results and applies to all regions of the electromag-

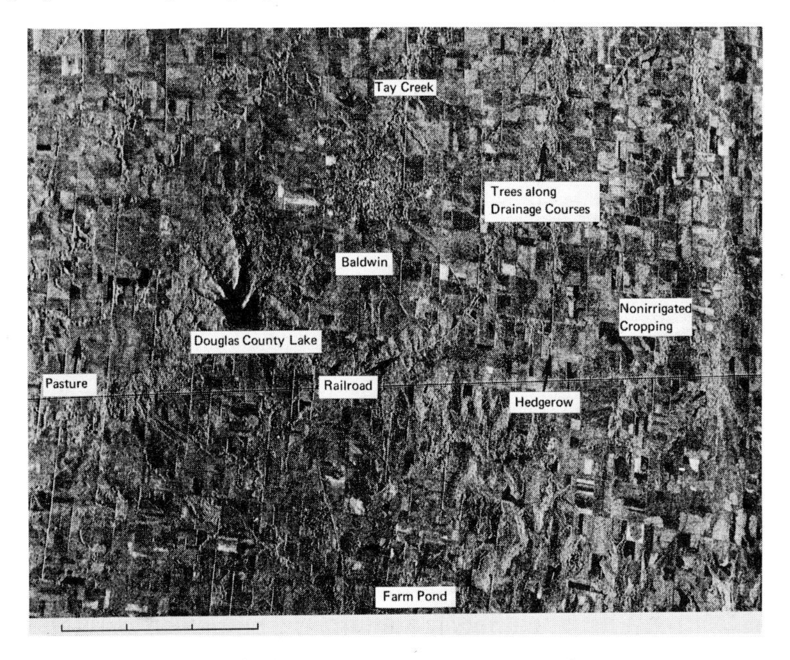

FIGURE 18 *K*-band imagery of Baldwin area, eastern Kansas.

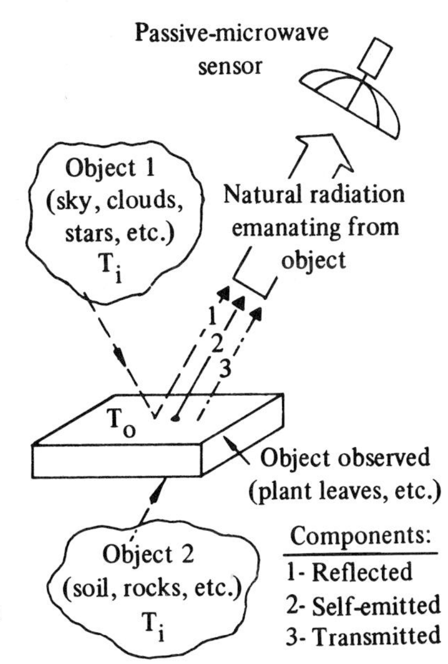

FIGURE 19 Natural microwave radiation emanating from an observed object.

netic spectrum. It states that the electromagnetic radiation given off by an object is a function of the object's absolute temperature and the wavelength (λ) of observation. In the microwave region, if we use the Rayleigh-Jeans* approximation to Planck's law,† the emitted radiant power, $W_{B\lambda}$, from a blackbody is

$$W_{B\lambda} = T_{B\lambda}{}^{-4}.$$

For real objects that are not perfect radiators, a radiation efficiency factor called spectral emittance, ϵ_λ, is used to describe the radiation efficiency of an object as a function of wavelength. This efficiency factor is defined as the ratio of observed radiant power, $W_{o\lambda}$, from an object to the radiant power from a blackbody at the same temperature and at the same wavelength:

$$\epsilon_\lambda = \frac{W_{o\lambda}}{W_{B\lambda}}, \qquad (T_o = T_B).$$

Effective target‡ temperature (or brightness temperature), T_T, a commonly used term, evolves from these basic radiation concepts. The radiation emanating from an object is, in general, made up of three parts: a self-emitted component, a reflected component, and a transmitted component. The self-emitted component, for instance, is proportional to the object's spectral emittance ϵ_λ and its temperature T_o. Thus, an effective temperature (T_{Te}) due to self-emission can be defined as the product of the object temperature and its emittance; i.e.,

$$T_{Te} = \epsilon_\lambda T_o. \tag{1}$$

Similarly, effective temperature contributions (T_{Tr} and $T_{T\tau}$) due to the object's spectral reflectance (ρ_λ) and transmittance (τ_λ) can be written in the forms

$$T_{Tr} = \rho_\lambda T_i \tag{2}$$

and

$$T_{T\tau} = \tau_\lambda T_i', \tag{3}$$

*These approximations, as well as a general discussion of the Planck law, can be found in most basic textbooks on modern physics.

†For a statement of Planck's law without approximation, see p. 130.

‡The word "target" is used here as a synonym for "object," and the identifying symbol is therefore the subscript T.

where T_i and T_i' are terms proportional to the radiation incident on the object. The effective target temperature T_T is thus the sum of these three terms; i.e.,

$$T_T = \epsilon_\lambda T_o + \rho_\lambda T_i + \tau_\lambda T_i'. \quad (4)$$

This is the temperature of the object as one would measure it with a remote microwave sensor having an ideal antenna. It would be the same as the object's actual temperature only if the object were a blackbody, in which case $\rho_\lambda = \tau_\lambda = 0$, and $\epsilon_\lambda = 1$. Thus for real objects, the remotely observed radiation intensity is dependent not only on the object temperature and the incident radiation but also on several other properties of the object. The emittance, reflectance, and transmittance are, in general, functions of the object material's absorption coefficient,[9,10] its bulk configuration or shape, the aspect at which the object is viewed,[11] and the surface structure.

Object-Detection Parameters

Three basic parameters can be used for identification purposes: the polarization, and the temporal and spectral characteristics of the object's radiation. Each of them is discussed below (see also Chapter 9).

From basic electromagnetic theory it is evident that the reflection properties of most surfaces are polarization-sensitive (polarization is defined on page 85).

Examples of surfaces that exhibit pronounced polarization effects at microwave frequencies are water, concrete, asphalt, and ice. Results of theoretical computations[12] of water reflectance and emittance as a function of angle (using Maxwell's field equations) are shown in Figures 20 and 21. From these values and a theoretical value for the angular dependence of the sky effective temperature T_i, the effective temperature of water $T_{T(H_2O)}$ can be computed as a function of angle:

$$T_{T(H_2O)} = \epsilon T_o + \rho T_i. \quad (5)$$

The surface emittance shown in Figure 21 is obtained by the relationship $\epsilon_\lambda = 1 - \rho_\lambda$, which is a form of Kirchhoff's law. It means that, for opaque objects where the transmittance is zero, the fraction of the incident radiation that is absorbed must equal 1 minus the fraction reflected; and for a body in thermal equilibrium with its surroundings and transferring power only by radiation, the power emitted must equal the power absorbed.

The results of computations of effective temperature for horizontal and vertical polarizations are shown in Figure 22. These calculations assume that the surface is perfectly smooth, and it should be kept in mind that surface roughness will affect the polarization characteristics of radiation from seawater. The use of this roughness relationship for monitoring sea state has been described in a recent publication.[13] The term "apparent" temperature is used in Figure 23. It applies to that temperature determined from the remote measurement. Apparent temperature is usually different from the effective target temperature T_T because the receiver's field of view is usually much larger than the size of the object. Also, the side lobes of the antenna response allow other radiation to "leak" through (see Figure 6). Perfectly planar surfaces (which are mirrorlike), such as the ones used in the calculations for Figures 20–23, are called specular reflectors. Surfaces that scatter radiation in every direction are said to be diffuse. Grass, bushes, and crops tend to be diffuse, and their polarization effects are not nearly so pronounced (Figures 23 and 24), although a great deal more experimental work is required to understand their radiation characteristics more fully.

Some objects exhibit a time-dependent (temporal) change in their effective temperatures. These changes can be seasonal or diurnal and can result from fluctuations in the actual object temperature, changes

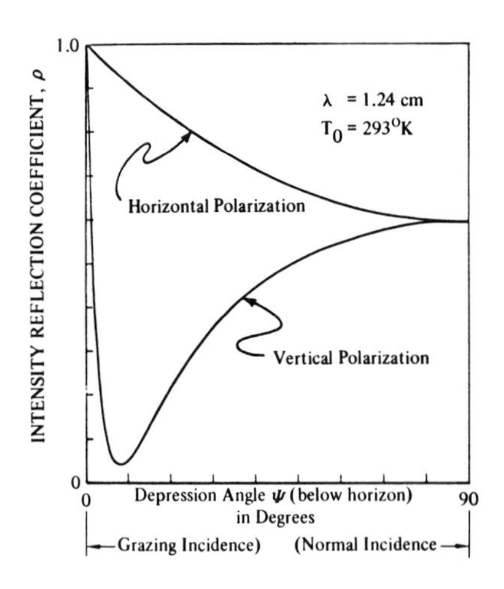

FIGURE 20 Calculated intensity reflection coefficients for a water surface.[12]

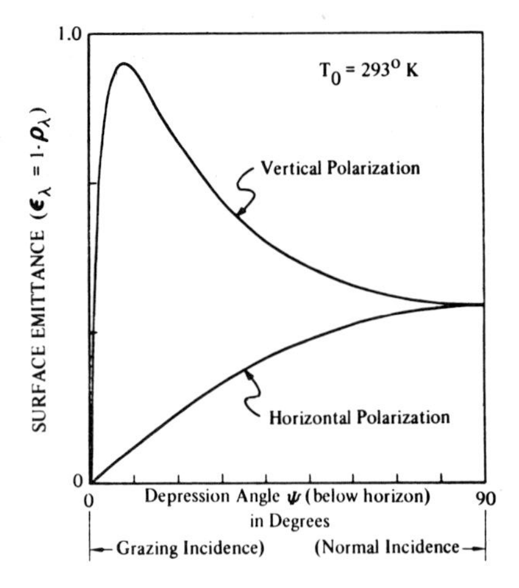

FIGURE 21 Calculated emittance for a water surface.[12]

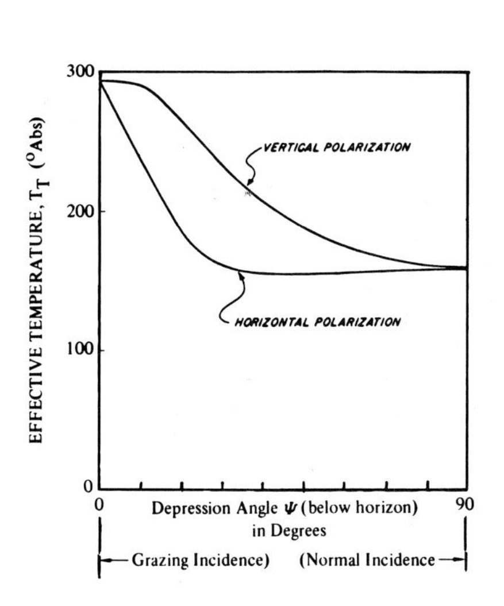

FIGURE 22 Calculated radiation from a water surface, including the reflected contribution from the sky.[10]

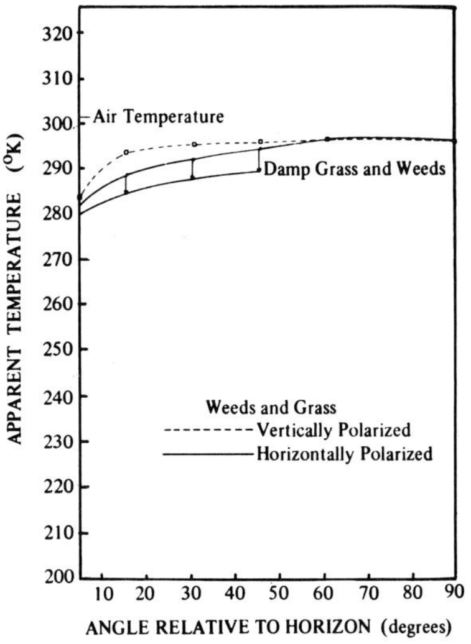

FIGURE 23 Temperature of grass and weeds.[14]

in sky temperature, or changes in the reflection and emittance properties of the object. Shown in Figure 25 are typical thermodynamic temperature ranges for various objects. Thermodynamic temperatures vary over an interval of 40° C for some objects. Comparable changes in effective target temperature will occur only for those objects that are highly emissive (see Equation 5).

The reflection characteristics of vegetation can change with seasons because of geometry, shape, and changes in terrain-covering ability during its maturation. In Figure 26, the normalized reflecting area* for wheat is compared at two times during its maturation. A maximum difference in echo occurs at an elevation angle of 10°; however, for the data shown, the differences at useful observation angles are much smaller. Unfortunately, a complete study of the temporal dependence of crop echoes on wavelength, polarization, and observation angles has not been made. Data of this kind are needed to make a judicious

*In microwave work, reflectivity is frequently expressed in an alternative but equivalent manner as the cross-sectional area of a perfectly reflecting sphere that would reflect the same power as the object being measured.

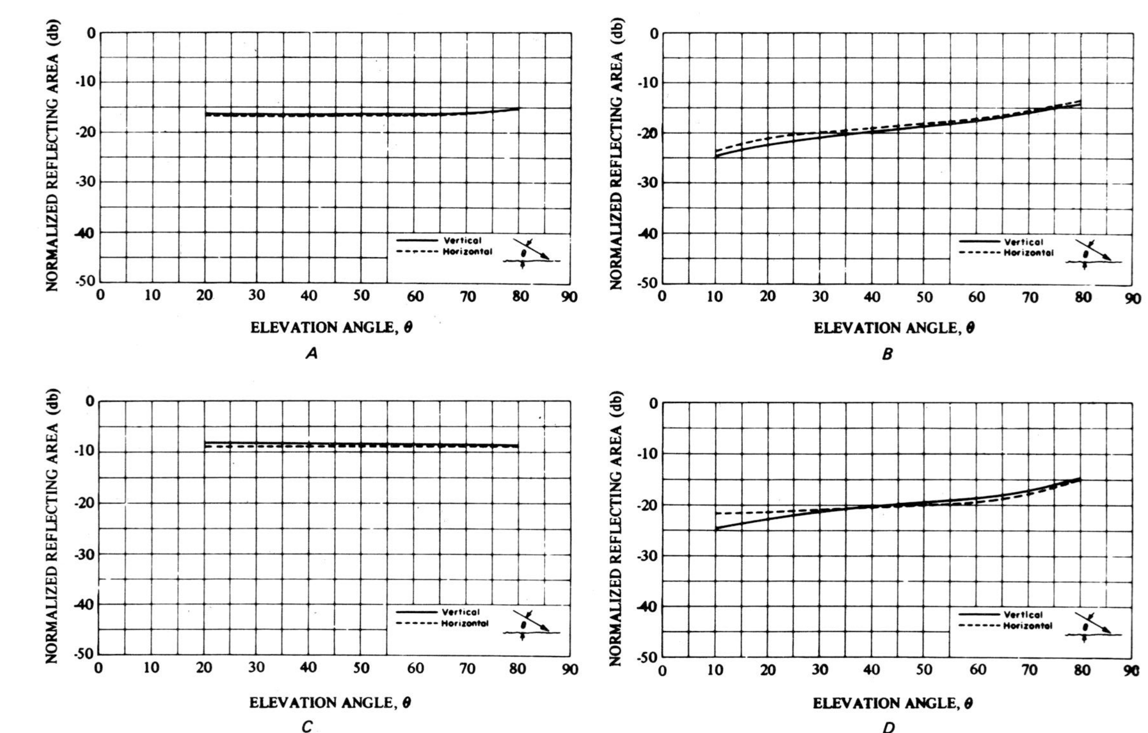

FIGURE 24 Normalized reflecting area of foliage. *Adapted from reference 15.* *A*: 8-in. grass flattened to a height of 3 in. at *K*-band (November). *B*: Soybeans at *K*-band (September). *C*: 6-in. alfalfa and grass at *X*-band (April). *D*: 2-in. wheat at *X*-band (April).

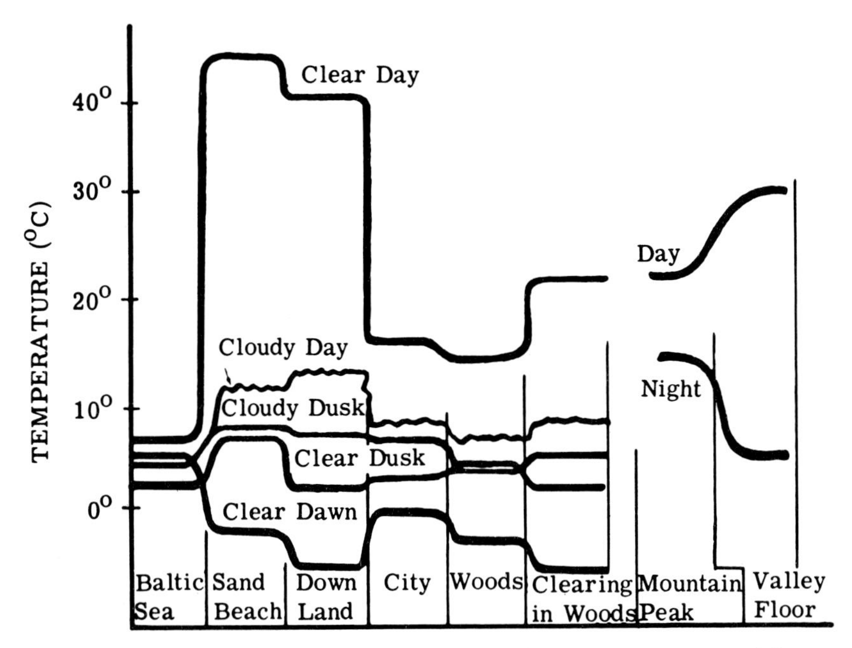

FIGURE 25 Typical gross temperature differences for various objects at different times of day.[16]

selection of the instrument parameters for obtaining maximum effects.

The spectral characteristics of an object are among the most useful parameters and have been exploited extensively in optical and infrared portions of the electromagnetic spectrum. This has not been the case in the microwave portion of the spectrum because of spectral band-pass constraints placed on conventional instrumentation. Some of the object data available show spectral characteristics (see Figures 27 and 28). With the advent of certain new hybrid instrumentation that incorporates bolometer and Putley-type detectors, thus combining[17,18,19] the advantages of both microwave and infrared measurement techniques, submillimeter spectral data are becoming available. The "target" of most studies to date has been the atmosphere.[18,20] As more submillimeter data become available, the feasibility of multicolor and other discrimination schemes currently applied at optical and infrared wavelengths can be evaluated for use at submillimeter wavelengths.

The spectral character of some materials can be expected to change because of the penetration capabilities of microwaves. For instance,

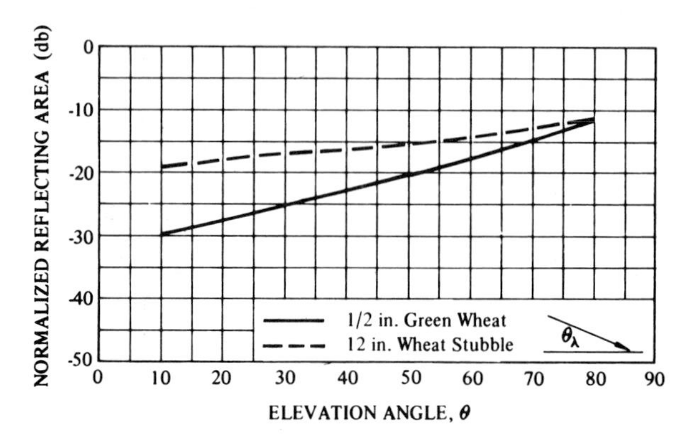

Figure 26 Seasonal variation of wheat, normalized reflecting area. *Adapted from reference 15.*

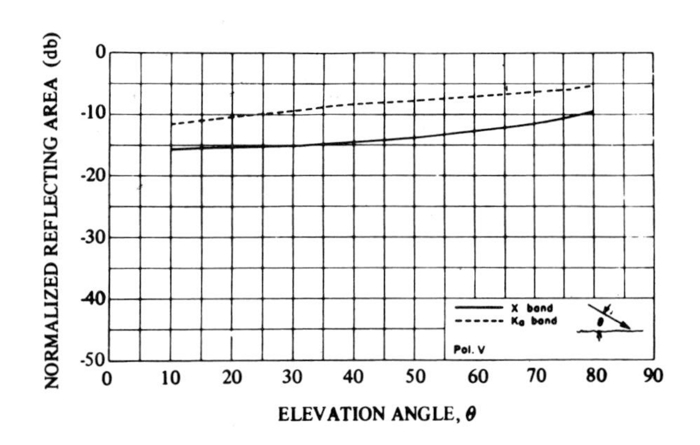

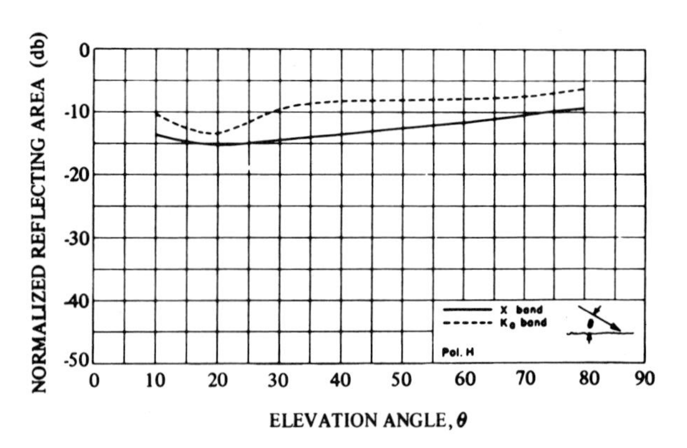

Figure 27 Normalized reflecting area of 4-in. wet soybean stubble. *Adapted from reference 15.*

the forest canopy, which is opaque at short wavelengths, becomes quite transparent at longer wavelengths. The depth of penetration in soils and snow is also a function of frequency.

Atmospheric Effects

Two parameters related to detection are affected by the properties of the atmosphere. First, radiation from the atmosphere itself is a source of illumination for all ground objects. Second, transmission of the atmosphere limits the ability to detect ground objects. In the discussion of radar imaging (page 80), it is noted that throughout all but the very-short-wavelength end of the region, molecular absorption by

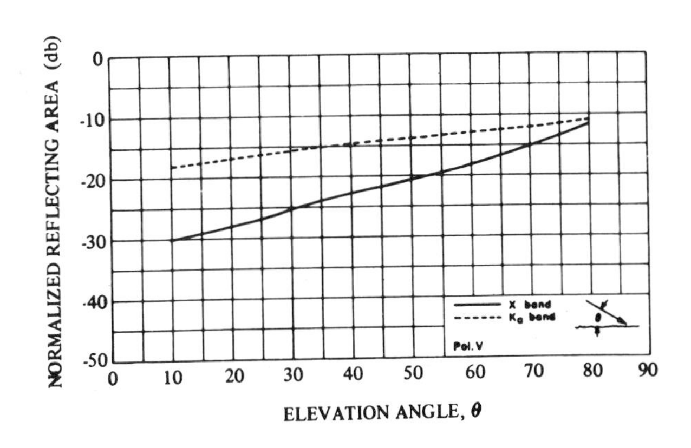

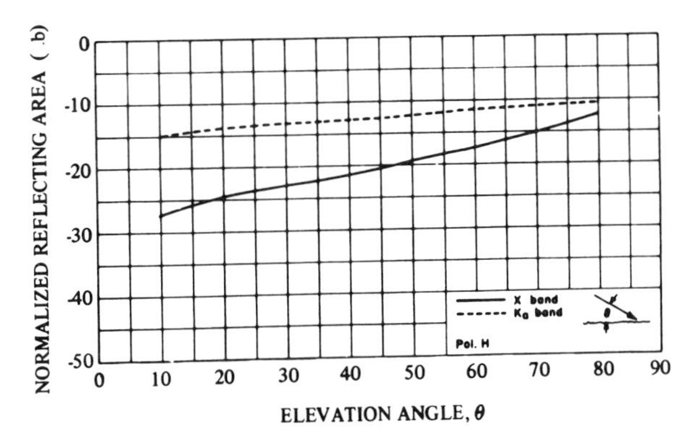

FIGURE 28 Normalized reflecting area of 0.5-in. green wheat. *Adapted from reference 15.*

atmospheric constituents does not play a significant role. Here in the passive-microwave region, molecular absorption begins to be significant and, as we will see later, becomes very significant at infrared wavelengths. Several "windows" in the microwave region are reasonably transparent during moderately poor to severe weather conditions. Figure 29 shows at a glance the locations of these windows. The opaque regions are due to water vapor and atmospheric-oxygen-absorption bands. At those wavelengths where the atmospheric transmittance is low, the atmosphere appears hot* (close to ambient;[18] i.e., the atmosphere has a high emittance at these frequencies and thus is a good emitter).

This correlation can readily be seen by comparing Figure 29 with Figure 30, which shows the atmospheric effective temperature due to clouds. It is therefore apparent that two reasons exist for operating in the windows. One is to minimize transmission losses, and the other is to maximize the effective temperature contrast. An example will illustrate this further. Suppose the object is a metal plate and the background is grass. The object at microwave frequencies is highly reflective, and the background is highly absorptive. Thus the effective temperature of the background is near its ambient temperature; however, the effective temperature of the highly reflecting object approaches the effective temperature of the illumination source, in this case the sky. To achieve a high contrast, the object must appear either much hotter or much colder than the ambient temperature. Since the sky as a source of illumination (excluding the sun) has an effective temperature between ambient and 0° K, it is apparent that the contrast occurs when the sky appears "cold," i.e., when the atmosphere has a high transmissivity.

It must be realized, however, that there are applications for which sensors operating in one of the nonwindow regions are possible solutions. For example, if it were the purpose of the instrument to monitor the amount of water content in the atmosphere, preferred operation would be in one of the water bands, e.g., at 23 GHz. The effective sky temperature at this frequency can be directly correlated with the amount of water-vapor content[24] (see Figure 30).

Until after about 1958, atmospheric transmission data at frequencies greater than 140 GHz were not available. Recent developments in microwave-receiver technology have made it possible to make atmospheric propagation measurements at higher frequencies. As a result,

*The illumination of the atmosphere, like the power received from any object at microwave frequencies, is expressed in terms of an effective temperature.

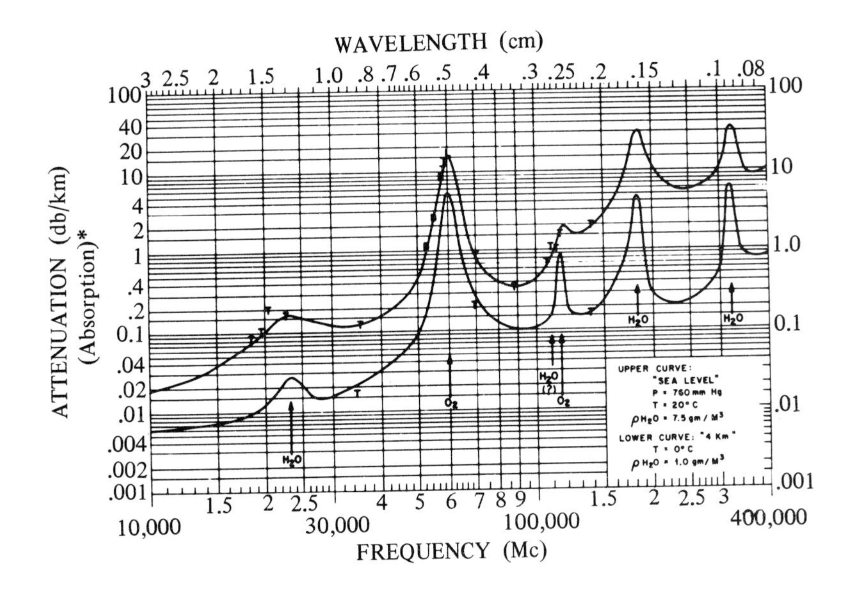

A

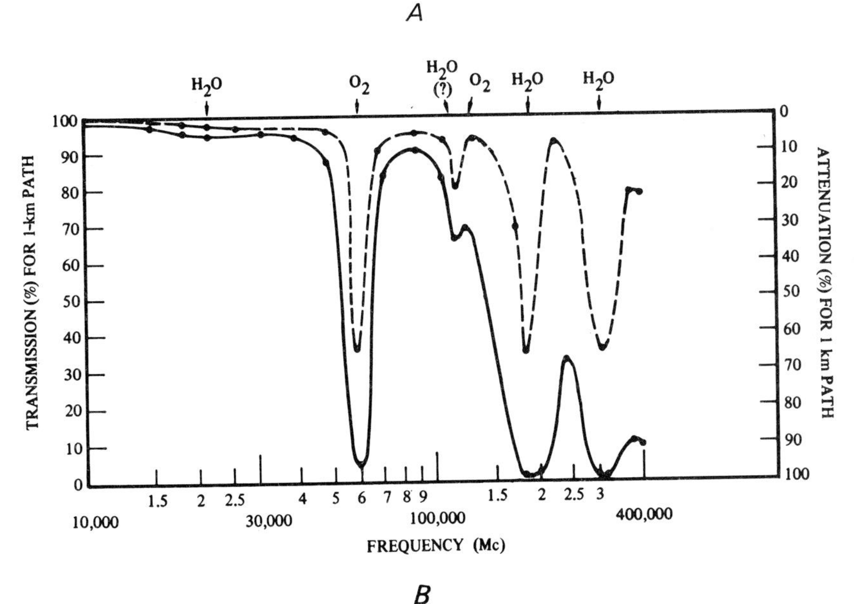

B

FIGURE 29 Attenuation (absorption) versus frequency. *A:* Logarithmic plot. *Adapted from reference 21. B:* Linear plot.

*The term attenuation is construed to include losses from a beam of radiation by all possible mechanisms, principally absorption and scattering by any of several modes. Since at these wavelengths scattering is relatively insignificant, for normal atmospheres absorption can be read for attenuation with negligible error.

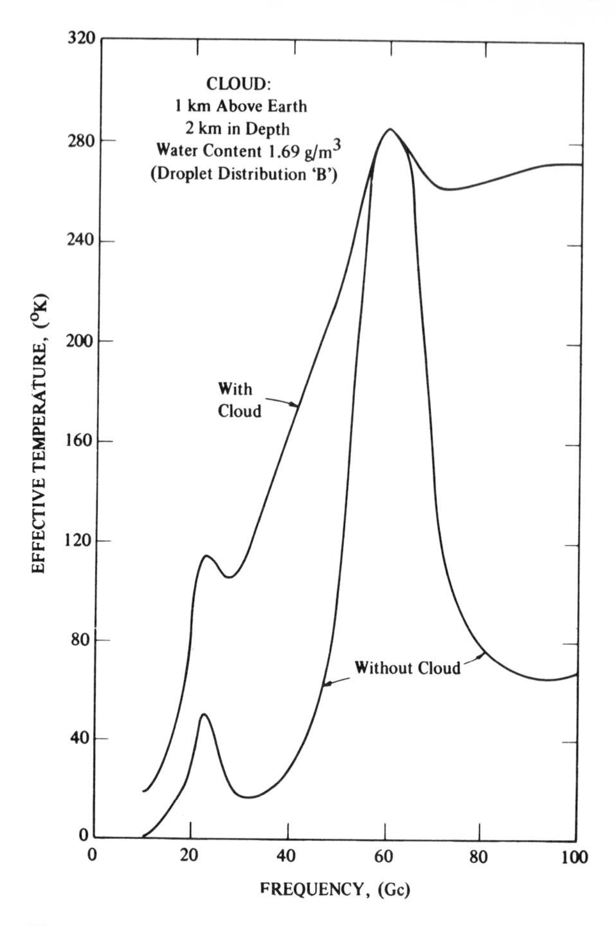

FIGURE 30 Zenith sky effective temperature with cloud and without cloud.[23]

some atmospheric transmission data are now available well into the submillimeter regions.[17,20,25,26] In the centimeter and millimeter regions, atmospheric effects in the water- and oxygen-absorption bands at the lower frequencies have been studied in great detail.[27,28]

Microwave Instrumentation

A microwave radiometer measures the radiant energy that emanates from a target within its field of view. The microwave instrumentation and the concepts developed to make these measurements are discussed in this section. Detection sensitivities of state-of-the-art instruments are summarized, the concept of apparent antenna temperature is reviewed, a basic radiometer is described, and finally, the scanning parameters are introduced, all to provide the reader with a basic understanding of microwave instrumentation.

The power received by the antenna of a radiometer and available at its output is commonly expressed in terms of an apparent antenna temperature.[29-31] This temperature is the thermodynamic incident temperature, T_i, of a Johnson noise source, which will deliver an equivalent power to the remainder of the system, the power being proportional to $kT_i \Delta f$, where k is Boltzmann's constant and Δf is the electrical bandwidth of the electronics.

This can be justified by the following approach. Assume that an antenna is enclosed by a blackbody at a thermodynamic temperature T (see Figure 31), and the antenna feed is terminated by a matched load, also at temperature T. From the second law of thermodynamics, it is known that the net power flow between the termination and the blackbody must be zero. Therefore, the conclusion can be drawn that the power delivered to the system is equal to that generated by the termination, i.e., the available Johnson noise power. These results are generally valid for wavelengths greater than 1 mm. For shorter wavelengths, the approximation errors become large and invalidate the results obtained.[29,30]

A number of radiometric system configurations have been developed in the past few years to measure accurately the apparent antenna temperature, T_A, and, more specifically, to measure the change in apparent temperature when a specific target enters the primary antenna pattern. Of these proposed configurations,[31] the input-modulated system proposed by Dicke[32] in 1946 has since been widely and frequently used. The discussions presented stem from this system concept.

A functional block diagram of the input-signal-modulated system, a Dicke radiometer, is illustrated in Figure 32. The input signal proportional to T_A and a known reference signal proportional to T_R are alternately sampled at the input switch. The resulting radio-frequency signal, which is amplitude modulated at the input switching rate, is amplified and detected. The modulation envelope is then sent to a coherent detection system that demodulates the signals. The filtered dc output signal then becomes a linear measure of the difference in signal between T_R and T_A. The sign of the signal and its relative magnitude are thus a measure of T_A.

The prime advantage of the Dicke system over others proposed is that output signal variations due to short-term gain instabilities of

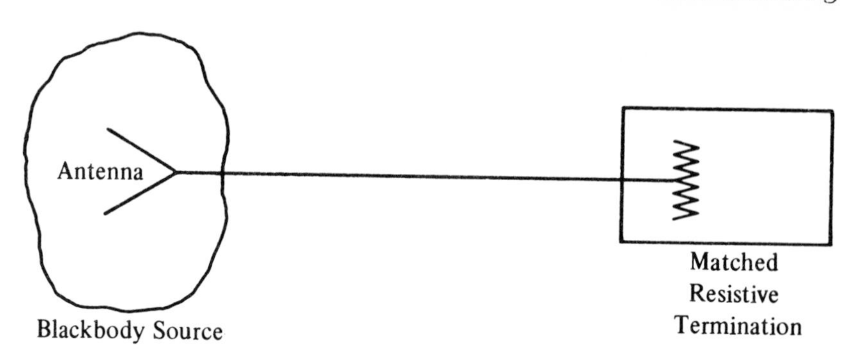

FIGURE 31 Blackbody and Johnson noise power equivalence.

the receiver can be greatly reduced. Thus, considerable improvements in system performance can be realized without employment of elaborate gain-stabilizing devices.

To evaluate the performance of any radiometer, a signal-to-noise (S/N) analysis of the system must be performed by use of standard noise theories and definitions.[22,23] For $S/N = 1$, the minimum detectable root mean square signal is given by

$$\Delta T = \frac{T_A + (F - 1)T_a}{(KB/B_o)^{1/2}} \quad (6)$$

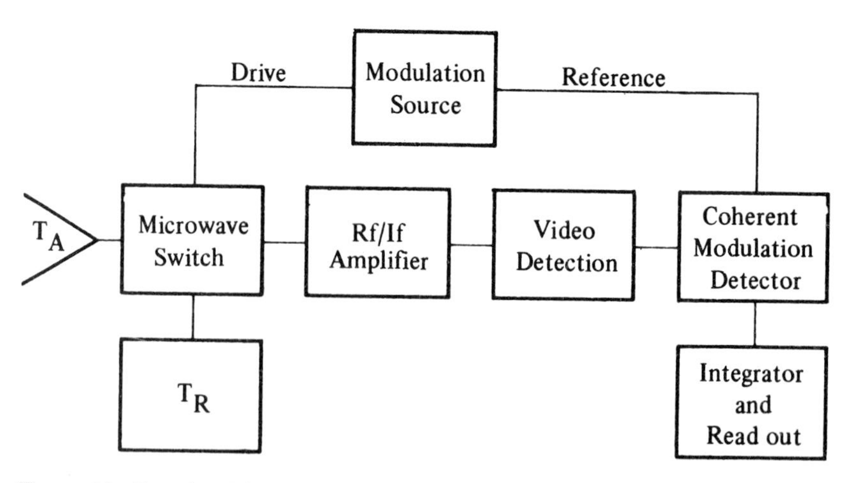

FIGURE 32 Functional block diagram of an input-modulated (Dicke) radiometer.

or as

$$\Delta T = K_1 \frac{(T_A + T_{\text{eff}})}{(B\tau)^{1/2}}, \tag{7}$$

where B = input signal (predetection) noise bandwidth
B_o = output signal (postdetection) noise bandwidth
K, K_1 = detection constants, determined by receiver configuration
F = receiver noise figure[31,33]
ΔT = minimum detectable root mean square temperature change
T_A = apparent antenna temperature
T_a = noise figure normalization temperature (290° K)[31]
T_{eff} = equivalent system input noise temperature
τ = integration time constant of the postdetection electronics (resistance-capacitance network is assumed)

Equations 6 and 7 are valid for any of the radiometer configurations described by McGillen and Seling.[31] The values of K_1 are given in Table 2, and the values of K can be computed, given that $\tau = 1/4B_o$.

From Equation 7 it is evident that to obtain optimum system performance, i.e., a low value for ΔT, a large predetection bandwidth (B) and/or a long integration period (τ) is desired. In addition, a low receiver-noise figure (F) is desirable. As would be suspected, a trade-off between bandwidth (B) and noise figure (F) is necessary for optimum performance since, in general, receiver noise figures increase with increased bandwidths.

To evaluate various types of receivers, i.e., those using superheterodyne techniques, masers, parametric amplifiers, and/or tunnel diodes for the first stage following the input signal switch (see Figure 32), a commonly used figure of merit, C, is given by

$$C = \Delta T(\tau)^{1/2} = \frac{K_1(290F)}{B^{1/2}}. \tag{8}$$

This is obtained by substituting Equation 6 for ΔT and assuming T_A is equal to T_a. Receivers for the microwave spectrum have been evaluated at The University of Michigan* by use of Equation 8.

*Specifically, the Infrared and Optical Sensor Laboratory of Willow Run Laboratories, a unit of the University's Institute of Science and Technology.

TABLE 2 Radiometer Constants for Various Circuit Configurations[31]

System Type	Modulation Waveform	Demodulation Waveform	Radiometer Constant (K_1)
Unmodulated	—	—	1/2
Postdetection modulated	Square wave	Sine wave	$2/\pi^2$
Signal modulated	Square wave	Square wave	1/8
	Square wave	Sine wave	$1/\pi^2$
	Square wave	([a])	$1/\pi^2$
	Sine wave	Sine wave	1/16
Cross correlation	—	—	1/4
Two-channel subtraction	Square wave	Square wave	1/4
	Square wave	Sine wave	$2/\pi^2$

[a] Assuming the signal has passed through a narrowband filter before demodulation. If no filter is used, then K_1 is 1/8.

The results are presented in Figure 33, which shows the current state of the art in microwave-receiver technology, including systems currently available.

From Figure 33 one can predict what system improvements will be of greatest significance. In general, the receivers at long wavelengths ($\lambda > 3$ cm) are wideband and are background limited when one looks earthward; i.e., the apparent temperature (T_A) exceeds the system equivalent input noise temperature (T_{eff}). Improvements in receiver noise levels will not significantly increase the radiometric performance capabilities at these wavelengths. At shorter wavelengths ($\lambda < 3$ cm), however, this is not the case, and receiver improvements can be expected since the basic physical limitations have not been reached. Currently, efforts are being made to develop low-noise receivers for the shorter wavelengths.

Finally, when one considers basic concepts of the signal-modulated system, another noise term must be included to obtain the final value of ΔT. This term is the product of the short-term receiver-gain instabilities and the difference temperature ($T_A - T_R$) between the signal and reference channel; i.e.,

$$\Delta T \propto \frac{\Delta G}{G} \mid T_A - T_R \mid, \tag{9}$$

where $\Delta G/G$ represents the short-term receiver-gain instabilities. This term[25] can become important when the temperature differential $| T_A -$

T_R | is large. This term can be negated by controlling the value of T_R such that

$$\frac{\Delta G}{G} \mid T_A - T_R \mid \ll \frac{K_1(T_{\text{eff}} + T_A)}{(B\tau)^{1/2}}.$$

As such, the control required, if any, is dependent upon the ultimate sensitivity required in ΔT.

To assess the scanning capabilities[13] of the radiometer, the integration time τ or the postdetection bandwidth B_o must be determined. Those parameters affecting the value of τ that must be evaluated include the antenna resolution β, the relative velocity V, the altitude H of the radiometer relative to the target, and the scan angle α of operation. It can be shown that, for an 87 percent rise in signal level while a point source traverses the main beam of the antenna, $\tau^{1/2}$ is expressed as

$$\tau^{1/2} = \frac{1.2\lambda}{D}\left(\frac{H}{2\alpha V}\right)^{1/2}, \tag{10}$$

where $1.2\lambda/D = \beta$ and is the Rayleigh resolution limit of the antenna system. (D is the aperture diameter and λ the wavelength of measurement.) Through substitution it is found that

$$\Delta T = \frac{CD}{1.2\lambda}\left(\frac{2\alpha V}{H}\right)^{1/2} \tag{11}$$

With this information, an evaluation of the system capabilities becomes possible.

To evaluate the usefulness of remote microwave sensors for agricultural applications, system constraints and trade-offs are now considered. These factors (parameters) include the physical constraints, the functional interdependence of the previously considered parameters, and the choice of operational wavelength.

Spatial resolution of the microwave system must be considered in order to evaluate its capabilities as a remote sensor. Diffraction theory implies that it is necessary to use both large apertures and short wavelengths to obtain good angular resolution; however, other constraints such as receiver performance figures (figures of merit), atmospheric transmission factors, and mechanical limitation must also be considered. Equation 11 implies that system spatial resolution and temperature resolution are inverse functions, and, therefore, a compromise

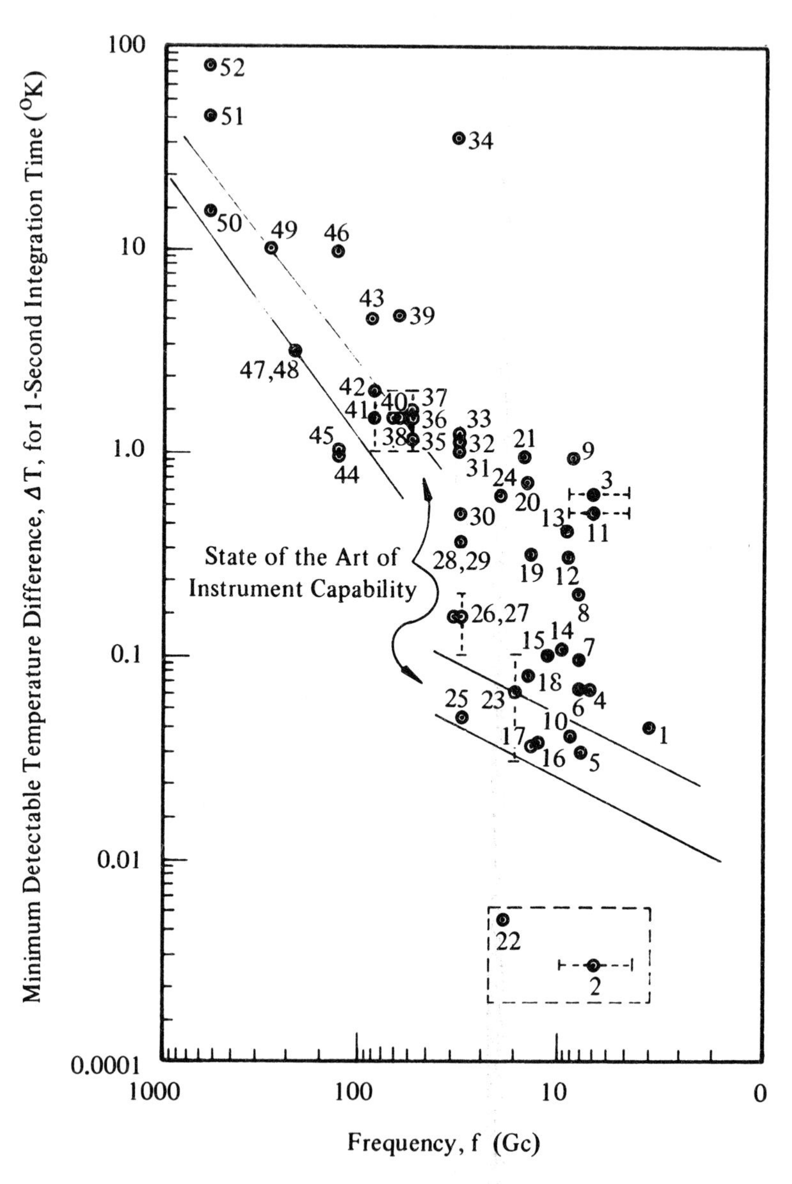

FIGURE 33 Review of passive-microwave state of the art. (*University of Michigan survey.*)

1. Airborne Instrument Laboratories parametric amplifier (in development).[34]
2. Autonetics, Inc. (C. Wiley).
3. North American Aviation (J. Hall).
4. Ewen-Knight Corp. (H. Ewen).
5. University of Michigan parametric amplifier tunnel-diode superheterodyne system (planned).
6. University of Michigan maser (operational).[35]
7. Airborne Instruments Laboratories.[36]
8. Naval Ordnance Laboratory, Corona, Calif.[37]
9. Raytheon (R. Porter).
10. Lear-Siegler, Inc. (D. Mathews).
11. North American Aviation.[38]
12. General Electric.[38]
13. Attainable using tunnel-diode amplifiers.
14. Sperry Microwave Electronics (Mr. Lazarchik).
15. Space General Corp. (T. Falco).
16. Sperry Microwave Electronics (Mr. Lazarchik).
17. Ewen-Knight Corp. (H. Ewen).
18. Collins Radio (W. Bellville).
19. North American Aviation (J. Hall).
20. General Electric.[38]
21. Raytheon (R. Porter).
22. Autonetics, Inc. (T. Falco).
23. Space General Corp. (T. Falco).
24. Nortronics.[37]
25. Airborne Instruments Laboratory.[37]
26. Space General Corp. (T. Falco).
27. Airborne Instruments Laboratory.[40]
28. Ewen-Knight Corp. (H. Ewen).
29. Sperry Microwave Electronics (Mr. Lazarchik).
30. Martin Co., Orlando, Fla.
31. University of Texas.[41]
32. General Electric Co.[42]
33. Sperry Microwave Electronics (Mr. Lazarchik).
34. University of Texas.[41]
35. Ewen-Knight Corp. (H. Ewen).
36. Collins Radio (W. Bellville).
37. Ewen-Knight Corp.[38]
38. General Electric Co.
39. General Electric Co.
40. North American Aviation (J. Hall).
41. Space General Corp. (T. Falco).
42. Aerospace Corp. (D. King).
43. Raytheon (R. Porter).
44. U.S. Army Ballistic Research Laboratories, Aberdeen (Md.) Proving Grounds (K. Richer).
45. Electronic Communications, Inc.[42]
46. Royal Radar Establishment.[43]
47. Electronic Communications, Inc.[41]
48. U.S. Army Ballistic Research Laboratories, Aberdeen (Md.) Proving Grounds (K. Richer).
49. Space General Corp.[38]
50. Department of the Army, Frankfort Arsenal.[20]
51. U.S. Army Ballistic Research Laboratories, Aberdeen (Md.) Proving Grounds (K. Richer).
52. Advanced Technology, Inc.[38]

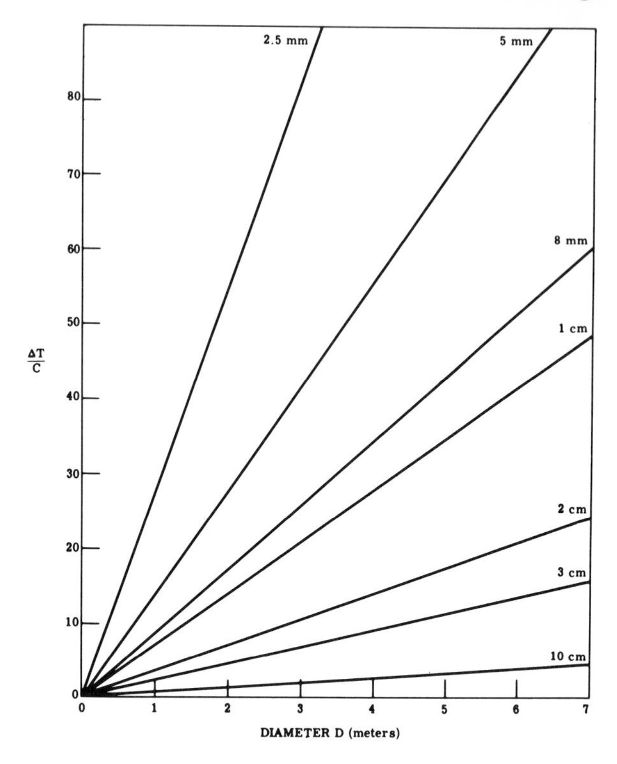

FIGURE 34 Plot of $\Delta T/C$ versus aperture diameter. This represents the operation that can be expected for a system at a velocity–altitude ratio of 0.02 and a scan angle of 10°.[13]

is necessary. The selection will be governed by both the spatial and temperature resolution desired as well as other physical limitations; i.e., for a given figure of merit and a given V/H, the temperature-detection ability can be improved only by degrading the spatial resolution or by reducing the scan angle. By assignment of values, then, a parametric plot of $\Delta T/C$ versus D, with wavelength as a parameter, can be established (see Figure 34). From this information, along with Figure 33, the system sensitivity (ΔT) can be obtained directly or by extrapolation if other parameters change.

The atmospheric losses encountered are a function of wavelength (see Figure 29) and must be considered in the selection of an operational system. In general, the losses are low at wavelengths longer than 2 cm and are relatively low in a window occurring at ~8 mm. The transmission losses have not been included in the equations and will degrade the operation to a varying extent. The degree of degradation will depend on the operating wavelengths chosen.

The mechanical constraints that must be considered apply primarily to the antenna configuration and size. For aircraft installations, maximum antenna apertures of 2 m are considered practical, and antenna surface tolerances can be maintained into the millimeter region. Several possibilities exist for space antenna structures. They may be designed into the surface of the vehicle; an antenna may be assembled outside the vehicle by a man; or the antenna may be an unfurlable or inflatable type. It is generally believed that the physical size of the collecting aperture should be 20 m or less. In general, the achievable size of the antenna is a function of wavelength; until 1970, practically achievable antenna diameters are expected to be as shown in Figure 35.

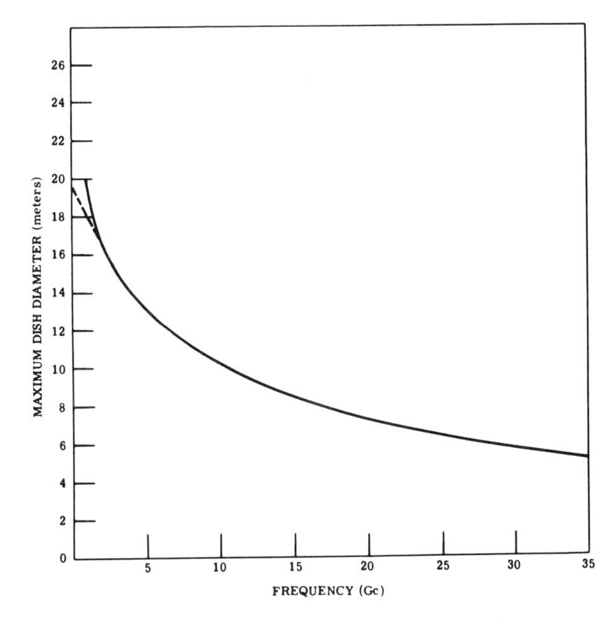

FIGURE 35 Attainable dish diameter versus frequency. For the 1970 time frame, the question of maximum dish diameter attainable in space with deployable antennas was explored at The University of Michigan in 1964 with these results.[13]

The size limits for an antenna are imposed by the surface tolerances that can be maintained. The angular resolution versus diameter of an antenna is plotted in Figure 36 with wavelength as a parameter. Also shown are the practical limits on a space installation that can be assumed for the 1970 time frame.

Selections of the wavelength for operation and the aperture size would be determined primarily by the resolution needed to map particular sources. Thus, if one assumes the use of the maximum aperture attainable at each wavelength, the wavelength to be used is determined by the resolution requirements. Then the system's temperature-resolution capabilities can be established, and the value of the system will thus be determined for each particular usage accordingly; conversely, since optimum detectivity implies a minimum aperture size (see Equation 11), resolution requirements dictate operation at the shortest possible wavelength. Therefore, the selection would generally be in the 8-mm region, necessitating slightly increased atmospheric losses.

Considering the system trade-offs discussed above, an airborne and a space platform for remote sensing of terrain features are examined.

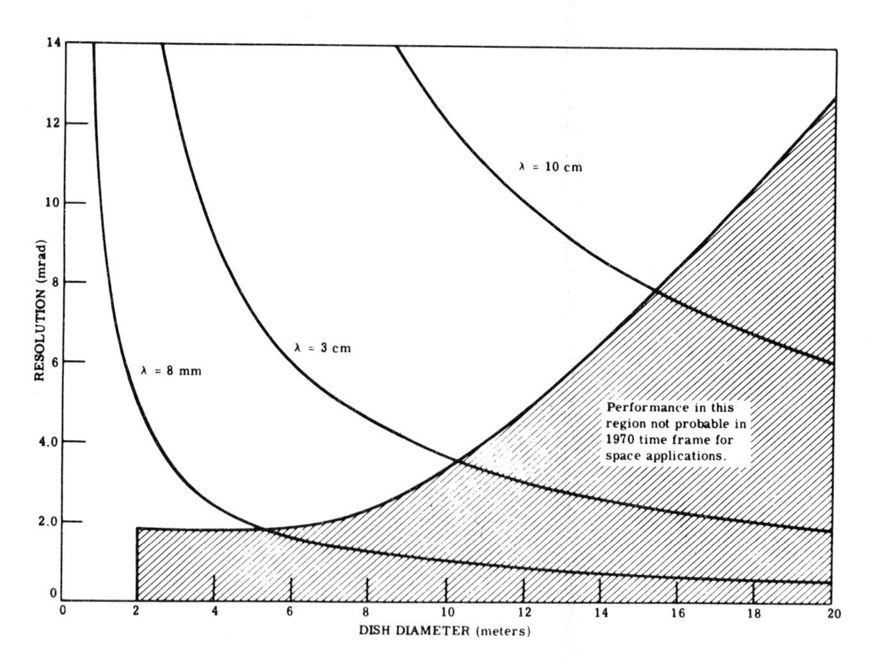

FIGURE 36 Plot of resolution versus antenna size.[13]

Typical airborne-system parameters for an antenna aperture diameter D of 1 m are:

$$\begin{aligned} \lambda &= 8 \text{ mm} \\ C &= 5 \times 10^{-2} \text{ (see Figure 33)} \\ V/H &= 0.02 \\ \alpha &= 10^\circ \text{ (0.175 rad)} \end{aligned}$$

With these parameters, a temperature sensitivity of 0.5° K and a spatial resolution of 10 mrads are achievable. These system capabilities are adequate for remote sensing of terrain features. Among the features that should be detectable are field boundaries, roadways, water–land boundaries, snow cover, and buildings.

Typical space-system parameters are identical to those for an airborne system, except that the antenna aperture size can be larger. For this example the maximum available aperture (D = 5.0 m) was assumed, with a resulting spatial resolution of 2 mrads and a temperature sensitivity of approximately 2° K. At a satellite altitude of 200 miles, the system's angular resolution implies that a minimum ground-patch resolution of 0.4 mile is achievable. These capabilities would seem to be inadequate for many agricultural applications. However, a number of agricultural problems are amenable to study with only coarse resolutions; among them are obtaining information on soil temperatures and soil moisture over extended regions and hydrologic information such as area of snowpack.

In general, it would appear that, from airborne platforms, instrument capabilities are currently adequate for agricultural applications. However, for operation from space platforms, instrument capabilities must be improved before feasibility can be shown.

Infrared Imaging

The infrared portion of the electromagnetic spectrum lies between the visible and the microwave regions (0.78 μ to 1,000 μ). The existence of this "invisible" radiation was discovered in 1800 by Sir Frederich William Herschel, who used a thermometer to detect the energy beyond the red portion of a visible "rainbow" produced by a prism. In 1861, Richard Bunsen and Gustav Kirchhoff established the principles underlying infrared spectroscopy, a powerful tool in the basic studies of molecular structure. After a century of further progress, the uses of infrared technology have been extended to a host of other applications. Among them is the use of infrared sensors for producing images of remote scenes. These devices (remote-sensing systems) have been used on the ground, in airplanes, and in space vehicles. They have been applied mainly to military problems, the original motivation for their development. Compared with their radar counterparts, these devices are small, lightweight, and require little electric power. Like radar, infrared systems operate under both daytime and nighttime conditions. However, infrared systems are hampered by clouds and rain—a problem radar systems do not have when sufficiently low frequencies are used. The two important features of infrared surveillance systems that give them an advantage over radar are the high spatial resolution achievable with relatively simple designs and the fact that they operate passively; i.e., they do not need to illuminate the ground scene artificially in order to record an image of it.

Basic Radiation Principles

The infrared radiation emanating from a scene is due to both self-emission from each object within the scene and reflected radiation from the object as a result of illumination from natural sources such as the sun, clouds, moon, stars, and aurorae.* Generally speaking, in the daytime the radiation is predominantly reflected in the spectral region from 0.78 to 3.0 μ and is mostly self-emitted at wavelengths longer than 4.5 μ. During daylight hours, roughly comparable amounts of reflected sunlight and emitted radiation will be present in the wavelength region 3 μ to 4.5 μ. The relative importance of reflected versus emitted radiation is shown in Figure 37. Here, the radiated power from an object is plotted as a function of wavelength. The upper and lower limits for the spread of radiated source in the short-wavelength band are for objects with reflectances ρ of 1.0 (a perfect reflector), 0.03, and 0.003. Limits for the long-wavelength band are for objects at a temperature of 300° K and with an emittance ϵ of 1.0 (a perfect emitter) and 0.1. The irradiance H_λ (with units of $\mathrm{W cm^{-2} \mu^{-1}}$), which is the radiation incident on the aperture of an optical sensor from an object in the remote scene, can be written as follows:

$$H_\lambda = N_\lambda \Omega, \qquad (12)$$

*Four textbooks on infrared radiation that the reader should refer to for a more extensive treatment of the fundamentals are references 42, 43, 44, and 45.

Where Ω is the solid angular field of view

N_λ, radiance, is defined as the radiation power per unit area, directed into a unit solid angle from the object; it is measured in $\mathrm{Wcm^{-2}\,sr^{-1}\mu^{-1}}$

In completely general functional form, H can be expressed as

$$H_\lambda = H_\lambda\,(\lambda; t; p; \theta_i, \phi_i; \theta_r, \phi_r; \Delta x, \Delta y; x, y; T).$$

This shows the parametric dependence of the important quantities that affect the incident radiation. The wavelength λ is perhaps the most important variable. The infrared radiation from objects depends on the angles at which the object is viewed and illuminated, (θ_r, ϕ_r) and (θ_i, ϕ_i), respectively (see Figure 38). The irradiance also depends upon the observed ground-patch coordinates x and y (i.e., upon the material under observation) and on the instantaneous ground-patch size $(\Delta x, \Delta y)$.

The irradiance H_λ will also vary with time (t) on both short- and long-term bases. Regions on the moon that have just passed from sunlight to shadow, for example, may exhibit significant variability, in milliseconds. Substances such as foliage exhibit diurnal and seasonal variations. Radiation emanating from an object may also be polarized

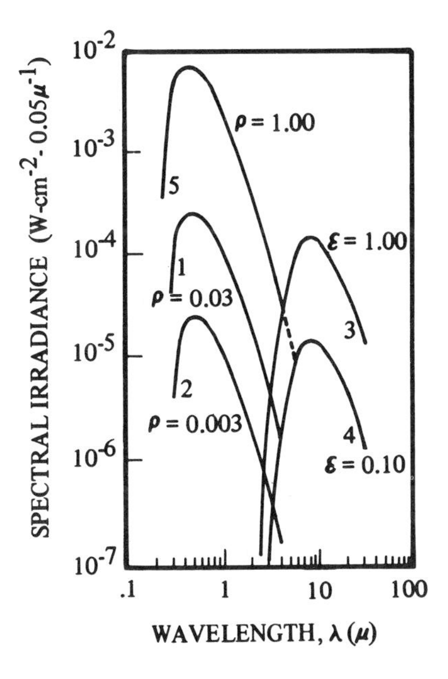

FIGURE 37 Comparison of reflected sunlight and emitted radiation. For many natural materials, the infrared emittances tend to be a larger fraction than the visible and near-infrared reflectances. Therefore, the wavelength of equal radiation will be shorter than indicated by the curves of equal emittance and reflectance. Emitted radiation curves were constructed from points calculated with a radiation slide rule. Solar irradiance data is taken from reference 49. Sonar irradiance is plotted at ground level for sun at zenith. Rayleigh and aerosol scattering and ozone attenuation are considered; absorption by other molecular species has been neglected.

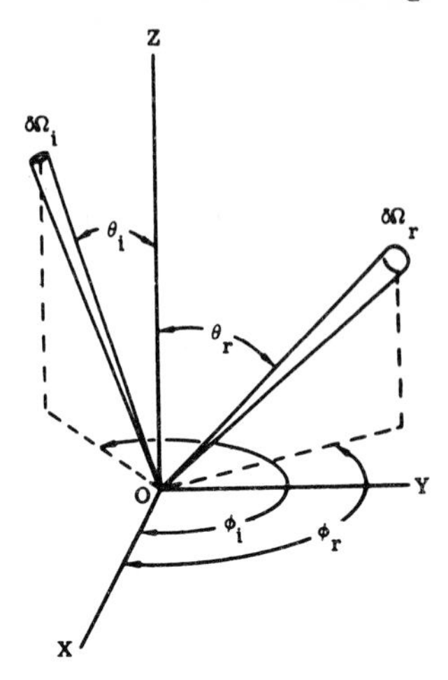

FIGURE 38 Diagram of directional reflectance.[45]

(p) to some degree, depending upon the surface topography and material constituents.

The change in irradiance in relation to λ; t; θ_i, ϕ_i; θ_r, ϕ_r; p; $\Delta x, \Delta y$; and x, y from one class of objects to another is the basis for detection and discrimination and is the starting point in listing the system specifications for purposes of designing remote sensors.

For blackbodies* and graybodies, the radiance N_λ due to self-emission will vary with λ according to Planck's law:†

$$N_\lambda = C_1 \epsilon_\lambda \lambda^{-5} \left[\exp\left(C_2/\lambda T\right) - 1\right]^{-1}, \tag{13}$$

where C_1 is a constant incorporating such quantities as π, the speed of light (c), and Planck's constant (h); C_2 is a constant with the form $C_2 = hc/k$ (k is the Stefan-Boltzmann constant); T is the temperature of the emitting object; and ϵ_λ is the spectral emittance, which is unity at all wavelengths for a blackbody and some constant between zero and unity at all wavelengths for a graybody.

Figure 39 is a spectral plot of the radiance from blackbodies of temperatures ranging from 100° K to 1,000° K. Note that the radiation peak moves to shorter wavelengths by increasing temperatures, and further, that the peak radiation for blackbodies at the earth's ambient temperature (approximately 300° K) lies at 9.7μ. For objects that are

*See discussion of basic radiation principles, page 106.

†As mentioned on page 106, it is possible at the longer, passive microwavelengths to use a much simpler approximation to Planck's law. At infrared wavelengths that approximation is not valid.

neither blackbodies nor graybodies, and almost all objects fall into this class, ϵ_λ can be a complicated function of wavelength. The effect of the temperature of the object on N_λ (and hence H_λ) is also shown in Equation 13. Higher temperatures mean more self-emission at all wavelengths, assuming that ϵ_λ is not inversely related to temperature at some wavelength.

The radiance N_λ due to reflected radiation from the object is expressed as

$$N_\lambda = C_3 \rho_\lambda H_\lambda', \tag{14}$$

where C_3 is a constant; ρ_λ is the spectral directional reflectance; and H_λ' is the spectral irradiance incident on the object from the natural or artificial source.

Equation 14 demonstrated the importance of the spectral shape of the source radiation. Substituting Equations 13 and 14 into Equation

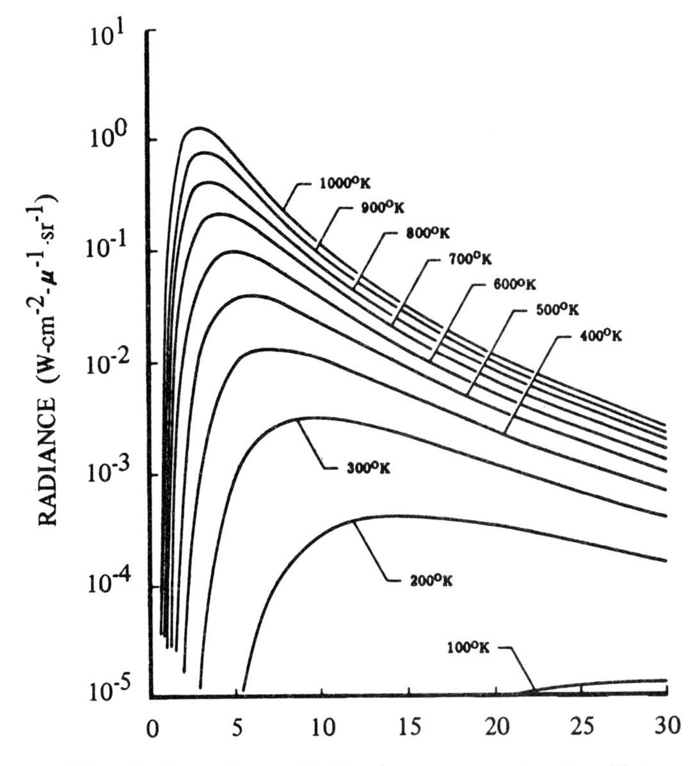

FIGURE 39 Plot of radiance from a blackbody versus wavelength, with temperature as a parameter.[45]

12 gives an expression for the radiant flux at an optical sensor, resulting from radiation reflected and emitted by the object.

An additional factor that must be considered in the design of infrared systems is the transmission of the atmosphere. As noted in the discussion of passive-microwave imaging (page 113), molecular absorption at microwavelengths by atmospheric constituents becomes significant. At infrared wavelengths, molecular absorption becomes very significant, and additional atmospheric gases play a role. Each of the various gases that comprise the atmosphere has its own characteristic infrared-absorption spectrum. The gases that have the most significant effect include CO_2, N_2O, H_2O, and O_3. For those portions of the infrared spectrum where one or more of these gases is strongly absorbing, the atmosphere is essentially opaque. The spectral intervals at which the atmosphere is reasonably transparent are (approximately, in microns) 0.7–1.35, 1.35–1.8, 2.0–2.4, 3.5–4.1, 4.5–5.5, 8–14, 16–21, and 750–1,000. These transparent spectral regions are called windows (see Figure 40). The detailed spectral transmission curve has a complicated structure, and it changes with season, time of day, viewing angle, etc. Thus, system designers must consider the spectral transmission function along with H_λ for the target and the background in the design of infrared-imagery systems.

Infrared Scanners

Infrared images can be obtained with a number of devices: infrared photographic film, image tubes, and optical scanners. Infrared photographic film is used in the same way as film in conventional photography. It is made by adding sensitizing dyes during the manufacturing

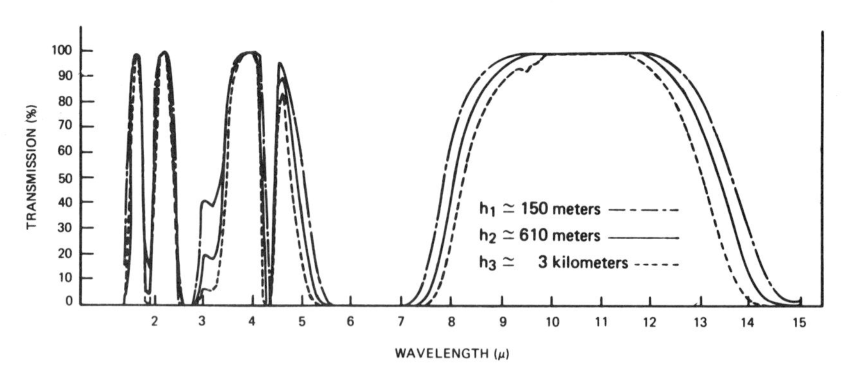

FIGURE 40 Plot of atmospheric transmission spectra as a function of altitude, *H*.

process to extend the spectral response to about 1.2 μ; however, aerial films extend the response to only 0.9 μ (see Chapter 2, Figure 2). As indicated in Chapters 5 and 6, the reflectance spectrum of green plants ranges in value from about 5 percent at 0.6 μ to about 45 percent at 0.75 μ; and further, this region, which contains a pronounced chlorophyll absorption band, is very sensitive to changes in the health of green plants. Because of this, infrared film has been used both by the military for camouflage detection and by foresters for detection of diseased trees in forests.

Image tubes are electronically scanned image-forming devices such as those used in television systems. No mechanical moving parts are required, although some of them use an electron beam to convert the invisible infrared image on the sensitive surface to a visible image on a phosphor screen. Infrared image tubes that are sensitive to wavelengths longer than 1.2 μ tend to have long time constants; i.e., the speed at which an image is formed is quite slow. Although this sluggish response is not a severe problem for some ground-based systems, it is so for airborne or space applications and therefore will not be discussed further here.

The most widely used airborne infrared surveillance device is the scanner. In systems of this kind, a single-element detector, or a one- or two-dimensional array (mosaic) is used to view small parts of the total field of view. This type of sampling is generally called optical-mechanical scanning or simply "scanning," as opposed to the electron-beam scanning associated with image tubes. There are, in principle, two types of scanning—image plane and object plane. The first type is so named because an image of the total object field is formed by a collecting telescope. This image is then sampled by a second optical system containing the infrared detector. In this case, the scanning device can be small and the optics for it easily designed. However, a high-quality image of the object field must be formed by the telescope. Designing and constructing the optics for such a telescope when the field of view is larger than 10° or 20° is a difficult problem. Because of this, image-plane scanning is not frequently used.

The object-plane scanner generally consists of a rotating mirror and a telescope that focuses the radiation from a small portion of the object plane. The rotating mirror causes the field of view to move smoothly across the object plane. When such a system is placed in an airplane, a strip map of the image of the ground scene is produced (Figure 41). Such devices have a very fine angular resolution ($\sim$1 mrad^2). This means an instantaneous ground resolution of 1 ft^2 for an aircraft altitude of 1,000 ft, or 6 $\times$ 6 in. for an aircraft altitude of 500 ft. For

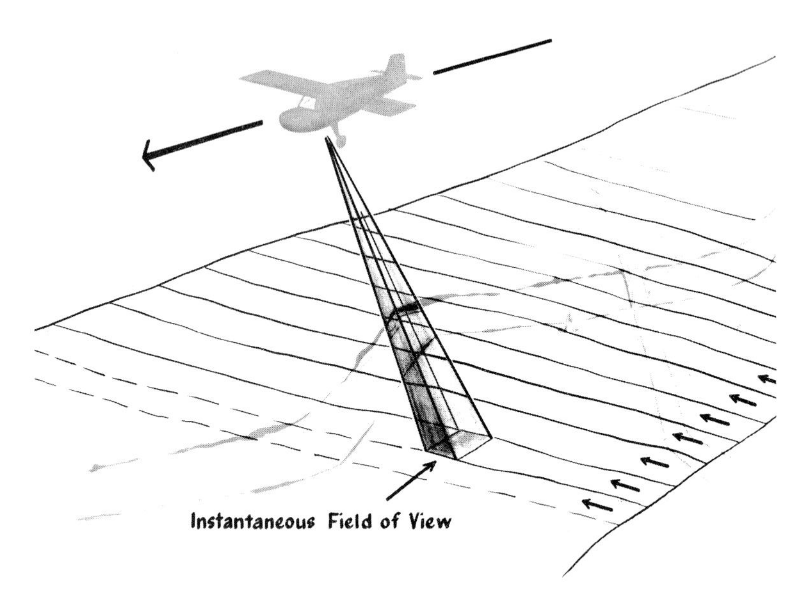

FIGURE 41 Scan pattern for a typical scanner.

satellites at altitudes of 100 miles, a scanner with 1-mrad^2 angular resolution would view an instantaneous ground patch of 600 $\times$ 600 ft.

Consider a surveillance aircraft traveling with speed V at altitude H, which carries a scanner having an instantaneous field of view (resolution element) of angular size β in both dimensions. To provide a good-quality image, β is typically quite small, on the order of milliradians. This instantaneous field of view is caused to scan through an angle (α) at right angles to the aircraft path by a rotating element in the scanner. The motion of the vehicle carries the scanner forward so that successive scans cover different strips of the ground. The portion of ground swept during a single scan through angle α is called a "line." If successive lines are not contiguous (if between the lines parts of the ground are not scanned), the condition is termed "underlap." This can occur if the vehicle speed is too high or if the rotating elements revolve too slowly. If successive lines scan partly over the same terrain the condition is called "overlap." Underlap is clearly undesirable, since information is not obtained between successive lines. Contiguous scanning is desirable in that no ground remains unscanned, and there is a certain economy in not scanning more than once over any part of the ground. Under certain circumstances, overlap is useful since the redundancies can be used to increase the signal-to-noise ratio.

A typical device for obtaining such a scan pattern is illustrated in Figure 42. A mirror in the form of a prism with n faces rotates at a rate r (revolutions per second) about an axis that is parallel to the flight path of the aircraft. Hence n lines are scanned per revolution when a single-element detector is used. Each face of the prism is inclined 45° to the axis of rotation. Normally, the detector acts as the field stop of the system, determining the angular dimensions of the instantaneous field of view. If the detector, instead of being a single element, consists of l identical elements arranged in a closely spaced linear array, then l contiguous lines will be swept out simultaneously by each face of the prism. Thus $n \times l$ lines will be scanned per revolution. The principal detector characteristic of significance here is the time constant τ. It will be assumed that the scanner must dwell on each resolution element for a time not less than $k \times \tau$, where k is a positive

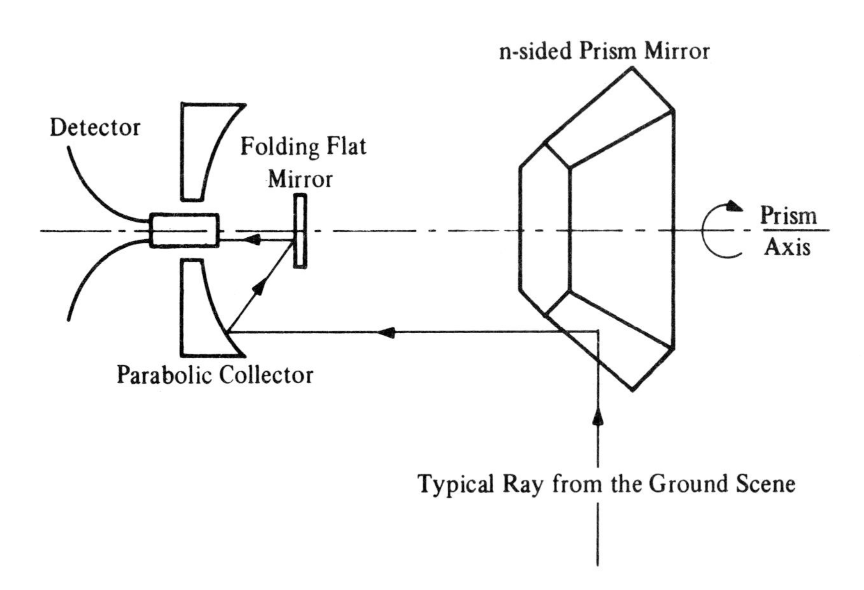

FIGURE 42 Diagram of a typical scanning system.

dimensionless number representing the dwell time in terms of the detector time constant.

The operation of such a scanner must meet two conditions: (1) The scanner must dwell on each resolution element for a time not less than $k\tau$. The number of resolution elements scanned per second is $2\pi r/\beta$, so the dwell time is $\beta/2\pi r$, where r is the rotational rate (rps) of the scan mirror. Hence, $(\beta/2\pi r) = k\tau$ (see reference 2). The scanner must be operated at such a rate that no underlap occurs (overlap may be permitted). In the direction of aircraft travel, the width of the ground strip scanned by each face of the prism is βHl; the width of the strip scanned each second is $\beta Hlnr$. If no underlap is to occur, this expression must be greater than or equal to the vehicle speed ($\beta Hlnr \geqslant V$). Rearranging these two expressions:

$$r < \beta/2\pi kr \qquad \text{and} \qquad r \geqslant V/Hln\beta,$$

in which all quantities are positive.

Thus, r has an upper limit set basically by the detector time constant and a lower limit set by the zero overlap requirement (by V/H). A third constraint upon r is also important: the maximum rotation rate permitted by mechanical considerations such as strength of materials, vibrations, and allowable distortions.

Eliminating r from the above inequalities, we obtain a constraint upon β:

$$\beta \geqslant \sqrt{(2\pi k/nl) \cdot (V/H) \cdot \tau} \cdot$$

Unlike r, β is constrained by a lower limit only. This constraint is imposed by a joint action of $V\beta H$ and r.

Under the limiting condition of contiguous lines, the inequalities become equalities, and the relations become

$$r = \sqrt{(1/2\pi knl) \cdot (V/H) \cdot (1/\tau)}$$

$$\beta = \sqrt{(2\pi k/nl) \cdot (V/H) \cdot \tau} \cdot$$

The individual terms in these expressions warrant some discussion. The designer usually will have little or no control over V, H, τ, and k. The first two are usually set by considerations relative to vehicle operation. The properties of the detector materials determine r, and these are not subject to change. Some small choice is available by selecting different detector materials. The amount of information degradation that is tolerable determines k, which is usually not less than 2. Thus, the only free variables are n and l, and to a lesser extent β in that it is constrained by a minimum value but not by a maximum value, except as the picture quality is degraded. As noted above, r is constrained by both greatest and least values as well as by mechanical considerations.

The heart of an infrared scanner is the detector, which transduces the incident infrared radiation into an electric signal. The early scanning systems developed in the 1950's used thermal detectors that were relatively insensitive by today's standards and had response times in the millisecond range. In thermal detectors, impinging infrared radiation heats the sensitive element, and a temperature-dependent property, such as resistance, is monitored. Generally, thermal detectors are slow because it takes time for the sensing element to lose its excess heat. Their chief advantage is that they respond to very broad portions of the infrared spectrum while operating at ambient temperatures. Today's scanners, because of their high-speed operation, require detectors with short time constants. Some modern scanners dwell on an instantaneous ground resolution element for 10^{-6} sec or less. Thermal detectors are 3 orders of magnitude too sluggish for systems such as these.

Photodetectors, on the other hand, are much faster and have better sensitivities. They differ from thermal detectors in that incident infrared photons react with bound electrons in the sensitive element, producing free charge carriers. Some electrical property that depends upon the concentration of these free carriers is then monitored. Detectors of this kind have been fabricated with time constants as short as a few nanoseconds and detectivities that are limited by the fluctuation in arrival rate of photons at the detector—an ideal situation. The two chief disadvantages of photodetectors when compared with thermal detectors is that they have a limited spectral response, and in most cases, they require cooling. Figure 43 is a plot of the figure of merit D^* (detectivity) versus wavelength for a few of the most commonly used infrared detectors operated at room temperature. Figure 44 is a similar plot of commonly used photodetectors that require cooling to 77° K. Figure 45 is the same kind of plot for detectors requiring cooling to temperatures below 77° K. Note that photodetectors sensitive to longer wavelengths require lower operating temperatures. For the most part, the data for these plots were taken from a survey paper by Potter and Eisenman.[46] At wavelengths longer than 25 μ, the indium antimonide detector, which is not illustrated, has a spectral response that is ad-

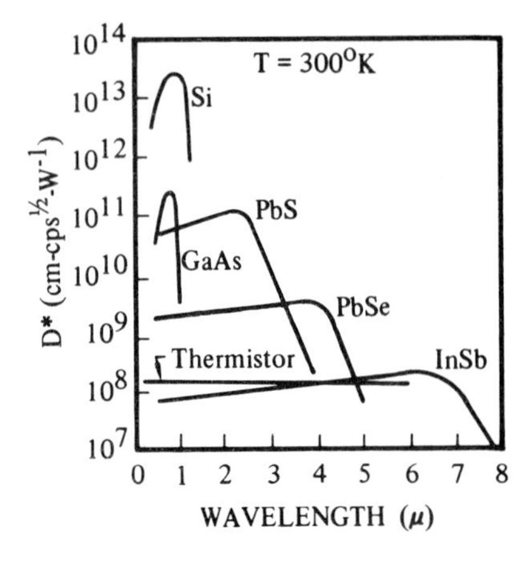

FIGURE 43 Comparison of the sensitivities of various commonly used photodetectors operated at room temperature. A typical thermistor thermal detector is included for comparison.

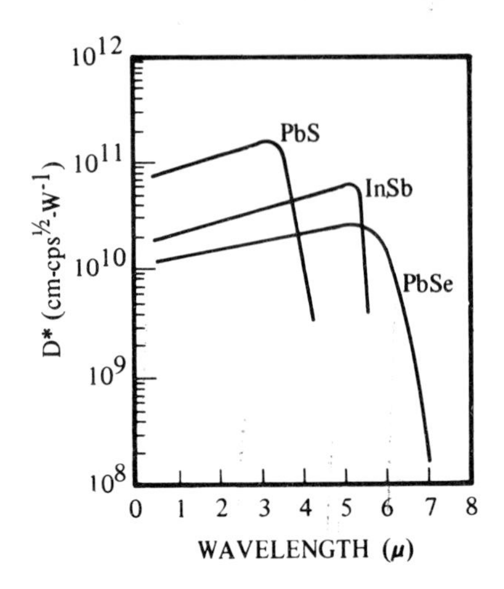

FIGURE 44 A comparison of the sensitivities of various commonly used photodetectors that require cooling to 77° K.

justable by varying the magnetic field strength used to produce the desired position of the energy levels in the sensitive element. This detector has been used at wavelengths as long as 8,000 μ.

The figure of merit D^* is related to the noise equivalent power (NEP) as

$$D^* = \sqrt{A \, \Delta f} \, (\text{NEP})^{-1},$$

where A is the detector's sensitive area

Δf is the electrical bandwidth of the preamplifier and amplifier combination

NEP is the amount of rms signal radiation power required to produce a ratio of root mean square signal to root mean square noise of unity

The sensitive elements of detectors most frequently used in infrared scanners today are made of indium antimonide, which is sensitive in the shorter wavelength atmospheric windows; and germanium : mercury, which is sensitive in the 8–14-μ window.

After amplification, the signal from the infrared detector is most commonly presented on a cathode-ray tube (CRT) scanned in synchronism with the infrared scanner. A permanent record is obtained by photographing the face of the cathode-ray tube. This type of display-and-recording system is essentially identical to that used in radar (see discussion of radar imaging, page 95, and Figure 12). Referring to Figure 12, note that only one line across the face of the CRT is used, each line of infrared and radar information being presented on the same part of the CRT in succession. This requires that between successive lines the recording film must be advanced by one line width, so the information is recorded one line at a time as the recording film moves along. The infrared-detector signal varies the brightness of the CRT spot.

It has recently become common practice to record the infrared-detector signal on magnetic tape that is later played into the type of display and pictorial recorder described above. Having the original signal available on magnetic tape permits a wide variety of electronic signal processing to be performed on the signal to achieve contrast enhancement, to present the data in other than pictorial form, e.g., total acreage of a specific crop, to identify specified crops automatically (as described in Chapter 9), etc. When the sensor is in an unmanned spacecraft, the infrared-detector signal will, of course, have to be trans-

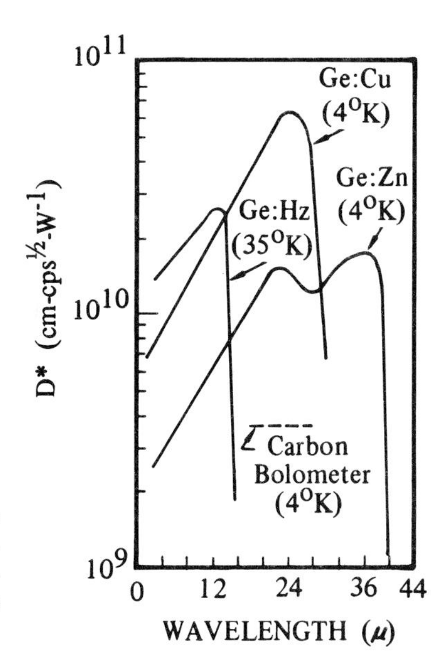

FIGURE 45 A comparison of the sensitivities of various commonly used photodetectors that require cooling to temperatures below 77° K. The temperatures are shown on the figure.

mitted to the ground, where it will then be recorded either on magnetic tape or on a CRT photographic-film system such as that described above.

Some examples of infrared imagery are described below.

Infrared Imagery

Plates 6 and 7 give examples of infrared imagery together with ultraviolet and visible imagery made simultaneously of the same areas; thus comparisons among these three spectral bands are easily possible. Only the infrared imagery is discussed in this section. For comparison with more conventional techniques, three types of photographic imagery of the same areas are also presented: panchromatic film, ordinary color film, and Infrared Ektachrome film. The remaining members of each set of imagery are scanner imagery in the narrow spectral bands indicated. These were produced by the type of scanning instrument just described, recorded on magnetic tape, and later displayed on and photographed from a CRT in the manner described above. In this section attention is directed to the images lying in the infrared region, namely, the seven narrow spectral intervals 0.72–0.80 μ, 0.80–1.0 μ, 1.5–1.8 μ, 2.0–2.6 μ, 3.0–4.1 μ, 4.5–5.5 μ, and 8–14 μ. These bands coincide with the regions in which there is good transmission through the atmosphere.

The areas shown in Plates 6 and 7 are located, as indicated on the accompanying map, in central Indiana near the Wabash River just west-southwest of the city of Lafayette. The areas are typical of Midwest farming regions.

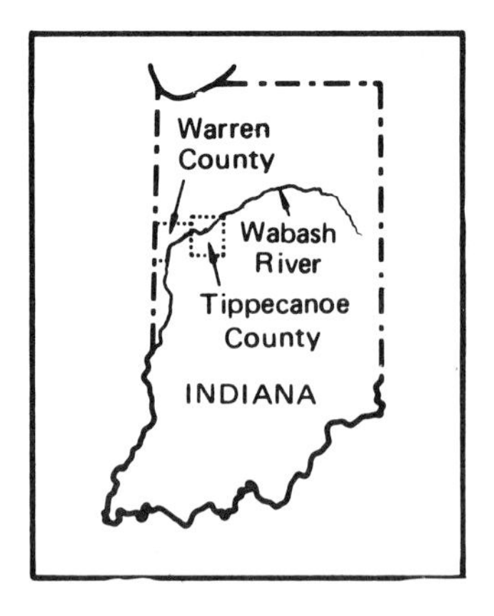

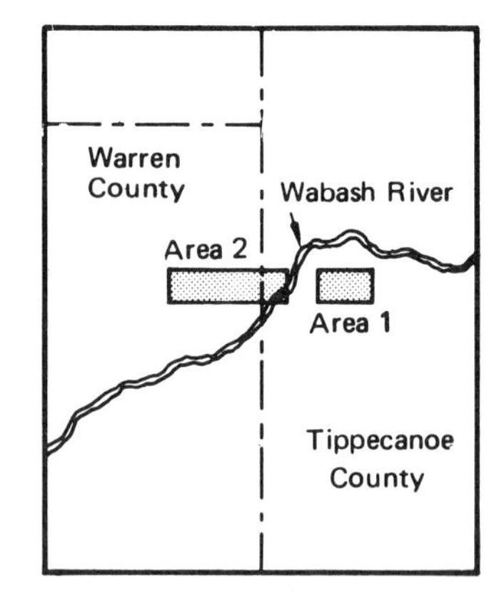

All the imagery was obtained at 2 P.M. on the afternoon of May 6, 1966, at an altitude of 3,500 ft above the terrain. The imagery in the 0.72–0.80-μ band was obtained with an RCA C70042 CP 1 photomultiplier detector with S-20* response. The 0.90–1.0-μ band was obtained with an RCA C70102 BP2 photomultiplier detector with S-1* response. The bandwidths were obtained by dispersing that part of the spectrum with a dense flint prism and placing apertures in the dispersed spectrum.[47] The several bands between 0.4 and 0.72 μ, discussed in the section on visible imaging (page 149), deal with the visible spectrum and were established with the same prism and similar apertures. The bands 1.5–1.8 μ, 2.0–2.6 μ, 3.0–4.0 μ, and 4.5–5.5 μ were obtained with InSb detectors cooled to liquid nitrogen temperature (approximately 70° K). The bandwidths were established by filters placed over the detectors. The band 8.0–14.0 μ was obtained with a mercury-doped germanium (Ge : Hg) detector. The long-wavelength band edge was established by the abrupt decrease in the sensitivity of this detector in the vicinity of 14 μ. The short-wavelength band edge was established by a filter placed over the detector. The detector was cooled to liquid helium temperature (approximately 4° K).

The ground area portrayed in the photographs does not coincide completely with that shown in the scanner imagery because the fields of view of the several available cameras and the scanner were not congruent. This will be true in general with multiple-sensor systems assembled from existing components and illustrates one of the several reasons behind the need to construct new systems and components for multiband remote-sensing research in agriculture.

Notice first that all the scanner imagery, although it does not equal, does approach photographic pictorial quality. This illustrates the progress that has been made in developing this type of sensor instrumentation during the last two decades. It is clear that it is now possible to produce imagery of quite good pictorial quality anywhere between 0.78 and 14.0 μ in the infrared part of the spectrum. The imagery also demonstrates that there is adequate reflective radiant power in the region 0.78 to 3.5 μ and adequate emissive radiant power in the region 3.5 to 14 μ so that these regions can be subdivided into much narrower subregions, each containing sufficient radiant power to permit the formation of acceptable imagery.

*See Figure 50 for curves of S-20 and S-1 spectral response.

Nothing even approaching a complete interpretation of these images can be given here because research in the use of these spectral intervals is just beginning; the general procedures for interpretation and, in general, even the spectral reflectivities and emissivities of these materials are not known. The imagery is presented to demonstrate that it can be obtained and that there is information contained in those bands that is not obtainable by conventional means, as evidenced by the shifting relative contrasts among the various materials as one progresses from band to band. Further than that, all that can be done here is to call attention to representative features of the imagery. Full realization of the potential must await the results of research yet to be done.

The materials present in the imagery of area 1 (Plate 6) are indicated in the accompanying labeled outline drawing of the field pattern. Notice that the relative contrasts in the 0.72–0.80-μ and 0.80–1.0-μ bands are quite different from those in the longer wavelength bands. There is less contrast between fields 1 and 3, which are plowed and sown, and wheat fields 2 and 11 than in the other bands. In the remaining bands there is strong contrast between the plowed and sown fields and the wheat. There is strong contrast between plowed and sown fields 1 and 3 and freshly plowed field 8. This contrast diminishes markedly in the longer wavelength bands. The wheat, and, in fact, almost all the vegetation, appears bright in the 0.72–0.80-μ and 0.80–1.0-μ bands and dark in all others. This indicates high reflectivity in these bands and either low reflectivity (in all bands below 4.0 μ) or low temperature or low emissivity, or both, in the bands longer than 3.0 μ. Oats in fields 9 and 12 appears in most cases intermediate in density between wheat and bare soil, probably because it is a young crop and a high fraction of soil is seen between the leaves of the young plants. The photography appears to confirm this. There are other relative contrast variations that are apparent but that are not discussed here.

In area 2 (Plate 7), the same relative contrast of wheat, red clover, and freshly plowed soil is evident in a number of instances. A new feature is the water present in the river, which appears uniformly dark toned in all wavelength bands. This is characteristic of water in the infrared region as well as the visible and ultraviolet. Sometime after dark, however, water will appear relatively bright in tone in the bands 4.5–5.5 μ and 8.0–14 μ.

Almost no other material in these images appears uniformly dark toned. There is one area, however, marked *anomaly,* that is uniformly dark toned. Although it is plowed earth, its contrast behavior differs from other plowed areas. No certain explanation is presently available. A likely speculation is that the soil is unusually moist and the plowing very recent.

There will, of course, be diurnal and seasonal changes that can furnish significant diagnostic information when time-lapse sets of imagery are compared. Diurnal effects are especially strong in the emitted part of the infrared, i.e., longer than 4 μ, because, as is well known, temperatures fluctuate diurnally. This has already been alluded to in discussing the river water in area 2.

To gain an impression of how strong diurnal effects can be, refer to Plate 8, which shows contact thermocouple measurements for a number of materials over one complete 24-hour period, near Ann Arbor, Michigan, in June 1963. A number of things are immediately apparent. First, just before dawn, a quasi-equilibrium has been established; i.e., the slopes of the curves are very small in most cases. After dawn this quasi-equilibrium is upset. All materials warm up, but some faster than others, and the curves cross. Peak temperatures are reached shortly after noon, and the reverse process occurs during the afternoon and evening, although somewhat more gradually than in the morning.

Notice particularly the temperature curve for water. It is distinctly different from the others in at least two strong features. First, the maximum temperature excursion is significantly less than any of the other materials shown. Second, it reaches its maximum temperature an hour or two after any of the other materials. This is explainable in terms of the absorptivity and emissivity of water and its very large heat capacity in comparison with most other materials. Because of this behavior, terrain temperatures will usually swing above water during the day and below water temperatures at some time during the night. During the winter, however, when the water is frozen, different behavior will be evidenced because ice has different thermal and radiative behavior than liquid water.

These diurnal (and seasonal) variations in radiative behavior clearly might provide a great deal of diagnostic remote-sensing information. This little-investigated group of phenomena represents one of the many areas in which research is needed to develop remote-sensing systems.

It is clear from the infrared imagery shown here and the comments above that many-band infrared imagery can provide new and more diagnostic remote-sensing information by capitalizing on wavelength variations, diurnal changes, and seasonal changes and that present capabilities will permit a great deal of shape information to be presented in the infrared. Polarization effects (see Chapter 9) are also present in the infrared, but they have been studied so little in agricultural applications that nothing can be said at this time about their possible utility.

Visible Imaging (Nonphotographic)

Photographic and television devices are the most common types of remote sensors used in the visible spectrum today. (For an extensive discussion of photographic systems see Chapter 2.) Photographic devices are ideal when a high-resolution image is desired and when the necessary provisions have been made for processing the film and returning it to a ground station. The NASA man-in-space programs are excellent examples. Since the missions were manned, simple cameras were used to bring back a wealth of valuable data recorded on film. However, for those space missions that are unmanned, such as Tiros and Ranger satellites, during which an extensive area below is to be surveyed, cameras are impractical and television systems, remotely (or automatically) controlled, are used since they produce electric signals that are easily telemetered to the earth station and then stored or displayed, or both. When high spatial resolution is required to cover a limited ground scene, complete photographic units can be built for satellites where the film-processing equipment, and film scanning, readout, and telemetering equipment are all in one package. This was done in the Lunar Orbiter II, where 211 pictures were taken of 13 possible landing sites on the moon.

However, where multispectral information both within and outside the visible region is required for discrimination purposes (see Chapter 9 for a detailed discussion of this), and extensive ground areas are to be studied from unmanned spacecraft, neither photographic nor television devices are optimum. For visible-region sensors used in conjunction with ultraviolet and infrared sensors in a multispectral system, all three must view an identical scene at the same instant and have the same dwell time and observation geometry. This can best be done if all the sensors have the same system configurations. Since photographic films are not sensitive to wavelengths longer than about 1 μ, the infrared devices are for the most part electro-optical line scanners. Thus, the visible and ultraviolet sensors must be electro-optical line scanners as well. The operating principles of line-scanner devices have been discussed in some detail earlier in this chapter. This section deals with the atmospheric properties in the visible spectrum, the radiation properties of materials of interest, the detection mechanisms used, and

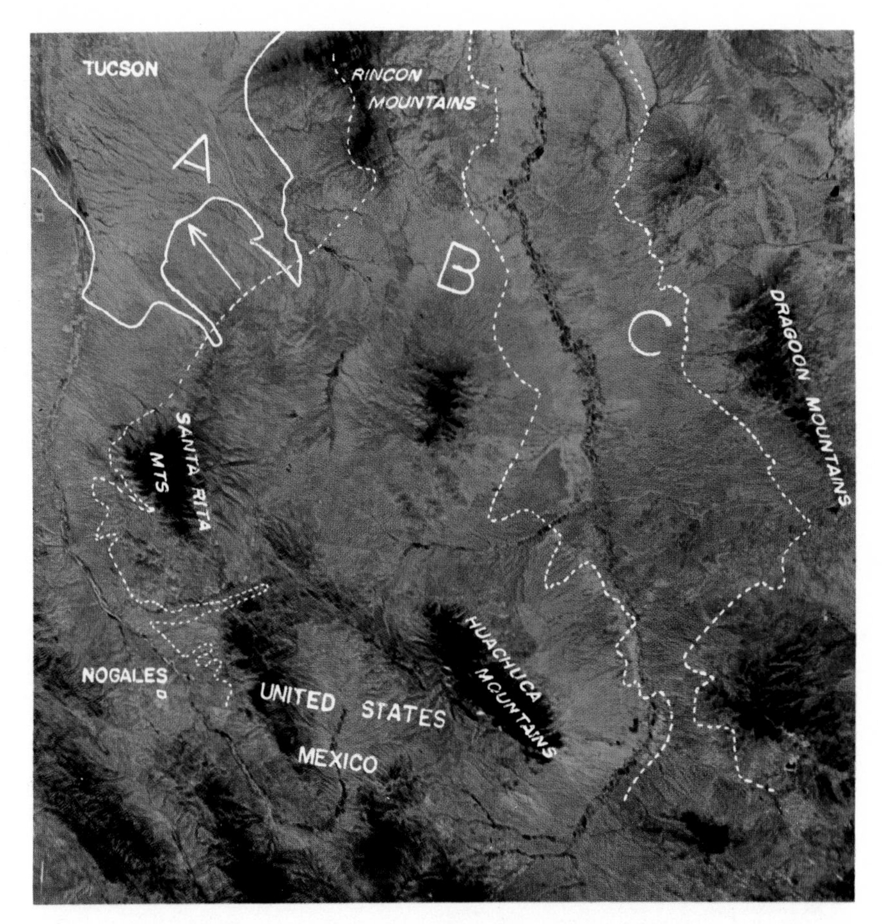

PLATE 1 Gemini IV photographic frame showing approximately the same area covered in Figure 5, Chapter 1. Area *A* shows the area of the desert scrub redefined (cf. Figure 5). Disparities between the dashed line derived from Figure 5 and the photo images indicate that proper interpretation of good-quality color imagery taken from space platforms will permit the refinement of vegetation-resource maps of arid regions. Limitations of time and funds for proper ground checking prevented redefinition of areas *B* and *C* from Gemini IV photography. (*Illustration and legend courtesy C. F. Poulton, Oregon State University.*) (For major discussion, see Chapter 1, p. 23.)

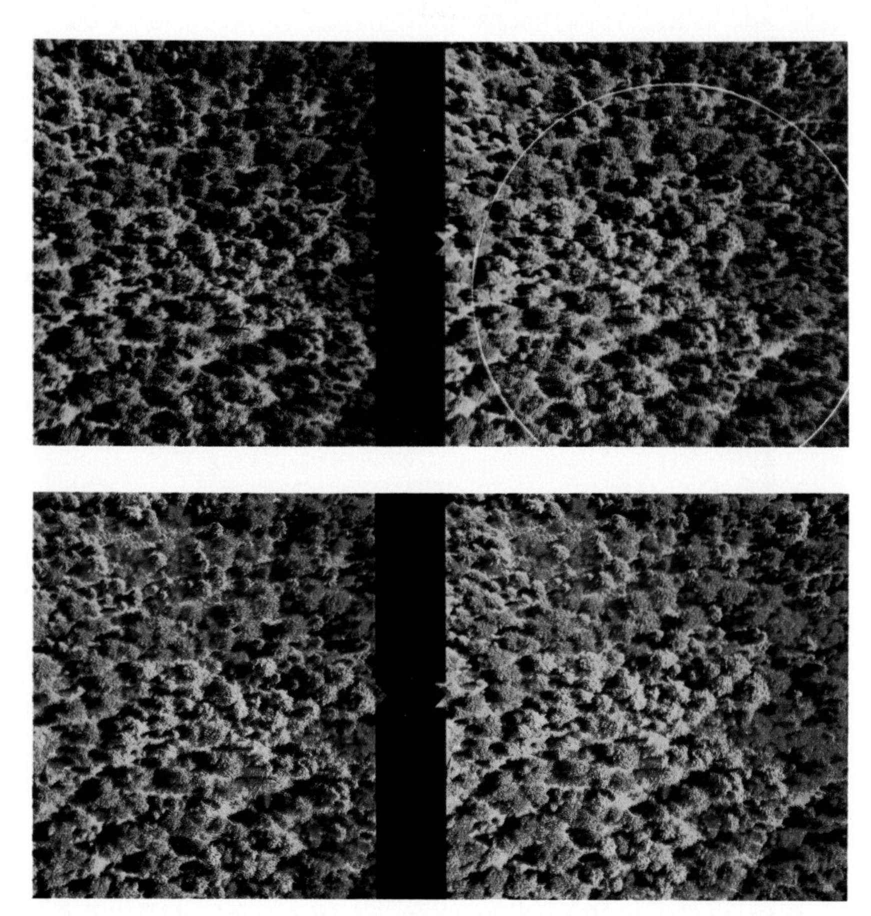

PLATE 2 Comparison of normal color film (Anscochrome D/200) and Kodak Ektachrome Infrared Aero film (E-3 process) for an area in which ponderosa pine (*Pinus ponderosa* Laws.) trees are dying from attack of Black Hills beetle (*Dendroctonus ponderosae* Hopk.). If the reader places a lens stereoscope over adjacent photos, he will get an enlarged, 3-dimensional view. Scale 1 : 1,584, taken July 1966. (*Courtesy U.S. Forest Service.*) (For major discussion, see Chapter 2, p. 48.)

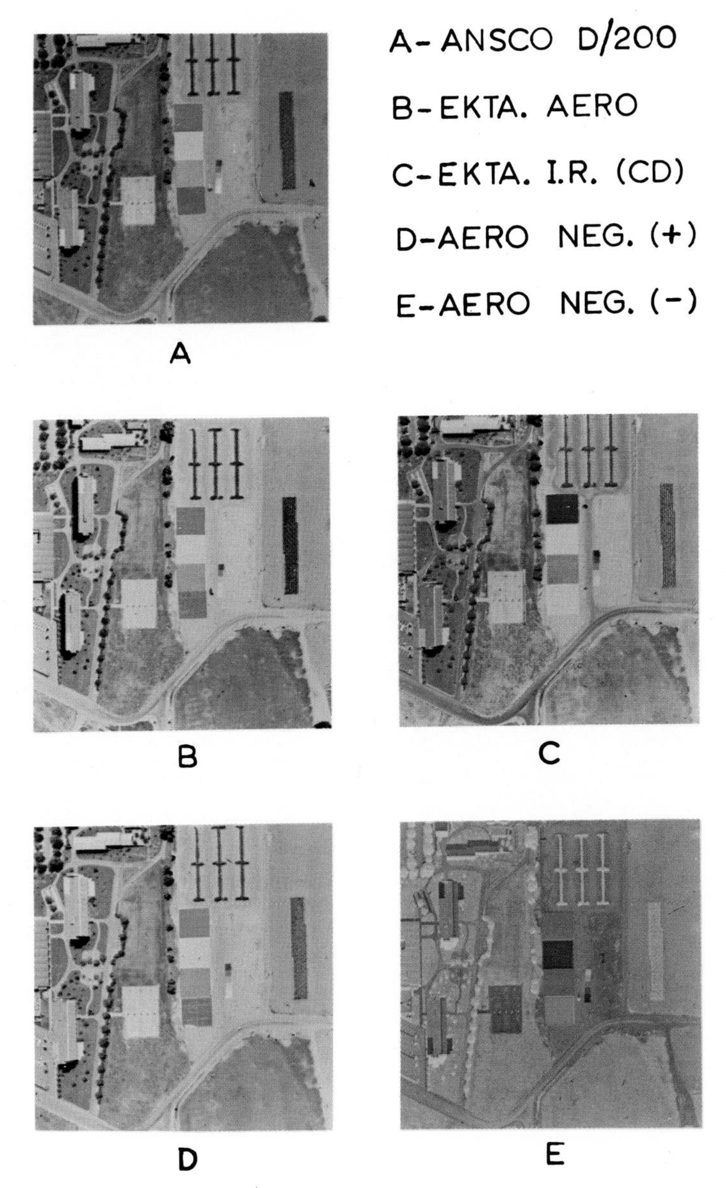

PLATE 3 Four aerial films are shown over a test site at the University of California campus at Davis, July 30, 1967. Approximate scale 1 : 6,000. *A,* Anscochrome D/200 with Wratten No. 1A filter; *B,* Kodak Ektachrome Aero film, no filter; *C,* Kodak Ektachrome Infrared Aero film with Wratten No. 12 filter; *D,* Kodak Aero-Neg film, no filter, processed to positive; *E,* Kodak Aero-Neg film, no filter, processed to negative. (For major discussion, see Chapter 2, p. 51.)

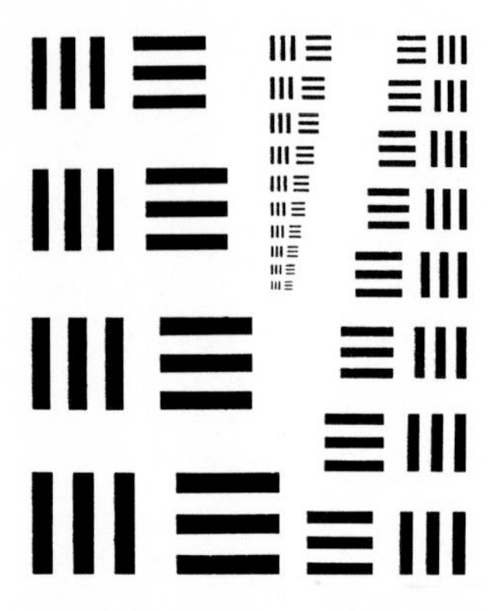

PLATE 4 U.S. Air Force bar target (upper left) is often used to assess resolution of a photographic system. Upper right photograph is made from an Anscochrome D/200 transparency. The color target shown here is used to check color fidelity and resolution capabilities of aerial color films. Lower stereo photograph is of resolution target shown in upper right photograph. (*Courtesy U.S. Forest Service.*) (For major discussion, see Chapter 2, p. 62.)

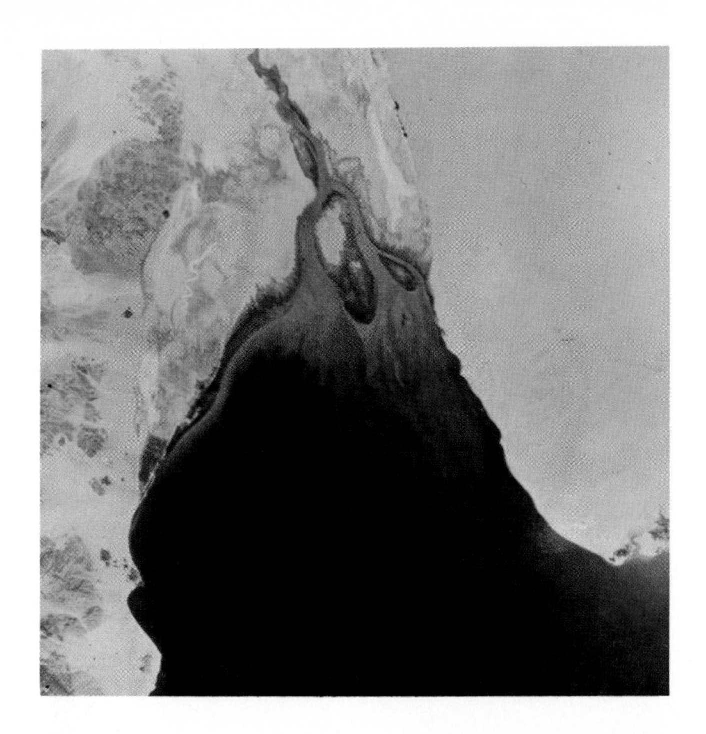

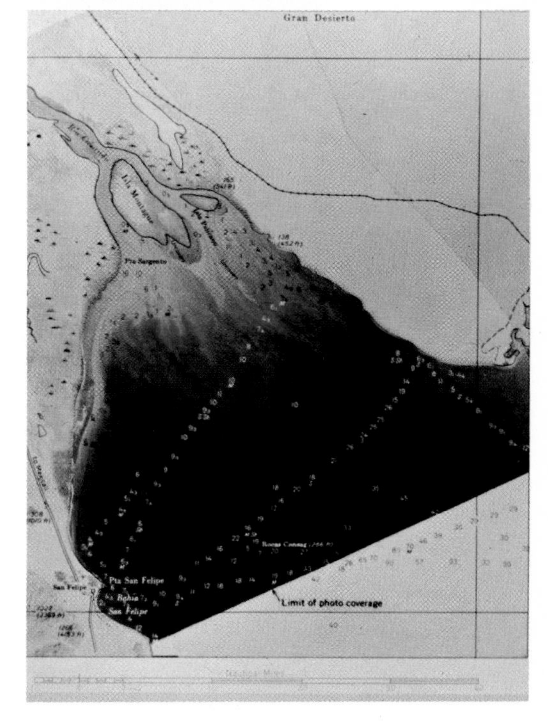

PLATE 5 Aerial photograph of Gulf of California taken from Gemini IV spacecraft (top). Experimental chartlet (U.S. Navy Hydrographic Chart No. 620) shows how information from one Gemini photograph could be used to identify water depth over 4,800 square miles (bottom). (*Courtesy* NASA.) (For major discussion, see Chapter 2, p. 63.)

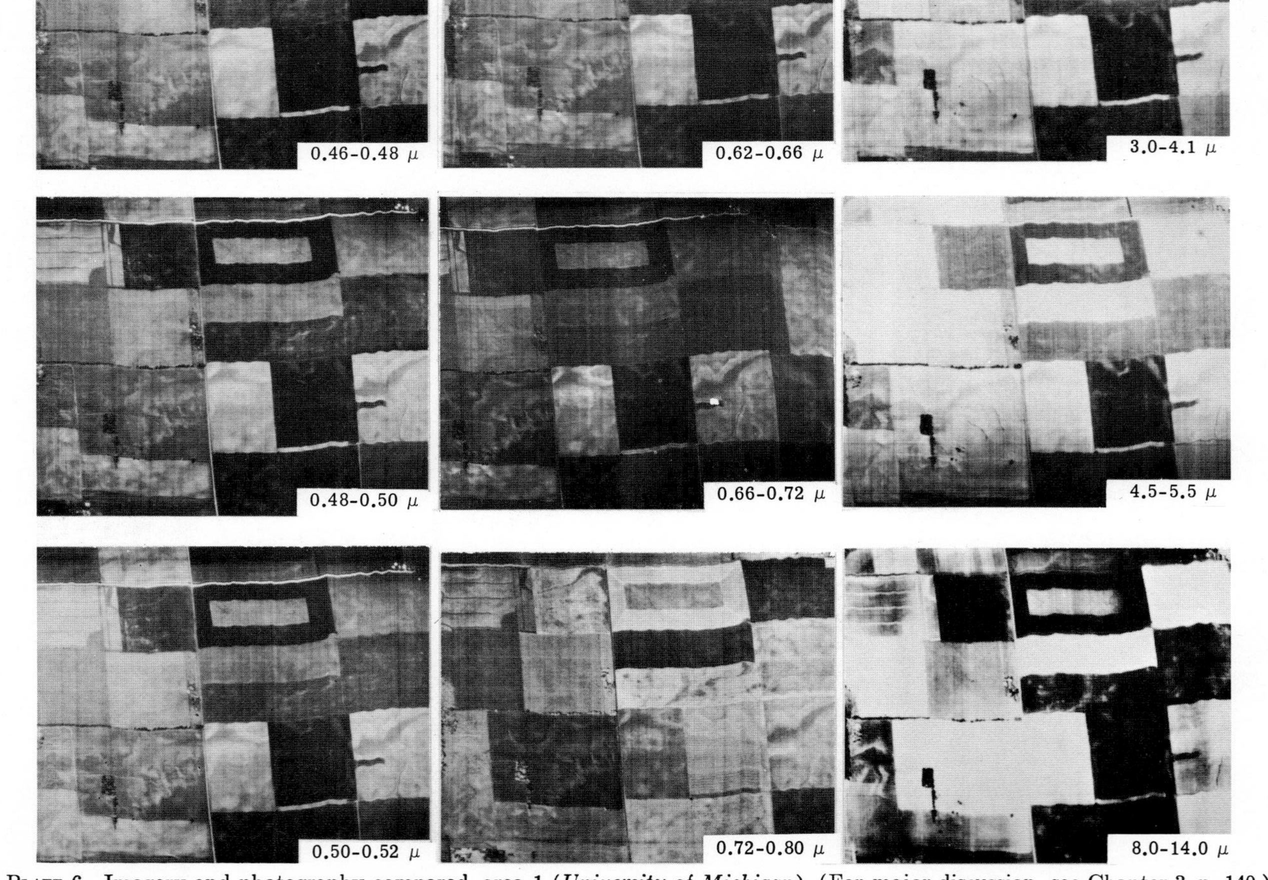

PLATE 6 Imagery and photography compared, area 1 (*University of Michigan*). (For major discussion, see Chapter 3, p. 140.)

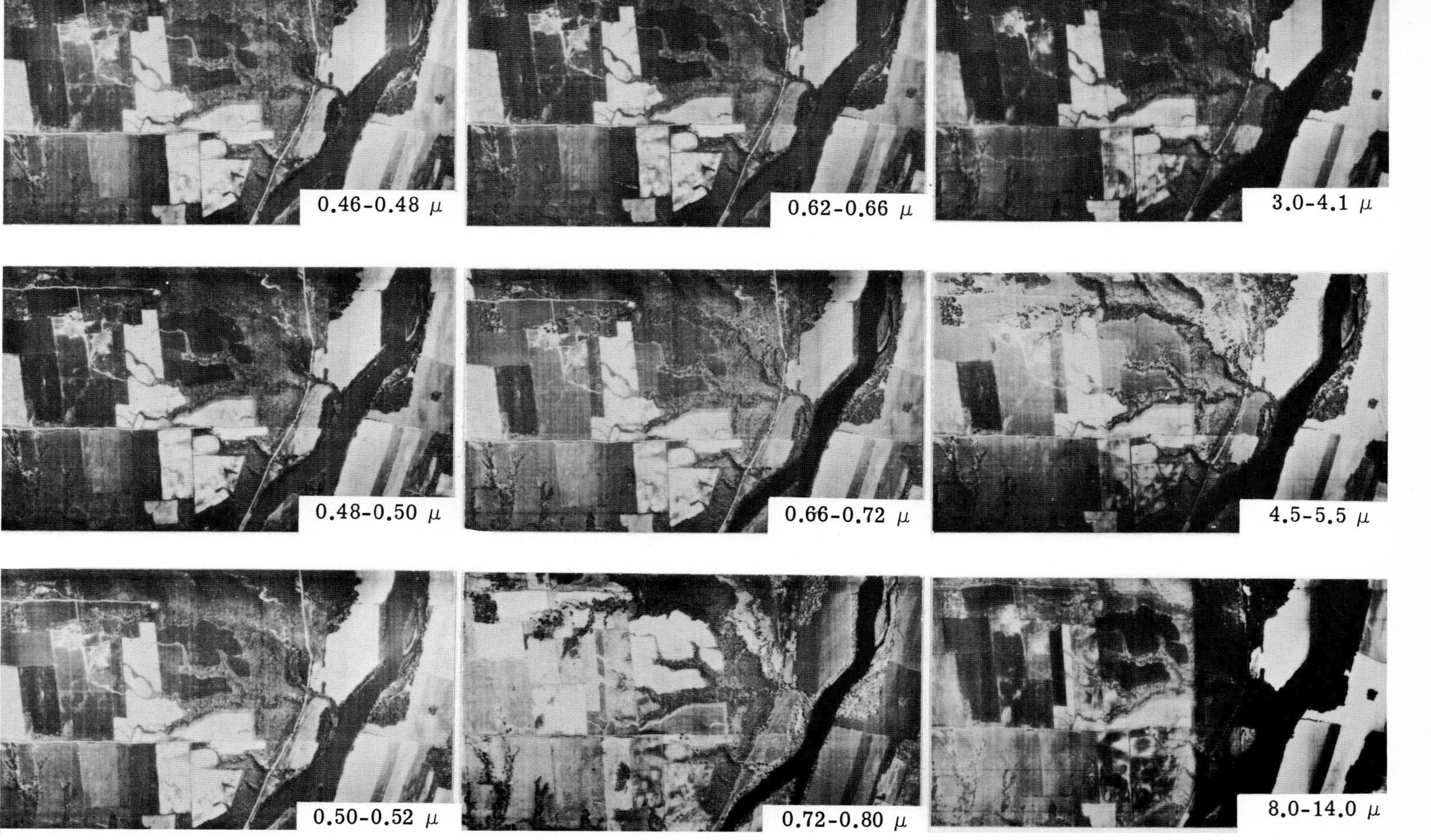

PLATE 7 Imagery and photography compared, area 2 (*University of Michigan*). (For major discussion, see Chapter 3, p. 140.)

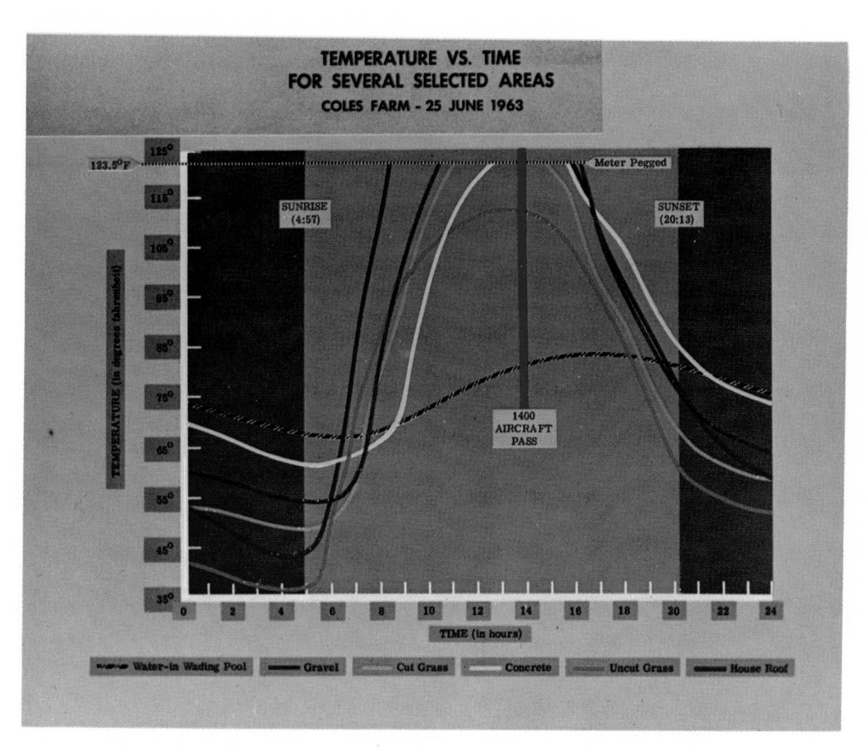

PLATE 8 Plot of temperature versus time for a farm near Ann Arbor, Michigan, June 25, 1963. (*University of Michigan.*) (For major discussion, see Chapter 3, p. 143.)

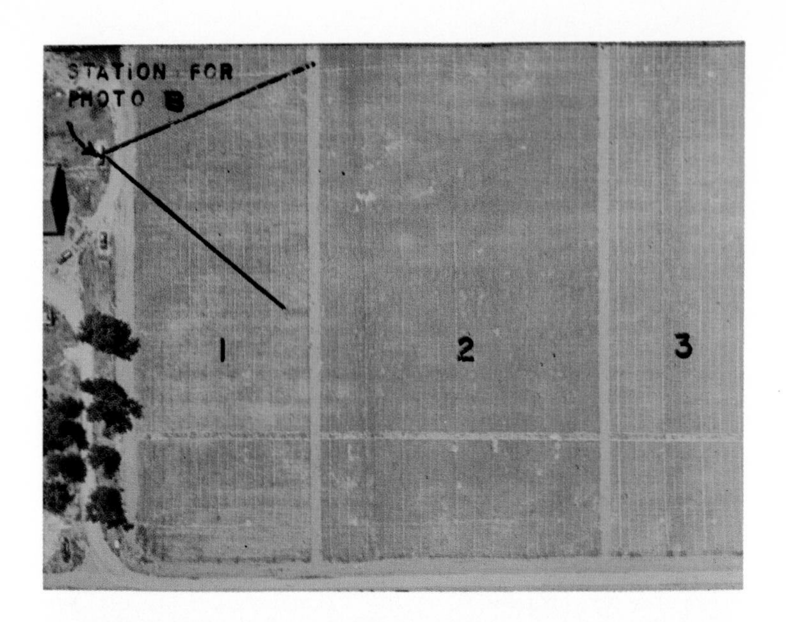

PLATE 9 Identification of crops by camera angle. *A,* Vertical photograph of an area near Davis, California, in which maturing wheat (blocks 1 and 3) cannot be differentiated from maturing oats (block 2). *B,* An oblique photograph of the same area shows a distinct tone or color contrast. (For major discussion, see Chapter 4, p. 168.)

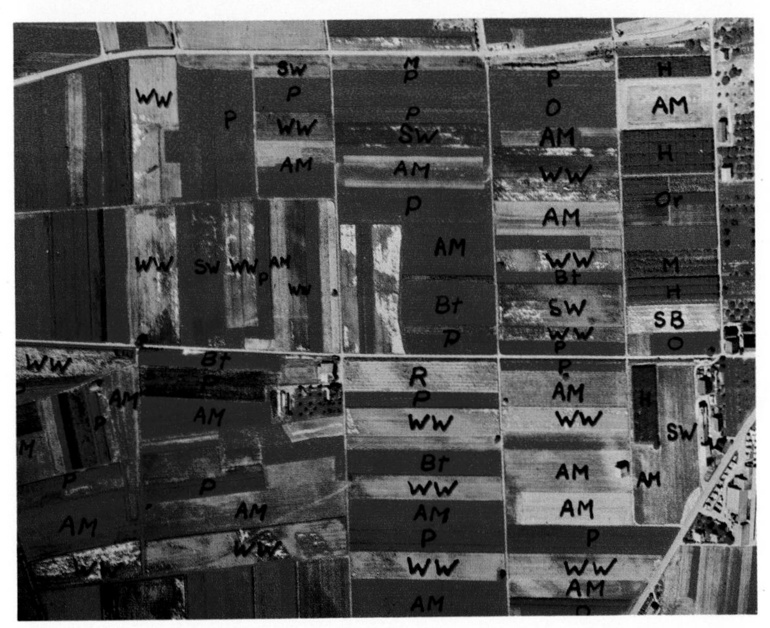

PLATE 10 Identification of crops by film type. A matched pair of Ektachrome and Infrared Ektachrome photos of an area near Zurich, Switzerland, in which Steiner is investigating the extent to which major crops can be identified on color aerial photos. On *B*, WW = winter wheat; SW = summer wheat; SB = summer barley; Bt = beets; P = potatoes; O = oats; M = maize; Or = orchards; AM = artificial meadow; R = Rape; H = hops. (For major discussion, see Chapter 4, p. 168.)

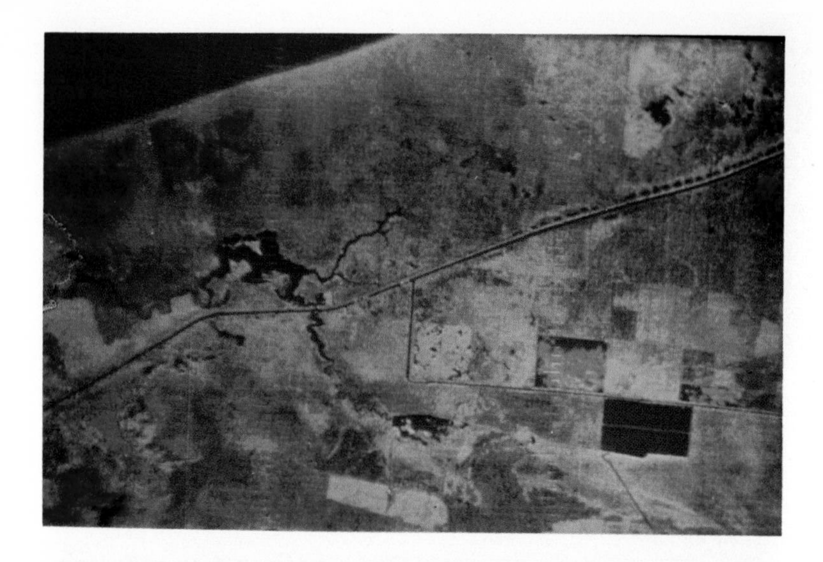

PLATE 11 Examples provided by Simonett, CRES, University of Kansas, of radar imagery obtained in areas in Kansas, using *K*-band radar. *A,* Brown and orange colors are areas of grassland and coastal marshland. The mottled appearance of the brown is caused by the pattern of surface vegetation and wet areas. Rectangular field patterns in mixed orange and green are rice fields. Reservoirs (lower right) are dark blue; two are partly empty, and bottom-growing vegetation produces a green pattern. Narrow blue line parallel to coast is the Intracoastal Waterway; blue spots are adjacent ponds. *B,* All yellow fields are sugar beets; all dark brown–blue fields are bare ground; all marshes are bright amber. Sorghum fields are more amber than wheat, and alfalfa fields are more green than sorghum fields. (*Courtesy Westinghouse Corp. and Earth Resources Survey Program of* NASA.) (For major discussion, see Chapter 4, p. 168.)

PLATE 12 Ektachrome (*A*) and Infrared Ektachrome (*B*) photographs of an experimental plot in which late blight of potato is present in specific subplots. On *B,* the subplots are readily detected by their dark blue to black color (healthy vegetation shows red). Photos taken on same day near Orono, Maine. (*Courtesy Manzer and Cooper, University of Maine.*) (For major discussion, see Chapter 4, p. 175.)

PLATE 13 Top pair: Matching Ektachrome (left) and Infrared Ektachrome aerial photos of an area in which some of the tree species are best identified by their unique colorations on the Ektachrome photo, and the remainder, by their unique colorations on the Infrared Ektachrome photo. It follows that a more complete identification of the many tree species in this area could be made through a study of both types of photography than from a study of either type alone. Bottom pair: Matching Ektachrome (left) and Infrared Ektachrome aerial photos of an area in the Sierra Nevada Mountains of California in which a bark beetle is causing heavy mortality to ponderosa pine. Note that the total mortality is far more readily discerned on the Infrared Ektachrome photo, in which the dead trees appear blue to black, than on the Ektachrome photo. (For major discussion, see Chapter 4, p. 197.)

some discussion of system operation. Television systems are described extensively in widely available literature and therefore are not discussed here.

Basic Radiation Principles

In the visible region of the electromagnetic spectrum, the irradiance at a remote sensor is determined primarily by the characteristics of the illuminating source, the reflectance of objects, the spatial distribution of the objects, and the atmospheric transmission. In general, the illuminating source may be a natural one such as the sun, moon, or stars, or an intense artificial one such as a laser or a mercury-arc lamp. At present the artificial sources do not have sufficient power for use in satellite systems: the long atmospheric path that the radiation must travel—from the sensor to the earth's surface and back again—causes serious attenuation. Pulsed lasers (i.e., those operated so that discrete pulses of radiation are generated with constant repetitive rates) that can generate high power for short time intervals have power-output levels that are marginal at best. Since the state of the art of artificial illumination sources for the visible spectrum is not sufficiently advanced for these systems, they are not covered here.

The spectral distribution of direct solar illumination incident on a plate 1 m^2 at sea level is shown in Figure 46. For comparison, curve *a* in the figure is the solar irradiance arriving at a point outside the earth's atmosphere. Curves *b, c, d, e,* and *f* are irradiance values when the sun is at 0°, 60°, 70.5°, 75.5°, and 78.5°, respectively, from the zenith. These values are typical for a clear day. Not shown in the figure is the irradiance on the plate due to radiation from other parts of the sky. Quantitative data of this kind are not readily available, but one can assume that the spectral distribution would be equivalent to the Rayleigh and aerosol scattering curves shown in Figure 47. The total natural illumination on an object at the earth's surface is the sum of these two components as well as the reflections from contiguous objects. It varies greatly with type and amount of cloud cover, haze level, and other environmental factors.

The next important factor that affects the visible radiation received at the remote sensor is the object's reflectance. The incident natural illumination, $S(\lambda)$, has a spectral characteristic, and if the reflectance, ρ, varies with wavelength, it will modify the spectral shape by a multiplicative factor: i.e.,

$$\text{reflected radiation} = K\rho_\lambda S_\lambda,$$

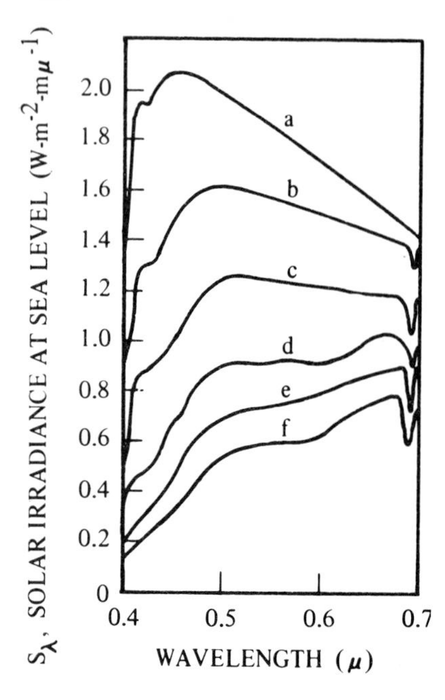

FIGURE 46 Plot of spectral distribution of solar irradiance at sea level.[48]

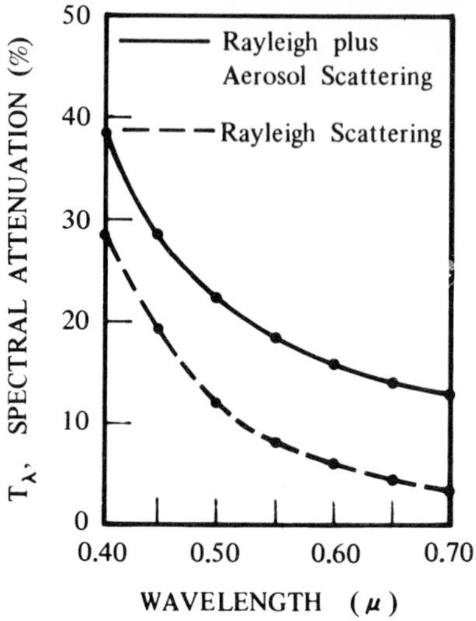

FIGURE 47 Plot showing percentage of attenuation due to Rayleigh plus aerosol scattering and Rayleigh scattering alone.

where K is a constant. The spectral shape is further changed by the transmission of the atmosphere, τ_λ, discussed in the following section. The spectral irradiance, H_λ, at a remote sensor is then proportional to

$$H_\lambda \propto \rho_\lambda \tau_\lambda S_\lambda. \qquad (15)$$

The reflectance function varies considerably from one object to another and for each object within a class. This variation is caused by a number of factors, including, in the case of vegetation, level of maturation, soil condition, and thermal history of the environment; and for other objects, the variation is caused by differences in the surface topography and the moisture content.

Spectral-reflectance curves for several plants and soils under various conditions appear in Chapters 5, 6, and 7.

The spatial distribution of plants is characteristic and therefore may be used to identify the species and perhaps to determine levels of maturation. Naturally occurring vegetation is usually randomly distributed as compared to crops that are planted in some orderly geometrical pattern. Since high spatial resolution in the ground is achievable with modern visible remote sensors, these spatial patterns can be determined. Further, the variation of the spatial pattern with level of maturation may also be determined. Unfortunately, the data that do exist today have not been examined for this, to the author's knowledge.

A judicious consideration of the spectral, temporal, and spatial characteristics of the objects of interest will be valuable in the design of remote sensors for agricultural and forestry purposes, and especially in the analysis and processing of the sensor data.

Atmospheric Attenuation

The attenuation of visible-region electromagnetic radiation through the atmosphere depends upon absorption due to gases and water vapor and scattering by atmospheric aerosols. Attenuation by absorption plays a minor role in this spectral region. An absorption band for molecular oxygen and water vapor exists at about 0.68 μ. This band can attenuate 10 percent or more for a vertical path through the atmosphere to satellite-borne sensors.

Attenuation due to scattering of radiation is the most significant factor in the visible spectrum. It is caused by an interaction between the radiation and small particles that range in size from molecular dimensions to as large as 10 μ (water droplets). The amount of scattering will vary with the relationship between the size of the scattering particle and the wavelength of the radiation being scattered. These parameters have three commonly recognized ranges. When the diameter of the scattering particles is significantly less than the wavelength of the radiation, the scattering is proportional to $1/\lambda^4$, where λ is the wavelength. This is called Rayleigh scattering since a scattering theory formulated by Lord Rayleigh applies here. When the diameter of the scattering particle is of the order of the wavelength being scattered, a theory formulated by Mie must be considered. In this region the amount of scattering will be a varying function of λ. Mie theory predicts scattering proportional to $1/\lambda^4$ for particles significantly smaller than λ (in agreement with Rayleigh theory), proportional to $1/\lambda^2$ for particle diameters equal to λ, proportional to $1/\lambda$ for particle diameters equal to $3\lambda/2$, and independent of λ for particle diameters equal to 2λ.[45] This last case is the third commonly recognized scattering region; i.e., when the scattering particles are significantly greater in diameter than λ, there is no wavelength dependence of the scattering.

An example of this is the white appearance of scattering from fogs of large-diameter water particles. Water droplets in the form of haze,

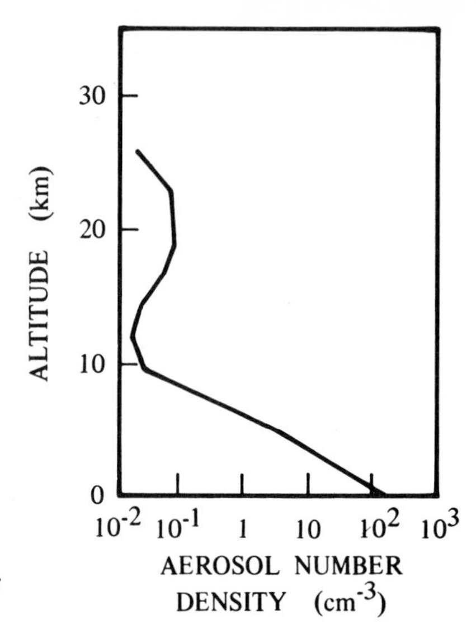

FIGURE 48 Plot of aerosol number density versus altitude.

fog, and clouds and dust particles constitute an aerosol that scatters radiation significantly in the visible and ultraviolet portions of the spectrum. The relative contributions of Rayleigh and Mie scattering are given in Figure 47, which presents the percentage of attenuation due to Rayleigh scattering alone as well as the percentage of attenuation due to Rayleigh scattering plus aerosol (Mie) scattering. These data were taken from a study by Elterman[49] and were calculated for a "Clear Standard Atmosphere" (CSA). This implies a relative humidity of 55 percent, a temperature of 23° C, and an aerosol consisting of a particle concentration of 215/cc at sea level with an altitude distribution shown in Figure 48. Aerosol scattering can and usually is much worse than this because of the presence of water droplets in clouds and large haze contributions, especially in urban areas. A substantial amount of data on aerosol scattering exists. However, the problem is complex, and reasonable, predictive, analytical models for other than clear days have not been calculated. In general, one can say that the earth as viewed from a satellite has 50 percent of its surface area obscured by clouds at any specific time.

Electro-optical Line Scanners

Except for the detector element, line scanners for use in the visible region are identical in design to those used in the infrared region. See the discussion of infrared scanners (page 133) for details.

The detectors used in scanners for the visible spectrum differ considerably from those used in the infrared. In the visible region they are photoemissive types. Here, incident photons cause electrons to be emitted by the photosensitive surfaces. These electrons are then collected by an anode, and the electric current through the circuit (Figure 49) is proportional to the incident radiation intensity. The photosignal is then read out as an output voltage across the load resistor R_l. These devices have been used extensively in the laboratory and in working systems for many years. Figure 50 gives the sensitivity of various commercially available photoemissive detectors.

In addition to the photoemissive detectors, the infrared photoconductive detectors described in the section on infrared imaging, page 137 (see Figures 43 and 44), are also sensitive to visible electromagnetic radiation and are often used in line scanners with spectral filters that limit the incident radiation wavelengths to those desired. Their sensitivity in the visible, however, is not as great as that of photomultipliers.

Visible-Region Imagery

Scanner imagery in a number of narrow bands throughout the visible region is illustrated in Plates 6 and 7. Ten visible spectral intervals are shown in the bands labeled 0.40–0.44 μ, 0.44–0.46 μ, 0.46–0.48 μ, 0.48–0.50 μ, 0.50–0.52 μ, 0.52–0.55 μ, 0.55–0.58 μ, 0.58–0.63 μ, 0.62–0.66 μ, and 0.66–0.72 μ. There are changes in relative contrasts between pairs of fields in going from one band to another. However, many of these contrast shifts tend to be somewhat subtle, so it is laborious to employ them for crop and soil identification and differentiation simply by viewing them. Full utilization of these subtle differences requires the

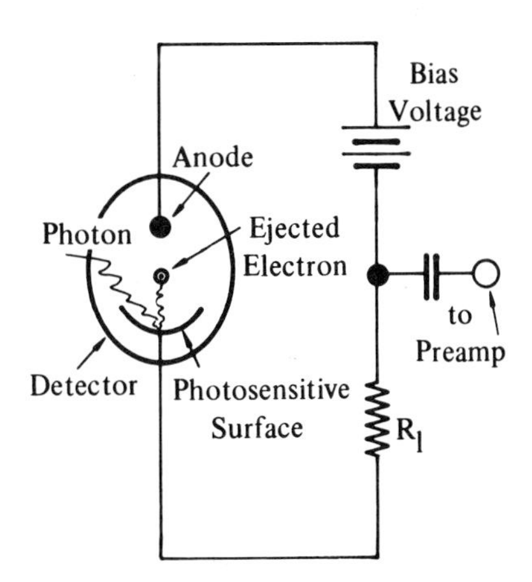

FIGURE 49 Diagram of a typical photoemissive detector circuit.

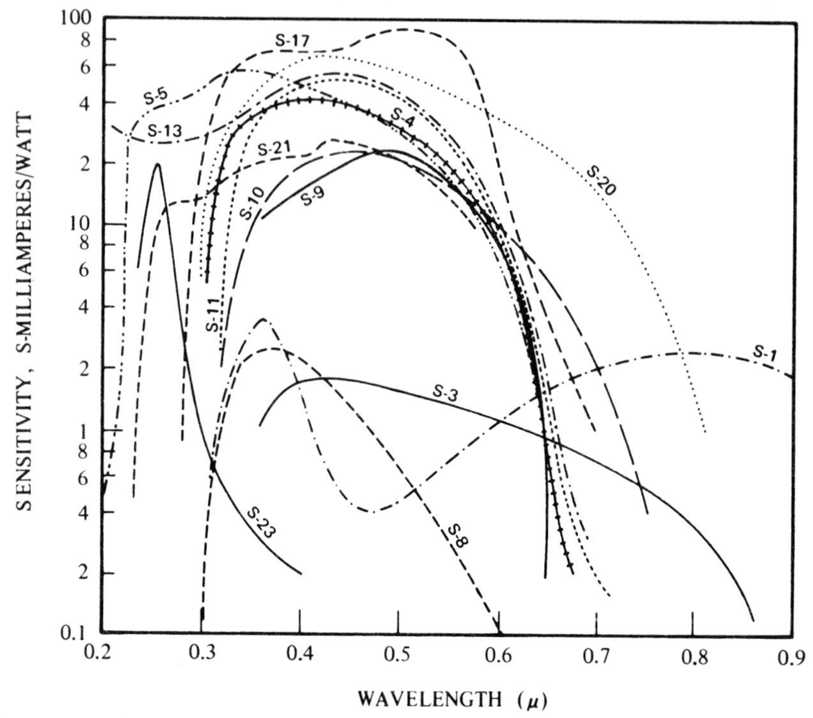

FIGURE 50 Comparison of spectral response of common photoemissive detectors.

more powerful method of automatic pattern recognition, which is described in Chapter 9.

In area 1 (Plate 6), notice the contrast between fields 9 and 10 and between 9 and the very narrow field just to the right of 9 in the band 0.66–0.72 μ. The contrasts almost disappear in the band 0.62–0.66 μ, reappear in the band 0.58–0.62 μ, and shift slightly throughout the remaining shorter wavelength bands. Also, notice fields 2, 6, 7, and 8. In the band 0.66–0.72 μ, they do not exhibit much contrast. In the band 0.62–0.66 μ, field 8 is quite different from the rest. In the bands 0.55–0.58 μ and 0.52–0.55 μ, fields 6, 7, and 8 are similar, but field 2 is darker. The remaining shorter wavelength bands show field 8 as lighter than the rest. Fields 3 and 4 do not exhibit strong contrast in any of the three types of photography, but do show easily noticeable contrasts in the ten visible-region narrowband scanner images. This illustrates the fact that merely observing in narrow visible wavelength bands can, on occasion, produce some stronger contrasts than any of the common photographic films. Other such contrast shifts with wavelength, and differences from photography can be recognized upon close examination.

In area 2 (Plate 7), notice that the river water is dark in all the visible bands as it was in all the infrared bands. This may be due to a high sediment load in the river water, as occurs in many rivers. It is well known that clear water is more or less transparent to most visible wavelengths. Notice that the wheat field labeled 1 is not sharply differentiated from the red clover immediately to its right in any of the photography, but it is differentiated from it in the scanner bands 0.52–0.55 μ, 0.55–0.58 μ, and again in the band 0.66–0.72 μ, but not so sharply in the other scanner bands in the visible region.

This scanner imagery in the narrow-band visible region illustrates, first, that such imagery of nearly photographic quality can now be obtained. Second, it demonstrates that some contrasts can be obtained in this manner that are not apparent in photography. Many of these contrasts are, however, subtle. Experience with imagery in so many bands indicates that direct viewing and interpretation of the imagery by a human being is very difficult and laborious. Human beings appear to be able simultaneously to intercompare not more than three bands at any one time with any reasonable facility. Therefore, when the information resides in the joint effect of subtle changes or differences among more than three bands, direct human interpretation appears impractical in all but very restricted instances. The real power in employing many narrow bands appears later (Chapter 9) when computer interpretation methods are discussed. Computers can intercompare a larger number of bands simultaneously than humans can. For computer applications, photographs are not useful because the information must be available in electronic form as it is in the scanner data.

Ultraviolet Imaging

The ultraviolet portion of the electromagnetic spectrum is, by definition, the region between 0.004 and 0.380 μ. It lies between the long-wavelength x rays and the violet portion of the visible spectrum. The ultraviolet region was discovered in 1801 by J. W. Ritter, who noted that in a prismatic spectrum the region beyond the violet caused a chemical action. He found that the rate of silver chloride blackening, which is caused by decomposition, increased when the silver chloride was exposed to this invisible energy. Experiments by Thomas Young in 1804, and later by others, showed that this invisible energy had, in fact, electromagnetic wave properties like the energy in the visible and infrared regions. A recent textbook by L. R. Koller[51] presents a basic survey of ultraviolet technology.

Radiation Phenomena

The sun is the most important source of ultraviolet radiation. On the Kelvin scale, its surface temperature is about 6,000°, and its calculated internal temperature is around 20 million degrees. The average value of solar radiation arriving outside the earth's atmosphere at the earth's mean solar distance is about 0.1396 W cm^{-2}. This quantity is called the solar constant, and to give the reader an idea of its magnitude, he can think of it as equivalent to 1,170 watts or 1.56 horsepower per square yard. About two thirds of this radiant energy actually reaches the earth's surface; the other third is lost by reflection, scattering, or atmospheric absorption. About 50 percent of this incident solar energy is concentrated in the visible spectrum, about 40 percent is in the infrared, and about 10 percent is in the ultraviolet. The spectral distribution of the ultraviolet portion of solar energy arriving outside the earth's atmosphere is shown in Figure 51.[48] This plot illustrates how rapidly the radiation intensity diminishes toward the shorter wavelengths. The value of irradiance is actually down to 2×10^{-6} W cm^{-2} at 0.14 μ. The earth's atmosphere attenuates strongly in the ultraviolet at wavelengths shorter than 0.28 μ, so much that the existence of solar radiation in this region was not experimentally verified until after World War II, when high-resolution solar ultraviolet spectroscopy was made possible by use of captured V-2 rockets. Since that time the region between 168 and 3,000 Å has been studied with photographic instruments[51] and the region from 120 to 1,300 Å has been studied by use of sensors with photoelectric detectors.* These measurements have indicated that the solar radiation in the region from 1,400 to 2,800 Å is continuous and approximately follows the blackbody law given by the Planck equation (see Equation 13, page 130). For wavelengths shorter than 1,400 Å, the major portion of the solar radiation appears as discrete spikes of spectral irradiance. These spikes are caused by atomic emission lines, which have as their source the chromosphere of the sun, where a great deal of atomic activity takes place.

Unlike the infrared region (for which remote sensors can be fabricated to observe the reflected radiation and the self-emitted radiation

*Ten thousand angstroms (Å) are equivalent to 1 micron. Angstroms are preferred usage with fractions of less than a tenth of a micron.

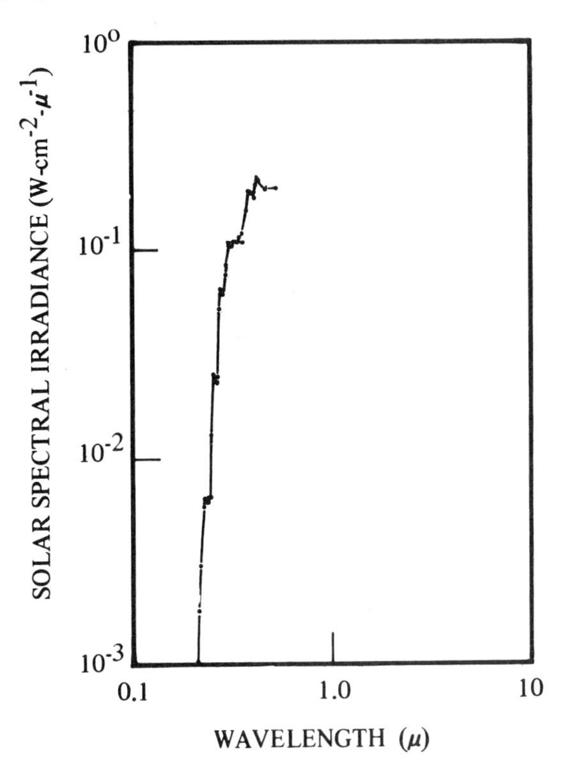

FIGURE 51 Plot of solar spectral irradiance at given wavelengths.

from an object, or both), remote sensors for the ultraviolet make use mainly of reflected radiation. This reflected radiation can have as its source the sun and sky during daytime, or some artificial lamp located at the remote-sensor platform at night. For sufficient self-emission in the ultraviolet region (see Figure 52), the object of interest must have a temperature of about 1,000° K or more. In order to detect the characteristics of agricultural materials such as crops and soils, which have temperatures around 300° K, it is reasonable to ignore the ultraviolet self-emission from the object because it contributes relatively little radiation in this part of the spectrum.

Four important properties must be considered in the design of ultraviolet remote sensors: the spectral characteristics of the radiation source, the effects of the intervening atmosphere, the spectral reflectance of the object of interest, and the spectral reflectance of surrounding objects that may be considered background materials. These prop-

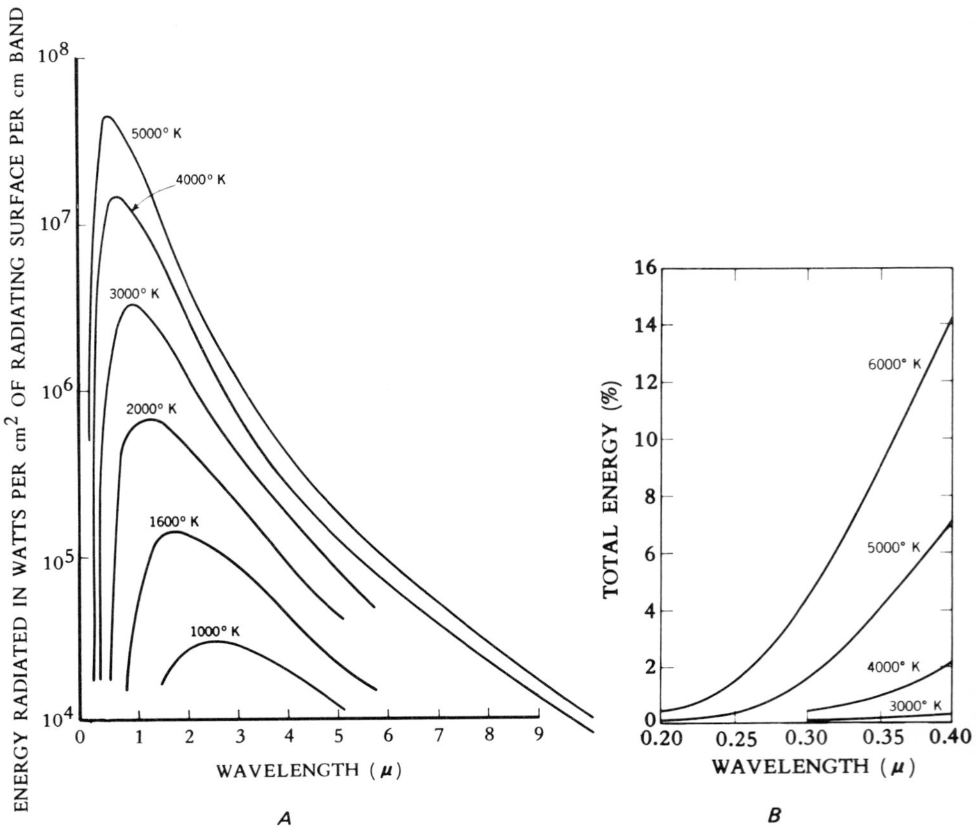

FIGURE 52 Blackbody energy. *A:* Energy distribution for a blackbody at various temperatures. *B:* Percentage of total energy radiation by a blackbody.[51]

erties are related by Equation 15 to the spectral irradiance, H_λ, arriving at the remote sensor:

$$H_\lambda \sim \rho_\lambda \tau_\lambda S_\lambda,$$

where ρ_λ is the spectral reflectance,
τ_λ is the spectral transmittance, and
S_λ is the source spectral irradiance.

The characteristics of solar radiation arriving outside the earth's atmosphere are described earlier in this section. Attenuation effects by the intervening atmosphere include primarily scattering and absorption. The fraction of radiation intensity transmitted through x kilometers of homogeneous air is given by $e^{-\sigma x}$, where σ is the attenuation coefficient, which is a function of wavelength λ. Values of σ at a given wavelength for two different samples of air can easily differ by a factor of 100. To illustrate further the variability, two samples of air having the same σ at one wavelength often have quite different values at another. The value of σ for air[52] depends upon three additive factors: (1) σ_A, Rayleigh scattering by air molecules; (2) σ_B, scattering and absorption by airborne particles and droplets, and (3) σ_C, absorption by gases. Thus, $\sigma = \sigma_A + \sigma_B + \sigma_C$. The various scattering mechanisms are described in the discussion of atmospheric attenuation (page 147). Rayleigh scattering is the dominant mode in the short-wavelength end of the visible region and in the ultraviolet regions. For this reason, the sky appears blue; i.e., solar radiation in the blue end of the visible spectrum is scattered about in the atmosphere and appears to arrive at the observer from all parts of the hemisphere. Scattering is quite strong in this portion of the electromagnetic spectrum, as evidenced by the fact that on a clear day the ultraviolet radiation due to scattered light from the sky, if observed as it falls on a horizontal plate on the earth's surface, may be greater in intensity than the amount directly radiating from the sun.

The primary causes of atmospheric absorption in the ultraviolet are ozone, O_3, and molecular oxygen, O_2. Ozone exists in layers a few kilometers thick about 30 km above the earth. This layer of gas strongly absorbs ultraviolet radiation between 0.2 and 0.28 μ. It has a weaker absorption band at 0.32–0.35 μ. Oxygen absorbs strongly in the region 0.13 μ to about 0.2 μ. In the region from 0.176–0.2 μ, the absorption spectrum of oxygen is made up of a number of strong absorption bands, whereas between the bands, oxygen is relatively transparent. Figure 53 presents some experimental results of measurements of σ versus wavelength on various days in Pasadena, California, in the year 1949.[53] To provide the reader with some quantitative understanding of the data, consider the day September 20, 1949. Figure 53 gives a value of 1 km^{-1} for σ at 3,200 Å. Since the percent transmittance is given as $e^{-\sigma x} \cdot 100$, we obtain a value of $e^{-1} \cdot 100$ for a 1-km atmospheric path (i.e., $x = 1$ km). The transmission is then about 37 percent. The theoretically derived Rayleigh scattering curve is also presented in Figure 53 to indicate that the observed attenuation is due to more than just molecular scattering.

For the most part, the ultraviolet-reflectance properties of agricultural materials and other materials that appear in natural scenes are not well known. Although laboratory spectrophotometers with attachments for measuring reflectance generally operate to wavelengths as short as 0.3 μ, reliable measurements at shorter wavelengths are virtually nonexistent. Figure 54, prepared at The University of Michigan, presents a sample of data in the 0.3–0.4-μ region. The data were ob-

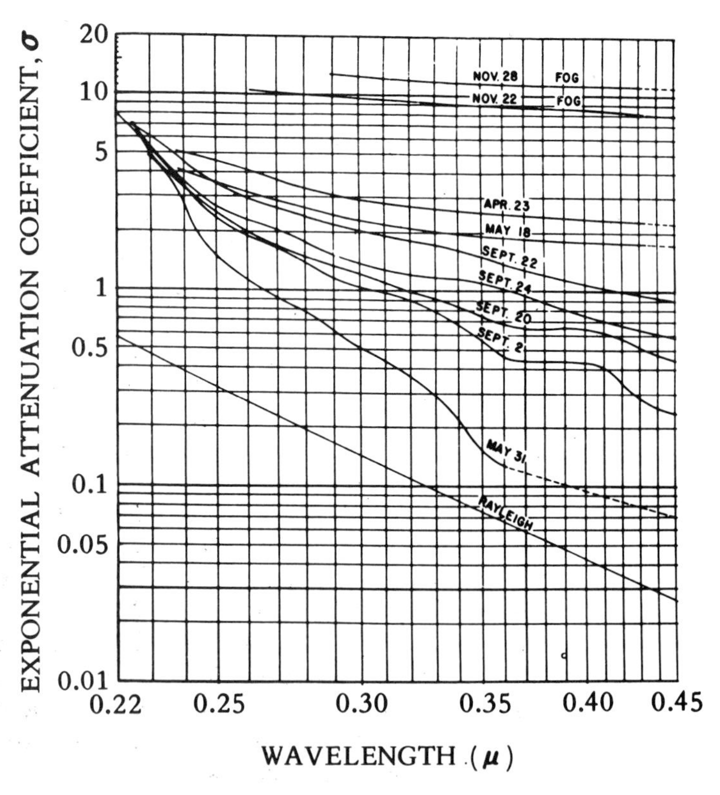

FIGURE 53 Typical attenuation curves illustrating the full range of atmospheric conditions, Pasadena, California, 1949.[56]

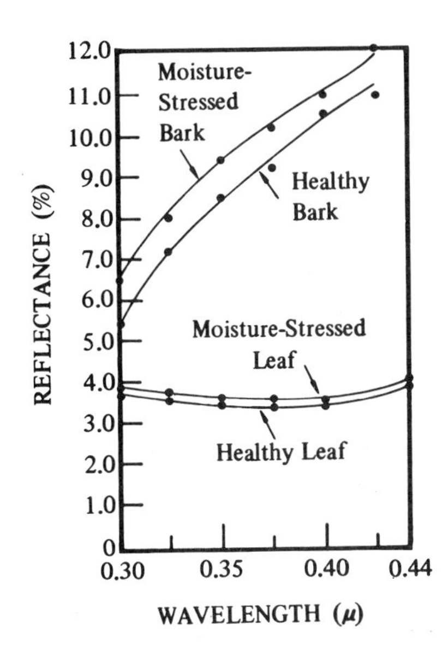

FIGURE 54 Spectral reflectance in the long-wavelength ultraviolet of bark and leaves of an apple tree. Data for both moisture-stressed and healthy materials are shown. (*University of Michigan.*)

tained with a Beckman DK-2 spectrometer with reflectance attachment. Included in the figure are data on the reflectance of leaves and bark from a healthy apple tree and from a moisture-stressed apple tree. Scientists at The University of Michigan have produced ultraviolet imagery of an apple orchard containing both healthy and diseased trees. This imagery (not shown here) shows noticeable contrast differences between healthy and diseased trees in consonance with the differences present in the spectrometer measurements. That is, the diseased tree had fewer leaves in the crown, so more of the bark was exposed, and the diseased tree crown appears lighter in tone than the healthy crown.

Ultraviolet Sensors

As compared to developments of radar and infrared nonphotographic sensors, relatively little has been done in the ultraviolet. Like infrared systems, ultraviolet systems employ such things as windows, mirrors and/or lenses, detectors, electronic amplifiers, and output-display devices. At wavelengths shorter than 0.3 μ, care must be taken in selecting a window material since many of them become opaque in this region. Some quartz glasses, quartz crystal, and several ionic salts such as lithium fluoride and barium fluoride are transparent at wavelengths as short as 0.12 μ. For wavelengths shorter than this, one is hard pressed to find a good window material.

For wavelengths longer than 0.28 μ, conventional lenses, as well as mirrors made of aluminum, or silver metal films are quite satisfactory for focusing the radiation onto a detector. At shorter wavelengths, however, the metal films become better absorbers than reflectors, and so refractive optics, i.e., lenses, may be more practical for focusing.

Ultraviolet radiation detectors are mainly photoemissive devices. This means that incident ultraviolet photons are absorbed by electrons in the sensitive film, imparting enough kinetic energy to the electron to allow it to escape from the surface. An electrically positive plate is placed sufficiently close to the sensitive surface to collect the emitted electrons. The electric current flow between the collector and the sensitive surface is monitored to provide a measure of the incident radiation flux. Commercial ultraviolet-sensitive detectors have been made by Westinghouse with platinum, tantalum, zirconium, or thorium as the sensitive element. Each has its own characteristic spectral response. In addition, sensitive surfaces composed of metal mixtures have been made by RCA and are commercially available. Relative spectral-response curves of these detectors are presented in Figure 55. Detectors operating at wavelengths shorter than 0.15 μ are research devices built for special applications. Apparently, the demand for detectors in this region has not been sufficient for industry to develop a commercial product.

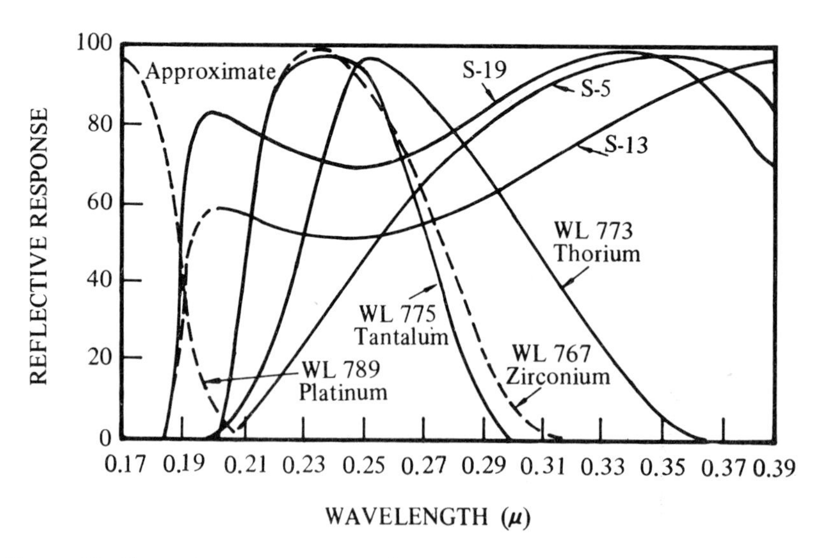

FIGURE 55 Spectral response curves of some ultraviolet-sensitive photoemissive devices.

In the part of the ultraviolet region of interest for remote sensing, i.e., where the atmosphere is transparent, scanning systems are of interest. These are identical to the system described earlier in this chapter (page 133), except that an ultraviolet rather than an infrared detector is used at the focus of the telescope.

It is worth noting that these scanning systems have an advantage over photographic methods in "seeing" through the high level of scattered radiation. In the scanning systems, part of the deleterious signal due to this scattered radiation can be subtracted electronically, permitting observation of more surface detail than can be recorded photographically.

Ultraviolet Scanner Imagery

In Plates 6 and 7, scanner imagery in the ultraviolet region is shown in the band 0.32–0.38 μ. This demonstrates that imagery of near-photographic quality can be obtained with scanners. For areas 1 and 2, there are no strong differences in contrast as compared with the adjacent visible band 0.40–0.44 μ. This lack of difference between the ultraviolet and the nearest visible bands, evidenced by the imagery of areas 1 and 2, is not true for all areas, as shown in Figure 56. That figure shows the Hayward fault area in California in two bands, the ultraviolet and the nearest adjacent visible band. A number of differences are immediately apparent. The road network contrasts more strongly with the surrounding terrain in the ultraviolet. The same is true for the human structures. Close examination will also reveal other differences between the two bands in the appearance of the terrain.

Conclusion

It has been shown that imagery can be generated in many narrow wavelength bands throughout the ultraviolet, visible, and infrared spectral regions, and the type of device required has been discussed. It has also been demonstrated that there are many contrast variations with changes in wavelength and that differences are observable that do not show in panchromatic, color, or Infrared Ektachrome photography. Full interpretation of this imagery has not been attempted, nor have the meanings of the contrast variations noted been specified since, for the most part, they are not yet known. The material presented does appear to indicate that these methods will provide a rich enough source of new information to warrant the research necessary to achieve a full understanding of the differing appearances of agricultural and construction materials and soils in the many available bands. The principal difficulty in using data of this sort, other than the present ignorance of its full meaning, i.e., the limitations of human interpretation, is shown to be surmountable when automatic recognition is discussed (Chapter 9).

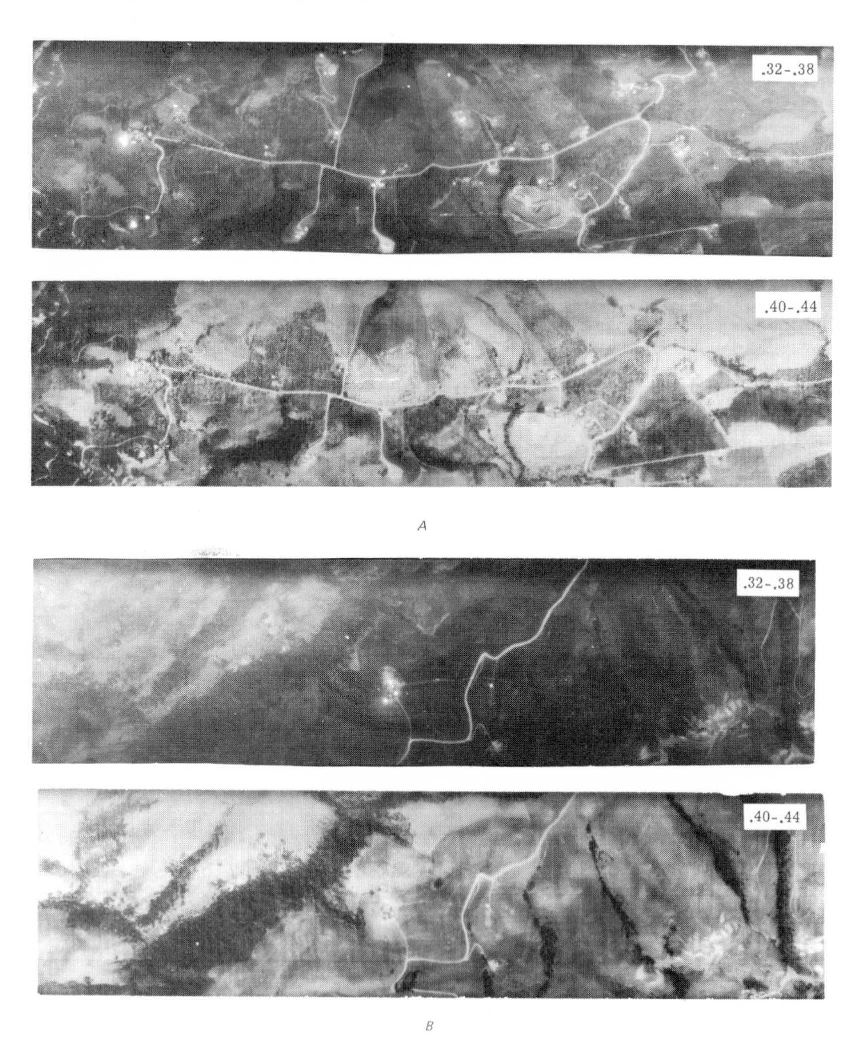

FIGURE 56 Imagery comparisons between ultraviolet (0.32–0.38 μ) and nearest adjacent visible band (0.40–0.44 μ), Hayward fault, California, May 26, 1966, 2,000-ft altitude, 2:40 P.M. (*University of Michigan.*)

REFERENCES

1. Beckman, P., and A. Spizzichino. 1963. The scattering of electromagnetic waves from rough surfaces. Macmillan, New York.
2. Cosgriff, R. L., W. H. Peake, and R. C. Taylor. 1960. Terrain scattering properties for sensor system design (Terrain Handbook II). Eng. Exp. Sta., The Ohio State University, Columbus. p. 16–18.
3. Carlson, N. L. 1967. Dielectric constant of vegetation at 8.5 GHz. Tech. Rep. 1903-5, Electroscience Lab., The Ohio State University, Columbus.
4. Morain, S. A., and D. S. Simonett. 1967. *K*-band radar in vegetation mapping. University of Kansas Center for Research in Engineering Sciences. CRES Rep. 6123. Lawrence, Kansas.
5. Medhurst, R. G. 1965. Rainfall attenuation of centimeter waves: comparison of theory and measurement. IEEE Trans. AP-13:550–564.
6. Bean, C. R., and E. J. Dutton. 1966. Radio meteorology. U.S. Nat. Bur. Standards Monog. 92, U.S. Govt. Printing Office, Washington, D. C. Chap. 7.
7. Cutrona, L. J., and G. O. Hall. 1962. A comparison of techniques for achieving fine azimuth resolution. IRE Trans., MIL-6, p. 119–121.
8. Planck, M. 1900. Ann. Phys. 4:553.
9. Gardon, R. 1956. The emissivity of transparent materials. J. Amer. Ceramic Soc. 39:278–287.
10. McMahon, H. O. 1950. Thermal radiation from partially transparent reflecting bodies. J. Opt. Soc. Amer. 40:376–380.
11. Nicodemus, F. E. 1965. Directional reflectance and emissivity of an opaque surface. Appl. Opt. 4:767–773.
12. Lewis, E. A., J. P. Casey, and A. J. Vaccaro. 1954. Polarized radiation from certain thermal emitters, Electronics Research Directorate, Air Force Cambridge Research Center, Air Force Research and Development Command. AFCRC Tech. Rep. No. 54-G.
13. University of Michigan. 1966. Peaceful uses of earth-observation spacecraft. Vol. III: Sensor requirements and experiments. Willow Run Laboratories, Rep. No. 7219-1-F(III). NASA Contract NASw-1084.
14. Straiton, A. W., C. W. Tolbert, and C. O. Britt. 1958. Apparent temperatures of some terrestrial materials and the sun at 4.3 mm. J. Appl. Phys. 29:776–782.
15. Cosgriff, R. L., *et al.* 1960. *Op cit.*
16. Conway, W. H., *et al.* 1963. A gradient microwave radiometer flight test program. Proc. 2nd Symp. Remote Sensing of Environment. Univ. of Michigan, Inst. Sci. Technol. Rep. No. 4864-3-X.
17. Chang, S. Y., and J. D. Lester. 1966. Radiometric measurement of atmospheric adsorption at 600 Gc/s. Proc. IEEE. 54:459–461.
18. Breeden, K. H., W. K. Rivers, and A. P. Sheppard. 1966. A submillimeter interference spectrometer: characteristics, performance and measurements. Georgia Inst. Technol., NASA CR-495.
19. Putley, E. H. 1965. Indium antimonide submillimeter photoconductive detectors. Appl. Opt. 4:649–658.
20. Williams, R. A., and W. S. C. Chang. 1966. Observation of solar radiation from 50 μ to 1 mm. Proc. IEEE. 54:462–470.
21. Meyer, J. W. 1966. Radio astronomy at millimeter and submillimeter wavelengths. Proc. IEEE. 54:484–492.

22. RAND Corp. 1963. Atmospheres, clouds, and spectra. The Application of Passive Microwave Technology to Satellite Meteorology: A Symposium, Sec. II. Santa Monica, Calif. Memorandum RM-3401-NASA.
23. Edison, A. R. 1966. Calculated cloud contributions to sky temperature at millimeter-wave frequencies. U.S. Dep. Commerce, Nat. Bur. Standards, Boulder Labs., Boulder, Colo. NBS Rep. 9138.
24. Cummings, C. A., and J. W. Hull. 1966. Microwave radiometric meteorological observations, North American Aviation, Inc., Columbus Div. Rep. No. NA 66H-371.
25. Hoffman, L. A. 1966. Propagation observations at 3.2 millimeters. Proc. IEEE. 54:449–454.
26. Frenkel, L., and D. Woods. 1966. The microwave absorption by H_2O vapor and its mixture with other gases between 100 and 300 Gc/s. Proc. IEEE. 54:498–505.
27. University of Texas. 1963. Attenuation and emission of 58 to 62 kMc/s freqencies in the earth's atmosphere. Elec. Eng. Res. Lab. Contract No. AF 33(657)-7333, Project 1002. AD 400 954.
28. Tolbert, C. W., A. W. Straiton, and R. A. Simons. 1964. The attenuation and emission in the earth's atmosphere of the complex of 60 Gc/s oxygen lines. Univ. of Texas, Elec. Eng. Res. Lab. Contract No. AF 33(657)-7333, Project No. 1002. AD 439 245.
29. Williams, R. A., and W. S. C. Chang. 1962. An analysis of the interferometric submillimeter radiometer. Ohio State Univ. Res. Found. Antenna Lab., Dep. Elec. Eng., Columbus. Rep. No. 1093-9. Grant No. NsG-74-60.
30. Ewen, H. I. 1967. State of the art of microwave and millimeter wave radiometric sensors. Int. Symp. on Electromagnetic Sensing of the Earth from Satellites. The Polytechnic Press, The Polytechnic Institute, Brooklyn, N. Y.
31. McGillem, C. D., and T. V. Seling. 1963. Influence of system parameters on airborne microwave radiometer design. IEEE Trans. on Mil. Electronics. MIL-7:(4).
32. Dicke, R. H. 1946. The measurement of thermal radiation at microwave freqencies. Rev. Sci. Instr. 17:268–275.
33. Sleven, R. L. 1964. A guide to accurate noise measurement. Microwaves. p. 14–21.
34. Moore, R. P., M. C. Hoover, J. S. Mendiola, R. L. Huchins, and C. A. Hawthorne. 1964. Microwave radiometric contrasts of metal targets against a terrain background. Naval Ordnance Lab. NAVWEPS Rep. 8198. Corona, Calif.
35. Microwave News. Microwave radiometry gains system interest. July 1965,
36. Nortronics Co. 1964. Radiometric data gathering program. Needham, Mass. Final Rep. No. AFCRL-64-170.
37. Arams, F., B. Peyton, and F. Haneman. 1963. Tunable millimeter traveling-wave master and 8-mm maser-radiometer system. Airborne Instruments Laboratories, Deer Park, N. Y. Final Rep. No. 8298-1.
38. Galloway, D. G., and C. W. Tolberts. 1964. Germanium bolometer detector of millimeter wavelength thermal energy. Rev. Sci. Instr. 33(5):628.
39. Cohn, M., F. L. Wentworth, and J. C. Wiltse. 1963. High sensitivity 100 to 300 Gc radiometers. Proc. IEEE. 51. p. 1227.
40. Meredith, R., and F. L. Warner. 1963. Superheterodyne radiometers for use at 70 Gc and 140 Gc. Millimeter Conf., Orlando, Fla.

41. Holter, M. R., S. Nudelman, G. Suits, W. L. Wolfe, and G. Zissis. 1962. Fundamentals of infrared technology. Macmillan, New York.
42. Kruse, P. W., L. D. McGlauchlin, and R. B. McQuistan. 1962. Elements of infrared technology: generation, transmission and detection. John Wiley, New York.
43. Jamieson, J. A., R. H. McFee, G. Plass, R. Grube, and R. Richards. 1963. Infrared physics and engineering. McGraw-Hill, New York.
44. Smith, R. A., F. E. Jones, and R. P. Chasmar. 1957. The detection and measurement of infrared radiation. Clarendon Press, Oxford.
45. Wolfe, W. L. (ed.) 1965. Handbook of military infrared technology. Office of Naval Research.
46. Potter, R. F., and W. L. Eisenman. 1962. Infrared photodetectors: a review of operational detectors. Appl. Opt. 1:567.
47. Work, E. A. To be published. A 12-channel multispectral imaging scanner, J. Opt. Soc. Amer.
48. Handbook of geophysics. 1961. Macmillan, New York.
49. Elterman, L. 1964. Atmospheric attenuation model, 1964, in the ultraviolet, visible, and infrared regions for altitudes to 50 km. Environmental Res. Paper #46. Opt. Phys. Lab. Project 767D, Air Force Cambridge Research Labs., Office of Aerospace Research, Hanscom Field, Mass.
50. Fairchild Camera and Instrument Corp. 1963. Du Mont multiplier phototubes, descriptions and specifications. 3rd ed. Du Mont Laboratories Div., Clifton, N. J.
51. Koller, L. R. 1965. Ultraviolet radiation. John Wiley, New York.
52. Johnson, F. S. 1954. The solar constant. J. Meteorol. 2:431.
53. Tousey, R. 1962. The extreme ultraviolet—past and future. Appl. Opt. 1:679.

4

Applications of Remote Sensing in Agriculture and Forestry

ROBERT N. COLWELL University of California, Berkeley, California
Contributing Authors: D. CARNEGGIE, R. CROXTON, F. MANZER, D. SIMONETT, D. STEINER

INTRODUCTION

As we consider applications of remote sensing, it is helpful to recognize that the intelligent management of resources entails three phases: (1) *inventory*—determining the quantity and quality of each resource in each portion of the area that is to be managed; (2) *analysis*—developing a plan based on this inventory, whereby each resource available in the area can be manipulated to provide maximum benefit; and (3) *operation*—the day-to-day implementation of the management plan. The objective of this chapter is to demonstrate, through specific examples, that aerial photographs and other remote-sensing data can be of benefit in each of these phases.

CULTIVATED CROPS

Despite the valiant efforts that reporting agencies have made from time to time to acquire information on crop acreages, whether from questionnaires sent to farmers or from direct on-the-ground surveys, it is often uncertain whether a satisfactory estimate has been obtained. Returns from the questionnaires have often been too few, too inaccurate, or too late. Returns from direct on-the-ground surveys, because of limitations of both time and funds, have sometimes constituted too small a sample of the vast agricultural area of interest for them to be used with confidence.

Aerial photos and other remote-sensing imagery can be of great value in helping to solve this problem, but it should not be inferred that the agriculturist can invariably determine the type of crop growing in each field merely by peering through a stereoscope or studying the readout from a magnetic tape. An approach that has proved helpful to the agriculturist confronted with the problem of quickly inventorying all of the crops growing in a vast agricultural basin is detailed here:

1. Develop a system whereby, from photo interpretation supplemented by a very limited amount of fieldwork, all agricultural crops in the area of interest can be accurately classified into a few major types. In many parts of the world the following six categories have proved to be both necessary and sufficient for this preliminary classification: (1) orchards, (2) vine and bush crops, (3) row crops, (4) continuous-cover crops, (5) irrigated pasture crops, and (6) fallow ground (Figure 1).
2. Compile a list of all crops encountered in each of these types within the agricultural area of interest.
3. From a preliminary study of aerial photographs that show representative examples of each of these crops, prepare a key, setting forth the photo-recognition features that differentiate each crop from all others and from non-cropland (Table 1). In this step, it may be found that some crops are virtually indistinguishable if reliance must be placed solely on the interpretation of so-called conventional aerial photography (vertical photography taken at a scale of 1 : 20,000 at high noon or thereabouts on a midsummer day, using panchromatic film and a Wratten No. 12 minus blue filter).* The crops may be readily distinguished, however, if (1) the photographs are taken with some other (economically feasible) specifications (the variant may be camera angle, scale, time of day, film–filter combination, or season in relation to the state of development of the crops that are to be identified); or (2) the photography is supplemented by other remote-sensing imagery in the thermal-infrared, microwave, or ultraviolet portions of the electromagnetic spectrum.

To illustrate this approach, let us assume that we are interested

*Types of film and filters named in this chapter are identified in more detail in Chapter 2.

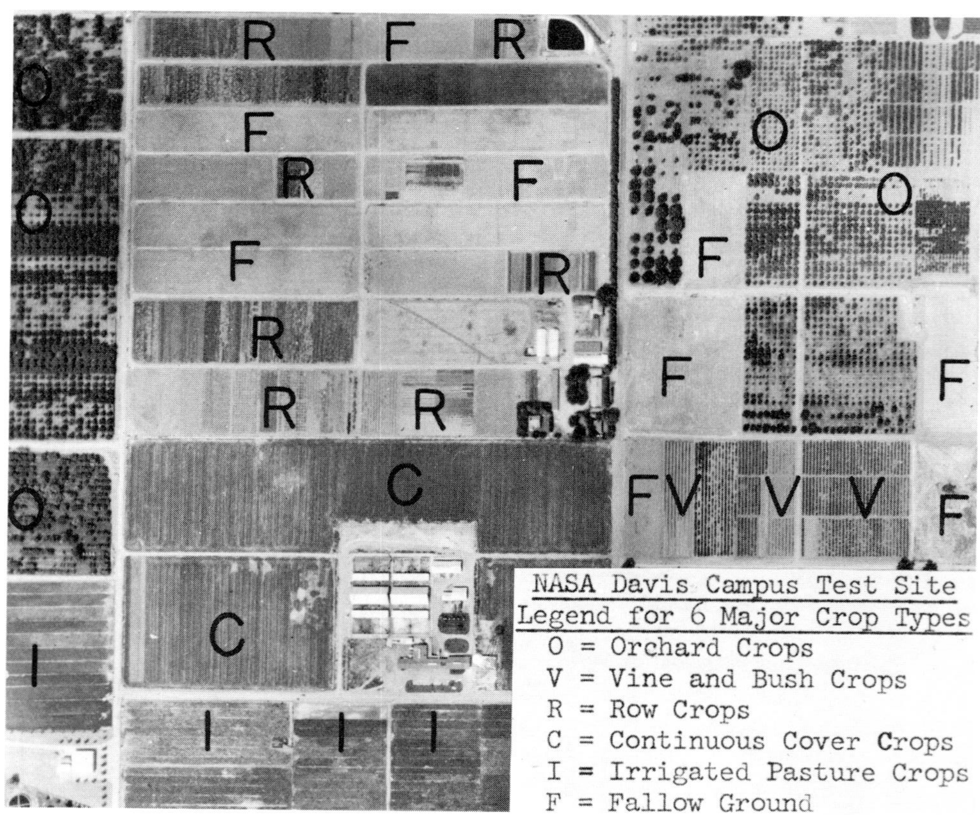

FIGURE 1 Panchromatic vertical aerial photograph of an area in California showing the six major cropland uses that are found in most of the world's agricultural areas.

primarily in detecting all fields devoted to the production of alfalfa and oats in a vast agricultural basin containing a variety of crops, including grapes, tomatoes, almonds, navel oranges, wheat, alfalfa, oats, and irrigated pasture crops. Without the key, we might have great difficulty in stating exactly how the photo images of alfalfa or oats differ from those of all other crops with which they might be confused. But merely by using the key (Table 1), we are readily able to distinguish continuous-cover crops (alfalfa, oats, and wheat are the only representatives in this example) from all other crops. This greatly simplifies our problem, for now, in attempting to identify alfalfa and oats we need seek only those photo-image characteristics that will separate wheat, oats, and alfalfa. This may still be a formidable task if reliance must be placed on conventional photography, but it is a very

TABLE 1 Aerial Photo-Interpretation Key to Major Crop Types and Land-Use Categories in Agricultural Areas of California. (For use in summer with panchromatic film, minus blue filter, scale 1 : 15,000.)

1. Vegetation or soil clearly discernible on photographs	See 2
1. Vegetation and soil either absent or largely obscured by man-made structures, bare rock, or water	*Nonproductive Lands*
2. Cultivation pattern absent; field boundaries irregularly shaped	See 3
2. Cultivation pattern present; field boundaries regularly shaped	See 5
3. Trees present, covering most of ground surface	*Timberland*
3. Trees absent or widely scattered; ground surface covered by low-lying vegetation	See 4
4. Crowns of individual plants discernible; texture coarse and mottled	*Brushland*
4. Crowns of individual plants not discernible; texture fine	*Grassland*
5. Crop vegetation absent	*Fallow*
5. Crop vegetation present	See 6
6. Crowns of individual plants clearly discernible	See 7
6. Crowns of individual plants not clearly discernible	See 8
7. Alignment and spacing of individual trees at intervals of 20 ft or more	*Orchards*
7. Alignment and spacing of individual plants at intervals of 10 ft or less	*Vine and Bush Crops*
8. Rows of vegetation clearly discernible, usually at intervals of 2–5 ft	*Row Crops*
8. Rows of vegetation not clearly discernible; crops forming a continuous cover before reaching maturity	See 9
9. Evidence of use by livestock present; evidence of irrigation from sprinklers or ditches usually conspicuous	*Irrigated Pasture Crops*
9. Evidence of use by livestock absent; evidence of irrigation from sprinklers or ditches usually inconspicuous or absent; bundles of straw or hay and harvesting marks frequently discernible	*Continuous Cover Crops* (small grains, hay, etc.)

simple task if we take the photos at the time when the oats and wheat are nearing maturity, using either Ektachrome film or panchromatic film with a light red Wratten No. 25A filter, and incline the camera sufficiently to take oblique photos rather than vertical ones. In the vertical view, the leaves of wheat and oats are most conspicuous, and since the leaves of wheat are of almost exactly the same color and tone as those of oats at each stage of maturity, the crops register in the same tone or color on vertical photos and are therefore indistinguishable from each other (Plate 9*A*). In the oblique view (Plate 9*B*), however, the brown heads of wheat, and correspondingly, the yellow panicles of oats, are most conspicuous as these crops near maturity, and hence are readily differentiated by the trained photo interpreter.

Plates 10*A* and *B* are Ektachrome and Infrared Ektachrome photos, respectively, of crops in Switzerland.[1] On the Infrared Ektachrome photos, the crops could be identified by their color characteristics with an accuracy of only about 30 percent. However, when crop height and other characteristics indicated by stereoscopic parallax were included, the accuracy more than doubled.

Plate 11 shows color-combined HH and HV radar images taken in the *K*-band (1–3-cm) wavelength range. The area is in Jefferson Parish, Louisiana. Simonett[2] found that the coastal location and low-lying terrain in this area make soil and vegetation sensitive to small variations in moisture. This type of color-combined imagery, despite its relatively low spatial resolution, is potentially quite useful for the identification of various kinds of crops and vegetation. Its value is greatest, however, when it is used in conjunction with conventional aerial photography, because more positive identification can be made when radar imagery and conventional aerial photography are used in concert than when either is used alone. It has the additional advantage of all-weather, day and night capability. A further example of the use of different wavelength ranges in identification is shown in Figures 15 and 16.

Figure 2 shows the relationship frequently encountered between topography and crop type. The three-dimensional study made possible by stereoscopy makes this relationship especially apparent.

Determining Crop Vigor

At present, the best possibility for determining the vigor of crops is usually offered through photographically recording foliage reflectance in the near infrared. The reason has not yet been ascertained, but a possible explanation follows:

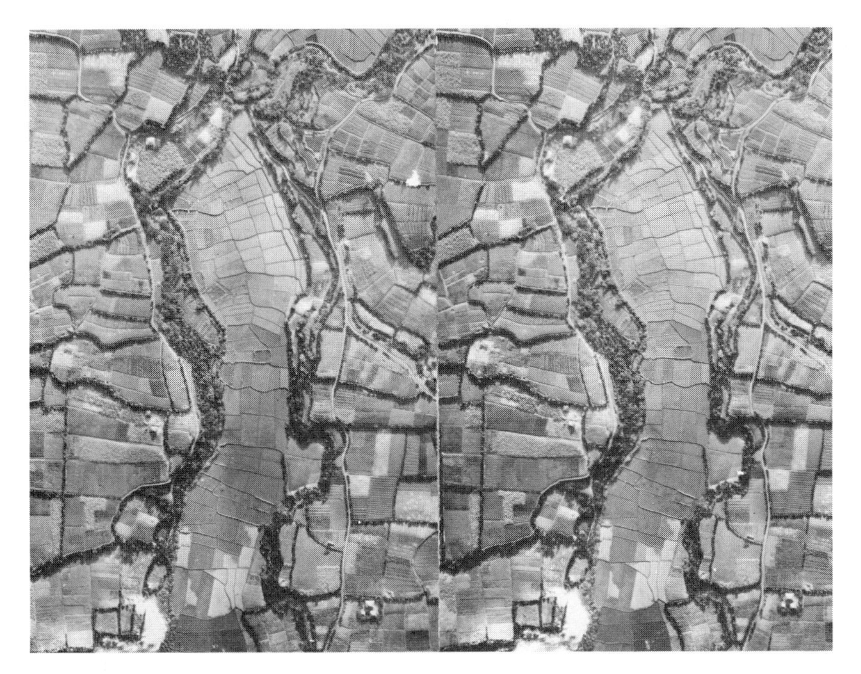

FIGURE 2 A stereoscopic pair of panchromatic vertical aerial photographs of an area in the Ryukyu Islands, illustrating the relationship that frequently exists between crop type and topography. Paddy crops such as rice and taro are seen to occupy most of the central portion, which is low, flat, and poorly drained with heavy (nonporous) soil. In marked contrast are the dry-land crops (mainly sugarcane and sweet potatoes) occupying most of the surrounding higher area.

The spongy mesophyll tissue of a healthy leaf, which is turgid, distended by water, and full of air spaces, is a very efficient reflector of any radiant energy and therefore of the near-infrared wavelengths. These pass the intervening palisade parenchyma tissue (which absorbs blue and red and reflects green from the visible). When its water relations are disturbed and the plant starts to lose vigor, the mesophyll collapses, and as a result there may be great loss in the reflectance of near-infrared energy from the leaves almost immediately after the damaging agent has struck a plant. Furthermore, this change may occur long before there is any detectable change in reflectance from the visible part of the spectrum, since no change has yet occurred in the quantity or quality of chlorophyll in the palisade parenchyma cells. (For a further discussion of plant physiology and radiation, see Chapter 5.)

To detect this change photographically, a film sensitive to these near-

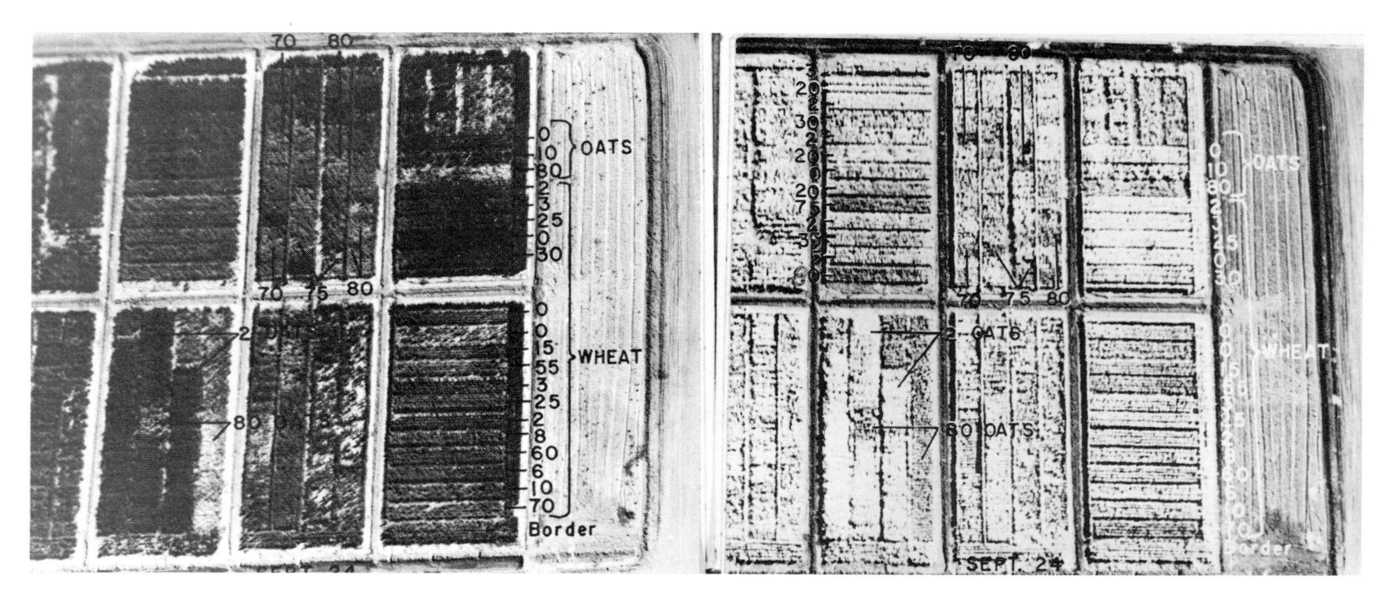

FIGURE 3 Vertical aerial photos, taken with panchromatic–25A (*left*) and infrared–89B film–filter combinations, of a cereal-crop nursery near Davis, California. Tone differences permit the photo interpreter to differentiate between diseased wheat, diseased oats, and healthy plants. Diseased wheat is dark in tone on the right photo but of normal tone on the left. Diseased oats are dark in tone on the right photo and light in tone on the left. Healthy plants, whether wheat or oats, are light in tone on the right photo but of normal tone on the left. The figures pertain to rust severity in percent.

infrared wavelengths is used. Infrared-sensitive films are also sensitive to much of the visible spectrum, yet to obtain maximum tone contrast between healthy and unhealthy foliage at this early date in the development of the disease, we should use only the near-infrared region where changes in reflectance have occurred. Consequently, a deep red filter (Wratten No. 89B) is commonly used in conjunction with the infrared-sensitive film, since this filter effectively prevents the unwanted wavelengths from reaching the film. On positive prints made with this infrared–89B combination, healthy plants consistently appear lighter in tone than unhealthy ones (see Figure 3). Panchromatic or conventional color photography (e.g., Aerial Ektachrome) taken at the same time, however, shows little or no tone difference between the healthy and the unhealthy plants, regardless of the filter used. (As the term panchromatic implies, this film is sensitive to all colors of the visible spectrum.) Corollary to the fact that at this early date we cannot obtain a tone difference between healthy and diseased plants on panchromatic or Aerial Ektachrome film, neither can we see a color difference with the naked eye. This leads to the rather startling conclusion, already borne out by numerous tests, that a loss of vigor in many plants can be seen more readily on infrared photography taken from an altitude of 2 miles or more above the earth's surface than by the expert on the ground as he walks through the fields.

Eventually, the loss of vigor may also lead to a reduction in the chlorophyll content of leaves and to an unmasking of the yellow pigments. By that time, the unhealthy plants exhibit a characteristic yellowish appearance, in contrast with the normal green plants; and this difference in spectral reflectance can be discerned on panchromatic photography taken with either a minus blue (Wratten No. 12) or a light red (Wratten No. 25A) filter (Figure 4).

In England preliminary trials in 1958 and 1959 confirmed that, for recording potato late blight, infrared film was far superior to panchromatic, giving much greater contrast between the diseased and healthy areas in the field.[3] The results indicate a most useful aid in the study of the development of blight epidemics. Manzer and Cooper[4] have demonstrated infrared aerial photography to be an effective tool both for basic research on potato diseases and for obtaining disease-survey information (Figure 5). They state:

> Periodic survey flights over potato fields sprayed by aircraft can, via infrared photography, show the aerial applicator and the grower when and where added fungicidal protection may be needed. Moreover the ability to obtain current in-

FIGURE 4 Panchromatic (left) and infrared, low-altitude, aerial oblique photos taken on two different dates of the area shown in Figure 3. Note that by October 15 all oats plants in the areas having the highest severity of black-stem rust have collapsed and fallen to the ground. Because of the early date at which these plants became infected, a very low yield was obtained from them. However, if either wheat or oats plants are infected at some later time in their development, much less yield reduction is suffered, even though the eventual "rust severity" (based on the proportion of the leaf and stem area covered by the pustules of the fungus) may become equally high in the latter instance. Hence, there is an obvious value to having repetitive cover of such an area on properly selected dates. Numbers refer to percentage of leaf area covered by the rust fungus in the plots indicated.

FIGURE 5 Late blight of potato is more easily discerned on infrared film (*B*) than on panchromatic film (*A*). (*Courtesy Manzer and Cooper, University of Maine.*)

FIGURE 6 Panchromatic vertical aerial photograph of orchards in the Santa Clara Valley of California. The light-toned circular spots indicate areas where the stone-fruit trees (mainly prunes and apricots) are suffering from attacks by the oak-root fungus (*Armillaria mellea*).

formation on disease incidence over a large area is useful not only in plant protection work but also in the making of reliable estimates of total crop production.

Manzer and Cooper also used Infrared Aero Ektachrome (camouflage-detection) film to detect differences in infrared reflectance. This film makes healthy plants, which are highly infrared reflective (0.7–0.9 μ in the near infrared), appear red, whereas the unhealthy plants do not (see explanation of false-color film in Chapter 2). Plate 12 is a matched pair of Ektachrome (*A*) and Infrared Ektachrome (*B*) images of the same potato field photographed on the same date. The diseased plants in plots shown in portions of this field are better discerned on the Infrared Ektachrome photo than on the conventional Ektachrome.

There is no universal approach to determining crop vigor from aerial photographs. There is considerable photographic evidence, however, that the first indication of loss of vigor due to black-stem rust on wheat or oats,[5] to root-rot disease on navel oranges, and to late blight of potatoes (Figure 5 and Plate 12) can be shown photographically in the near infrared (0.7–0.9 μ).

Areas of decidedly low vigor can usually be discerned with little difficulty, although the photo recognition may vary with the crop. Figure 6 shows stone-fruit trees infested with root rot. If most of the trees in an orchard have uniformly large crown diameter, but in one area this gradually diminishes from the periphery to the center of a circular area, some harmful condition is certainly present.

Similarly, in a row crop or continuous-cover crop, a loss of vigor may be indicated by a general sparseness of the stand, as compared with surrounding areas. Photographically, this usually will be manifest as a tone difference. The soil is usually of different tone than the foliage; consequently, the low-vigor area may appear abnormal in tone because more soil and less foliage are registered in that area than in the surrounding healthy area (see Figure 7).

Crop production can sometimes be used as an indication of vigor. A good example is shown in Figure 12, where the number of trays of drying grapes is the indicator.

Figure 8 is another example of crop vigor and associated yield, with a pair of stereo photographs as the "diagnostic" aid. This study of lodging in rice fields subjected to fertilizer and irrigation tests is aided by sequential photographing during crop development. Yield estimates are supplemented by limited on-the-ground observations.

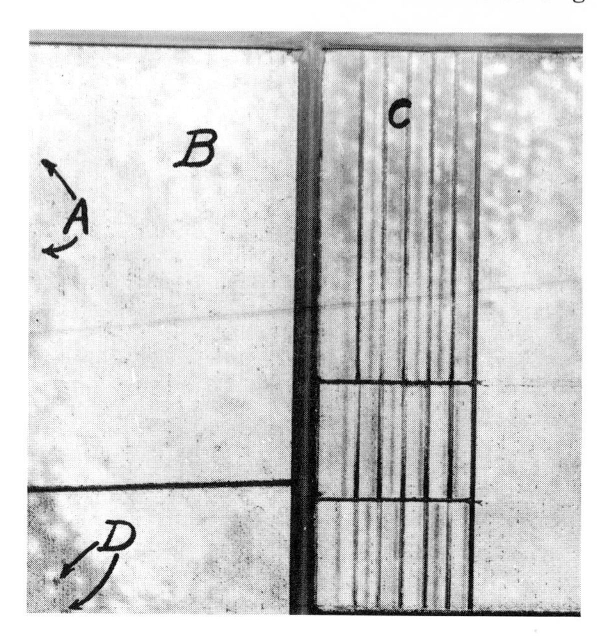

FIGURE 7 Vertical aerial photograph (infrared film, dark red filter) showing areas at *C* where a sparse stand of cotton is due primarily to adverse conditions of soil moisture and soil fertility. In attempts to estimate crop yield, it is important to recognize such areas and to make appropriate reductions. "Seedling gaps" in the cotton field at *A* also necessitate the application of appropriate yield-reduction factors, as compared with an area such as *B* where yield is uniformly high. Clumps of dark-toned weeds, which also can reduce crop yield, appear in several spots, including those labeled at *D*.

Determining the Agent Responsible for a Loss in Crop Vigor

The loss of near-infrared reflectance can be caused by a number of damaging agents, including pathogens, insects, sunscald, frost injury, mineral deficiency, mineral toxicity, drought, and flooding. The problem of determining the agent responsible for a loss in crop vigor usually is not as hopeless, however, as the foregoing statement might imply. In many cases, a given crop growing in a particular locality experiences almost all its loss in vigor as a result of a single damaging agent. Once the identity of that agent has been established by careful on-the-ground checks, further mapping of vigor losses by means of photographic tone differences seen in that crop is tantamount to mapping the locations of the single agent.

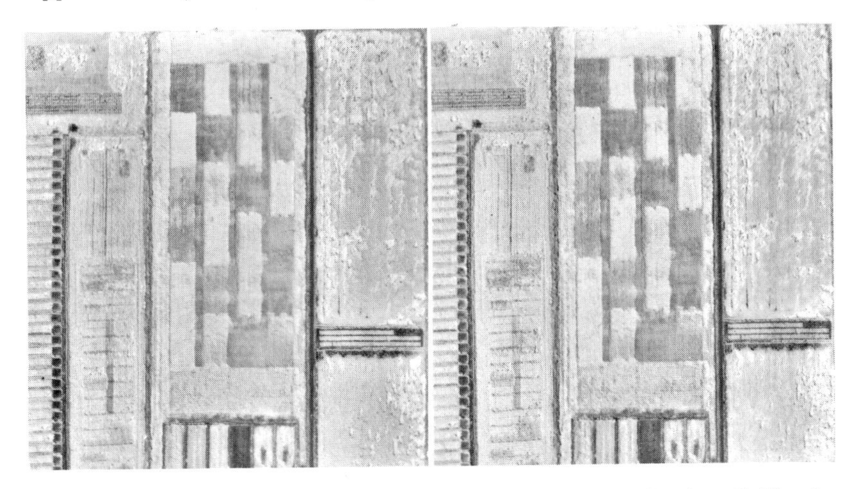

FIGURE 8 Stereo pair of part of the Biggs Rice Experiment Station, California, showing juxtaposition of natural and experimental plots. (*Courtesy Robin Welch, Airview Specialists Corp.*)

In a few instances, the first photographic evidence of a loss of vigor in plants consists not of a decrease in infrared reflectance but of an increase in yellow and red reflectance, as indicated in Table 2. Furthermore, in at least one instance there is initially an increase, rather than a decrease, in infrared reflectance. While this may complicate the problem of detecting a loss in crop vigor, in any local area it may facilitate both the identification of the crop and the determination of agents responsible for the vigor loss.

Figure 9 illustrates the possibility for employing spectral analyses as an aid to selecting the optimum film–filter combinations for identifying a particular damaging agent from its "tone signature" on multiband photographs. Figure 10 shows an example of pictures taken with a valuable research tool, the nine-lens camera, by means of which the identifying tone signature for a given crop condition may be arrived at empirically rather than through spectral analyses. Multiband photography is discussed further under "Estimating Crop Yield," page 184.

Estimating Crop Acreage

A vertical aerial photograph of flat terrain is truly a map. On it acreages can be accurately measured once the scale of the photograph is known. Although aerial photographs are rarely vertical (an average of about 1° tilt from the vertical is common), and although agricultural land is rarely perfectly flat, experience has shown that, even by ignoring these complications, the acreages of agricultural fields commonly

TABLE 2 Use of Multiband Tone Signatures in Identifying Crop Conditions and Types

		Panchromatic Light Red Filter (25A)		Infrared Film Dark Red Filter (89B)	
Crop	Condition	Tone	Probable Explanation	Tone	Probable Explanation
Wheat Barley Corn Alfalfa Potatoes Orange trees	Healthy	Dark	Large amount of chlorophyll present; it absorbs red light	Light	Turgid spongy mesophyll is highly reflective
Wheat	Rust infected	Dark	Same as above during early stages of infection	Dark	Spongy mesophyll not highly reflective; its cells have lost vigor
Potatoes	Downy mildew infected	Dark	Same as above	Dark	Same as above
Orange trees	Root rot infected	Dark	Same as above	Dark	Same as above
Barley	Powdery mildew	Dark	Little or no chlorophyll present; hence highly reflective carotenoid pigments are unmasked	Very light	Spongy mesophyll remains turgid and reflective; surface hyphae also highly reflective
Corn	K-deficient	Light	Same as above	Light	Spongy mesophyll remains turgid and reflective
Alfalfa	K-deficient; *Cercospera* infected	Light	Same as above	Dark	Spongy mesophyll is not reflective because cells have lost turgor; early loss of chlorophyll
Oats	Rust infected	Light	Same as above	Dark	Same as above

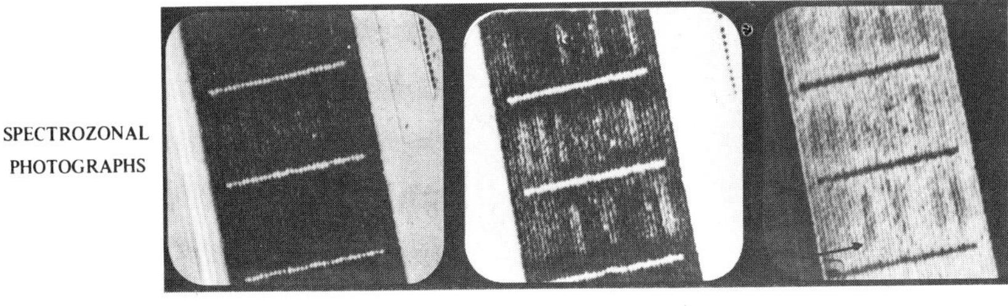

Band 1 Band 2 Band 3

Feature to be Identified	TONE ON POSITIVE PHOTOGRAPHIC PRINT		
	Band 1 0.38-0.46μ	Band 2 0.64-0.72μ	Band 3 0.80-0.90μ
Bare Soil	Light	Light	Dark
Healthy Vegetation	Dark	Dark	Light
Recent Infestations	Dark	Dark	Dark
Old Infestations	Dark	Light	Dark

Light
Tone on Positive Print
Dark
Bare Soil
Old Infestations
Healthy Vegetation
Recent Infestations
Band 1 0.38-0.46μ
Band 2 0.64-0.72μ
Band 3 0.80-0.90μ

FIGURE 9 Three-band photography of a sugar-beet experimental area on the Davis Campus of the University of California. Information on crop vigor and disease-causing agents can be interpreted from three bands and could not have been interpreted from any one or two of the bands.

can be measured to within 1 percent of their correct areas on conventional aerial photography (see Figure 11). Even if the agricultural land has a great deal of relief, and even if the photos have more than a few degrees of tilt, true maps can readily be prepared from the photos through the use of modern stereoscopic plotting equipment. Acreages can then be accurately measured from the maps.

The acreages within irregularly shaped fields, once the fields have been outlined either on the photos or on a map, can be determined by use of any of the following:

1. A polar planimeter, the pointer of which can be made to trace out

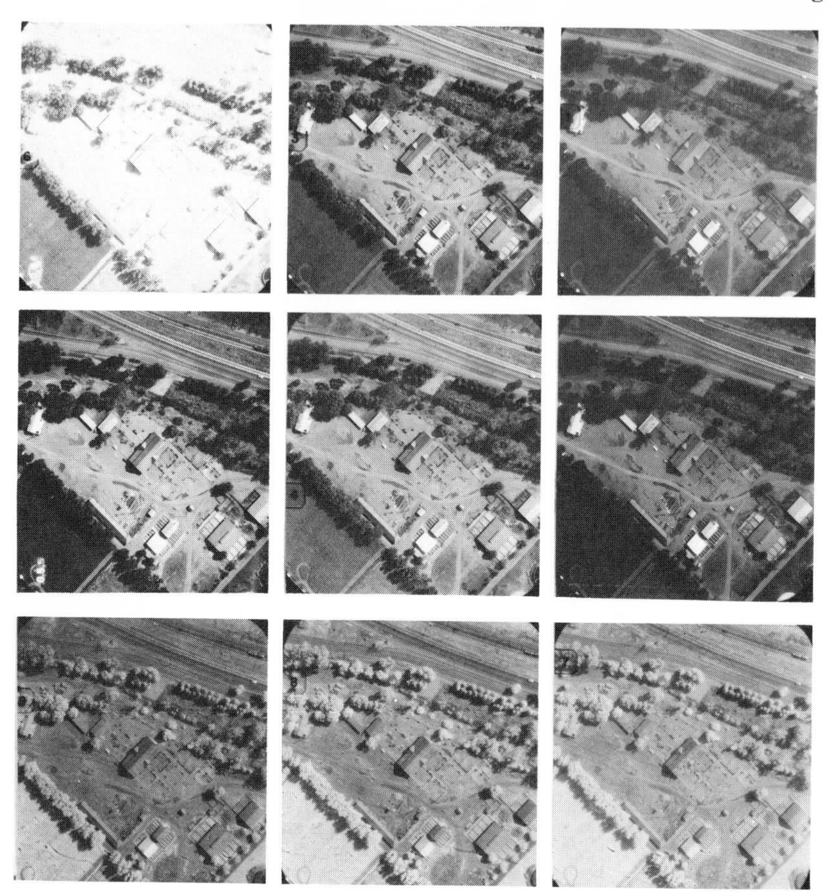

FIGURE 10 Example of multiband photography taken with a nine-lens camera. (*Courtesy Itek Corp.*)

the perimeter of the field and thus actuate a dial on which area can be read directly.

2. A suitable transparent plastic template as an overlay to the photos or map. The template contains either a dot-grid or a system of line-transects so constructed that each dot on the grid or each unit of length of line on the transect represents, say, one tenth of an acre at the scale of the photos or map. By laying this template randomly over the photos or map, and by then counting all the dots or measuring all the line transect units within the field boundary, the area of that field is readily determined.

3. A scissors or a razor blade to cut out each field as portrayed on the photos or map, and then to determine its weight. Once the scale of

the photos or map is known, a square representing a convenient area, (e.g., 10 acres), is cut out of the same type of paper and weighed. This provides a conversion factor for computing the field acreages from the paper weights.

Estimating Crop Yield

The crop yield in a field is ordinarily determined photogrammetrically by computation from the yield per unit area and the acreage. If the crop in a field appears on aerial photographs to be uniformly healthy and vigorous, it probably also has a uniformly high yield. One simple procedure for estimating the total yield for such a field is to visit the field on the ground as the crop nears maturity, measure the yield per unit area in each of several small sample plots randomly located in the field, calculate from these measurements the average yield per acre in the field, and multiply the last figure by the total acreage as measured photogrammetrically.

Ordinarily, ground checking must be used to obtain a reliable measure of the crop yield per unit of field area. In special instances, however, yield estimates can be reliably made by direct photo interpretation. An example is the vineyard area of the San Joaquin Valley of California (Figure 12). Here, both raisins and wine are commonly produced

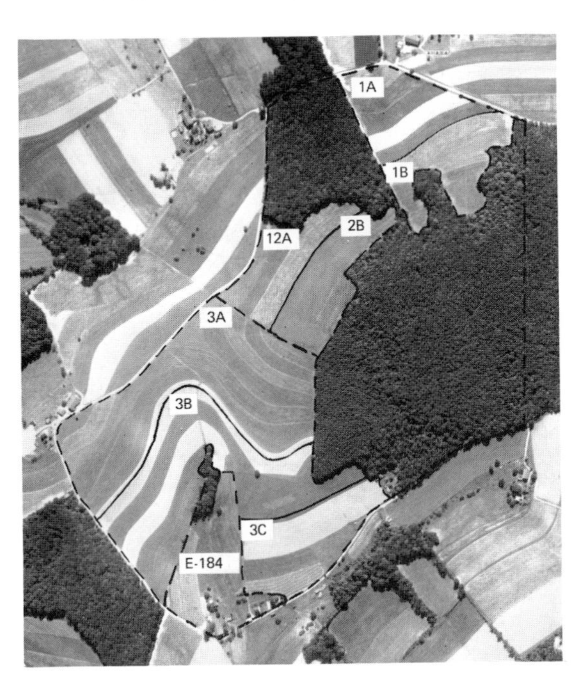

FIGURE 11 Photo interpretation and field inspection enable farm reporters to delineate field boundaries in preparation for area determination by planimeter. Where current field lines do not show on the photographs, observations are made for exact plotting. *(Courtesy U.S. Department of Agriculture.)*

FIGURE 12 Portion of a photograph, greatly enlarged, of raisins drying in the vineyards of the San Joaquin Valley, California. The dark spots between the rows of vines are raisin trays, which the photo interpreter counts.

from the same variety of grape. The raisin grapes are harvested and laid out for drying in the vineyards on trays that hold 20 to 25 pounds each; this is done several weeks before the wine grapes are harvested.

The purposes of the raisin-lay survey are: (1) to determine raisin crop acreage; (2) to determine the total fresh weight of the raisin grapes; (3) to determine the drying rate for comparison with a predetermined optimum rate; (4) to determine crop vigor; (5) to forecast unofficially the size of both the raisin-grape and the wine-grape crops; (6) to estimate the balance between the two crops and determine when to cut off the raisin-grape harvest. Growers use the survey information to improve their efforts to meet the needs of their customers and to operate their industry more stably and more profitably.

Tray counts* from aerial photographs have replaced the slow, costly, and inaccurate ground observations of the area (approximately 1,000 square miles). Surveys are made on each of seven suitably spaced dates during the raisin-harvest period, so that the time to stop the raisin harvest can be decided. The photos are taken at an altitude of 17,000 feet during 4-hour periods beginning at about 10 A.M. The film is processed and interpreted promptly, so the survey results are available by the following morning. In one recent year an expenditure of about $50,000 for surveys was credited with having saved growers about $5 million.

There are conditions under which crop-yield estimates can become somewhat complicated:

1. When the crop in the field is not uniformly healthy or vigorous. The remedy is to delineate each significant vigor class on the photos, determine the acreage in each vigor class photogrammetrically, make field measurements of yield by vigor classes, and from these data compute the yield in each vigor class. Then, by mere summation, the yield for the entire field is readily calculated.

2. When the crop is not accessible for field measurement of yield. One remedy is to obtain similar aerial photography of a limited number of similar fields in an accessible region, measuring yields in the accessible fields, and by photo-image comparison, applying these yield

*Sometimes, if the photography is of somewhat poorer quality than that shown in Figure 12, the photo interpreter merely determines, for each vineyard interpreted, whether trays were present at the time of photography. Sample tray counts are then made on the ground.

figures, crop by crop, and vigor class by vigor class, to the inaccessible fields.

3. When the estimate of crop yield must be a forecast made several weeks or months before the crop matures. The U.S. Department of Agriculture and the Crop and Livestock Reporting Services of the various states are among the agencies that commonly require this information. One long-range approach is as follows: On the approximate date when the yield forecast must be made each year, aerial photographs are taken, year after year, of selected test areas in which several representative fields of each crop of interest can be found. Also year after year, as these crops mature, measurements are made of the yield actually obtained in each field. Each yield figure is eventually correlated with weather history and the corresponding photo image of the field as seen on the aerial photography made earlier in the same season. By this means, the photo interpreter is able to compile, over a period of years, a "key" consisting of photographic examples of fields as seen on the date when a forecast must be made, and also showing for each such example the yield per acre actually obtained from the field later in the same year. Then, in attempting to forecast yield in any subsequent year, the photo interpreter has aerial photography of the fields for which he is forecasting taken to the same specifications and on approximately the same day of the growing cycle as that for which he has prepared his key. By matching each field, as seen on this photography, with one of roughly comparable appearance as seen in the key, and by knowing the weather history and yield per acre that was obtained in the key example, he computes the probable yield for the field in question. As the season progresses, weather data alone can revise estimates.

For most crops, the accuracy with which yield estimates can be made photogrammetrically is improved if two things are known: the agent responsible for vigor loss and the time during the development of the crop at which each portion of the field first suffered a vigor loss. As an example of the importance of the latter, black-stem rust on wheat was found[5] to cause a 90 percent yield reduction on portions of a field infected three weeks before first heading, but only a 10 percent yield reduction on portions of the field that were infected only one week before first heading.

One means of determining when each portion of the field has become infected is by flying several photographic missions over the fields of interest while the crops are developing. Another means, suggested by

Figure 9, is provided by the previously mentioned technique of multiband reconnaissance. In this example, the concept has been extended to encompass three spectral bands instead of two. From a study of tone signatures provided by two of these bands (0.64–0.72 μ and 0.80–0.90 μ), we can distinguish three crop conditions: (1) healthy, (2) recently infected, and (3) previously infected. Thus we may be able to obtain from one photographic mission vital information (for yield estimation) that might otherwise have required at least two photographic missions. By use of the third band (0.38–0.46 μ) we can make another important distinction—one that is not determinable when only the other two bands are used—diseased foliage from bare soil. In predicting yields of sugar beet fields, the latter distinction is important since bare soil usually results not from disease but from poor planting and cultivating techniques. "Cultivator blight" (as this last category has been called) does not spread through the field with the passage of time; disease does. Hence the need for distinguishing between the two in order to predict crop yield for a large area is obvious.

One possible technique suggested is to employ four collateral pieces of information: (1) temperature minimums and maximums, (2) relative humidity, (3) rainfall, and (4) number of daylight hours. These factors influence both the emergence of insects and the onset of disease and thus give the analyst information on the probability of the onset or spread of pests.

In several examples of multiband reconnaissance that have been given, two bands were found to be better than one, and three bands were found to be better than two (Figure 9). The question then arises whether still more information can be obtained if more than three bands are used. The nine-lens camera that was used to make Figure 10 was built by Itek Corporation partly in an attempt to answer this question. Numerous tests have been made to determine how many bands might be useful to the agriculturist. The results of each test are meaningful only when specific details are given as to the test objectives. However, one generalization that appears to be warranted is that, although four bands sometimes are better than three, and occasionally five bands are better than four, a still further increase in the number of bands yields little additional information. These statements are based on the presumption that only the spectral range in which photographic films can ordinarily be used directly as the sensors ($\sim$0.35–1.0 μ) is under consideration. There is evidence that additional information can be obtained when sensing is also done in the thermal-infrared or microwave regions.

Other Current Applications

Some additional applications of aerial photography and other remote-sensing imagery currently being made by agriculturists are listed below.

1. Detailed photo-interpretation studies to determine areas that need erosion control, weed control, fencing, soil amendment, or other remedial measures.
2. Studies to determine areas in which farmers have already taken remedial measures and thus have qualified to receive federal benefit payments. Since these payments are commonly made on a per acre basis, the acreage of each area in which a remedial measure has been taken also is determined from the photographs.
3. Land-use studies that will be of interest to agricultural economists and agricultural geographers.
4. Farmland appraisals for use by taxation authorities.
5. Rapid and accurate damage assessments following such natural disasters as floods, hurricanes, tornadoes, fires, and severe epidemics of insects and pathogens.
6. Determining the adequacy of an existing irrigation system for uniformly wetting an entire field (see Figure 13).
7. Surveys in large agricultural areas of the total farm animal populations. Such surveys may estimate the number of each type of animal, the breed, the ratio between the sexes, and also the general shape, fatness, etc., of each animal as a measure of its vigor, nutritional state, and marketability. Livestock surveys are discussed further later in this chapter (see page 205).

Future Applications

As indicated by some of the preceding examples, certain kinds of information desired by agriculturists are best obtained by remote sensing in one particular band of the electromagnetic spectrum, and other kinds are found in one of the other bands. It has been shown, in consequence, that a greater amount of information is obtainable if two or more aerial cameras or other remote-sensing devices are used, each especially adapted to sensing in its own spectral band, than if reliance is placed on either of the sensors alone.

The task of comparing tonal values on each of several photographs, however, can be very wearisome. Aerial color photography, by automatically integrating tonal values in each of three spectral zones,

FIGURE 13 Aerial photographs can be used effectively in determining the distribution of irrigation water throughout a field. Dark-toned areas are those that have been reached by the irrigation water in these two recently harvested alfalfa fields. White lines in some of the dark-toned areas are irrigation pipes.

presents to the interpreter a composite image, the hue, value, and chroma characteristics of which greatly facilitate tonal comparisons. Agriculturists have made only limited use of aerial color photography to date. However, in terms of future applications, its potential value is very great.

Color images can be made by various color-combining processes in addition to the one exemplified by Plate 9. Specifically, by proper choice of black-and-white film and of accompanying filters, three separate black-and-white negatives can be obtained in spectral zones corresponding to those for which the three dyes of a color film have been sensitized. The tonal densities of an image on the three black-and-

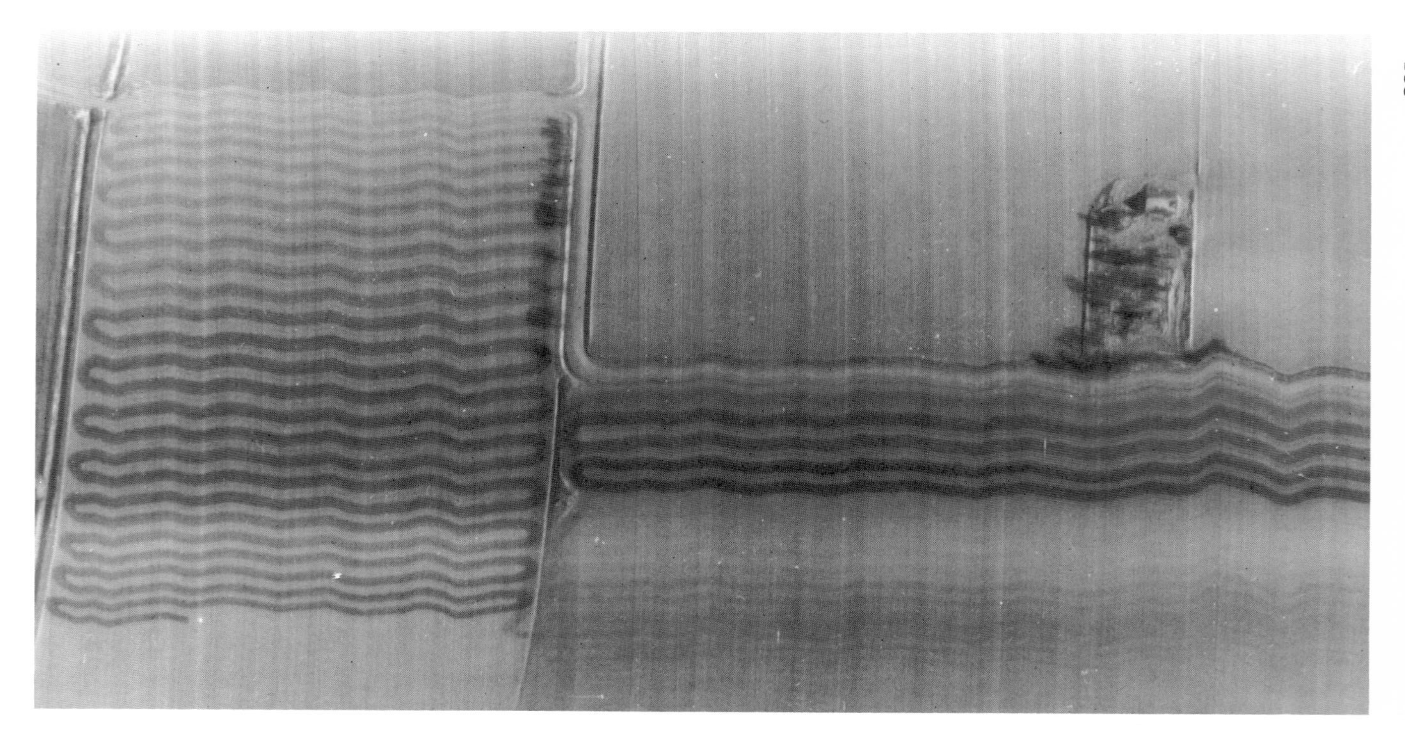

FIGURE 14 Portion of a thermogram taken by a Bendix scanner on the Davis Campus, University of California, in the 8–14-μ wavelength region. (Waviness of the plow lines is caused by lack of gyrostabilization of the scanner.) In the lower left, at the end of one of the plow lines, a light dot indicates the hot engine of the tractor. The dark tone of the freshly turned soil indicates that it is moist and cool. The longer the time that has elapsed since the soil has been exposed to the sun, the lighter the tone. Since the rate of warming depends on the soil moisture content, this offers a means, after suitable calibration, of estimating soil moisture in recently plowed fields. (*Courtesy Cartwright Aerial Surveys and Bendix Corp.*)

white negatives can then be used to govern the intensities of corresponding colors on the composite color image.

Imagery made from sensors utilizing the thermal-infrared* (Figure 14), microwave, and ultraviolet regions of the spectrum has great potential in agriculture (see Chapter 3). It has already been demonstrated, for example, that the earliest remotely sensed evidence of disease in some crops appears on ultraviolet imagery, that the level of grain in a silo is detectable only on thermal-infrared imagery, and that the depth to a water table or aquifer is sometimes detectable on imagery obtained from microwave bands but not from any other spectral region.

Radar imagery of a quality equal to or surpassing that shown in Figure 15 should soon be routinely obtainable (see also Plate 11). This kind of capability could be of tremendous value when periodic crop inventories must be made of rather large areas and on a quite rigorous time schedule.

Possibilities for using multiple polarization effects in radar imagery as an aid to crop identification are supported by the data plotted in Figure 16. On this graph the average gray-scale values (for each polarization) for approximately 400 fields are plotted. The separation shown between sugar beets and corn is surprisingly good. The major contributor to this separation is the HH (horizontal transmit and receive) polarization. Corn is also well separated from all other crops. Of 25 fields of corn, 23 are seen to fall in the compact zone between the 2 hyperplanes. The HV imagery contributes significantly to this separation.

The lowermost category in Figure 16 is dominated by bare ground (120 cases). It will be noted that only one bare field lies outside this region. Fields of emergent wheat, wheat stubble, and weeds consist overwhelmingly of bare ground; hence, a number of the most recently planted wheat fields with a thin stand of sprouts is included in the bare-ground category. As the growing season advances, the wheat fields will change in value, but the bare ground will remain unchanged. Alfalfa, of which there are six cases in this study, and grain sorghum (four cases) are the major inadequate separations.

The final group shown in Figure 16, while dominated by grain sorghum (102 cases), is mixed with emergent wheat (38 cases), wheat

*When thermal-infrared energy is used to form a photo-like image (as in Figures 14 and 24), the resulting imagery is known as a "thermogram." See also Chapter 6, Figure 19.

FIGURE 15 An example provided by Simonett, CRES, University of Kansas, of K-band radar imagery taken of agricultural fields near Garden City, Kansas, in the 1–3-cm wavelength range. Light-toned fields are sugar beets. (*Photo taken by Westinghouse Corp. under contract to* NASA *and the U.S. Army Signal Corps.*)

stubble and weeds (24 cases), and alfalfa (24 cases). Note that, whereas grain sorghum completely overlaps the other crops in the lower portion of its range, the upper third (37 cases) is overlapped by only eight fields in different crops.

It may be feasible to improve discrimination in this region by using multiple radar frequencies, but there has been no test of this possibility. It may also be that the addition of one or two channels in the visible and near-infrared regions would have made possible virtually complete separation of these crops.

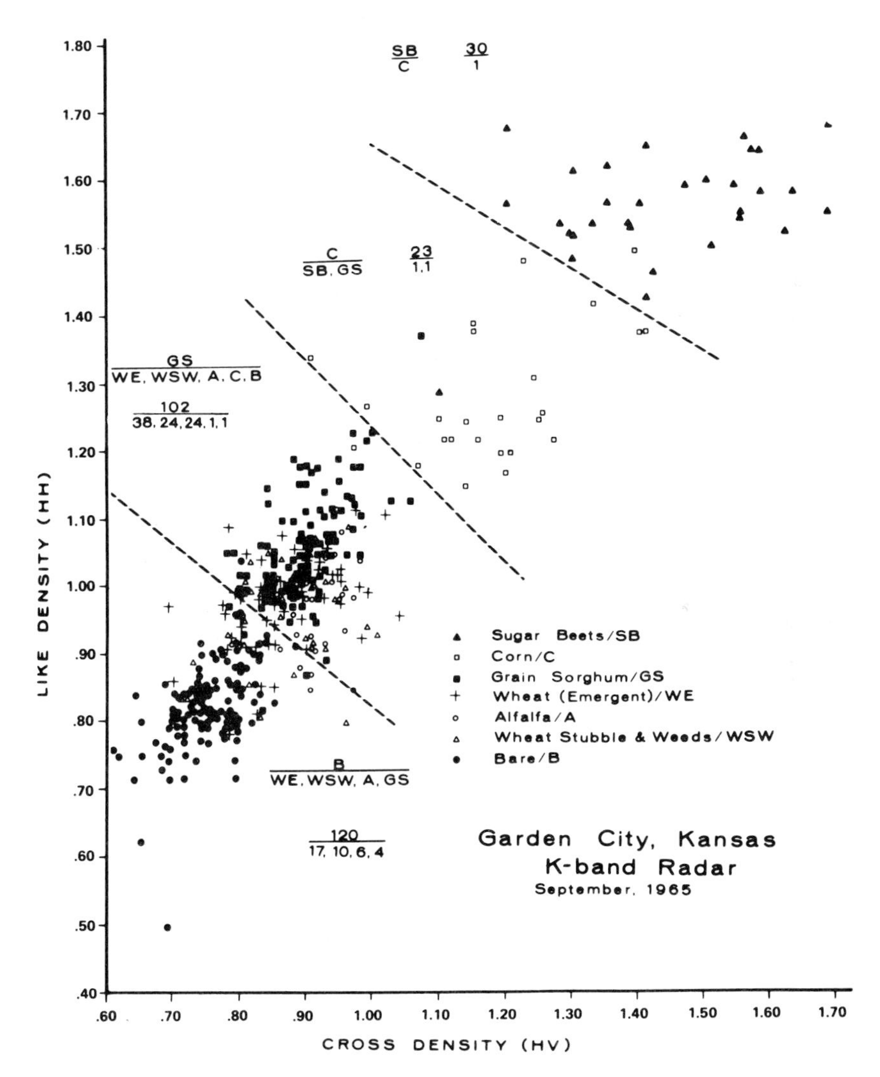

FIGURE 16 This graph clearly shows possibilities for discriminating between crop types by multiple polarization *K*-band radar imagery. (*Courtesy* CRES, *University of Kansas.*)

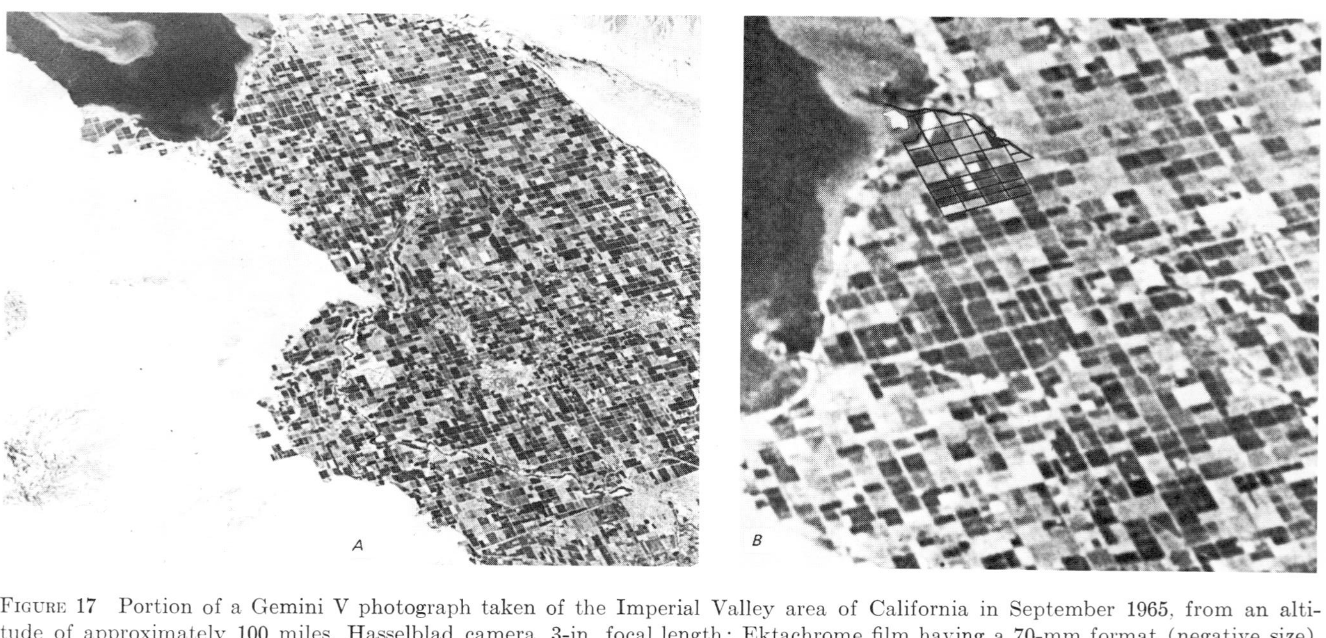

FIGURE 17 Portion of a Gemini V photograph taken of the Imperial Valley area of California in September 1965, from an altitude of approximately 100 miles. Hasselblad camera, 3-in. focal length; Ektachrome film having a 70-mm format (negative size). *A*, Fields as small as 40 acres are clearly seen on the original color transparency, and tone or color differences that are indicative of crop type and plant density also can be discerned. *B*, Enlargement of a portion of *A*. From field checking, a few of the individual field boundaries have been outlined to indicate the extent to which this might have been done directly from the Gemini photo. Tonal variations within certain of the outlined fields are indicative of differences in plant density and crop vigor. Most of the fields here are devoted to the production of barley and other small grains.

Satellite Photography

The most exciting achievements of the past decade are probably those relating to man's newly acquired ability to launch instrument-carrying satellites and to insert them into prescribed space orbits. The aerial camera carried aloft on such vehicles offers a great variety of possibilities for gathering valuable information about the physical universe. Figure 17 provides strong support for this statement. For further evidence, see Colwell.[7]

Figure 18 provides an excellent illustration of how the photo interpreter might aid the agriculturalist and engineer through utilization of a space photograph. Of course, the photo interpreter's recommendations must be checked on the ground before final decisions are made. In this case, the interpretation has been independently confirmed by agriculturalists and engineers making ground studies. The interpretation was, furthermore, quick and easy because the entire area was imaged on a single photograph.

Finally, space photos taken at intervals (e.g., Figure 19) are useful in recording the sequential conversion of wildlands to agricultural lands and agricultural lands to urban developments.

WILDLAND CROPS

A number of specific applications of aerial photography and other remote-sensing data have proved to be of primary value in the management of wildland crops.

Determining Species Composition of Forest Stands

The photo-interpretation approach to tree-species identification is much the same as that already discussed for crop-type identification. However, in two notable respects, the problem is more difficult for the forester.

A given piece of farmland is usually occupied by a single type of crop, and its photo identification is facilitated by exploiting the uniform mass effect that this single crop produces on the aerial photographs. In contrast, a given piece of forestland is often occupied by a random mixture of many tree species, making both a photo identification of the component species and a classification of the timber into meaningful categories most difficult. Most crop plants on a farm have

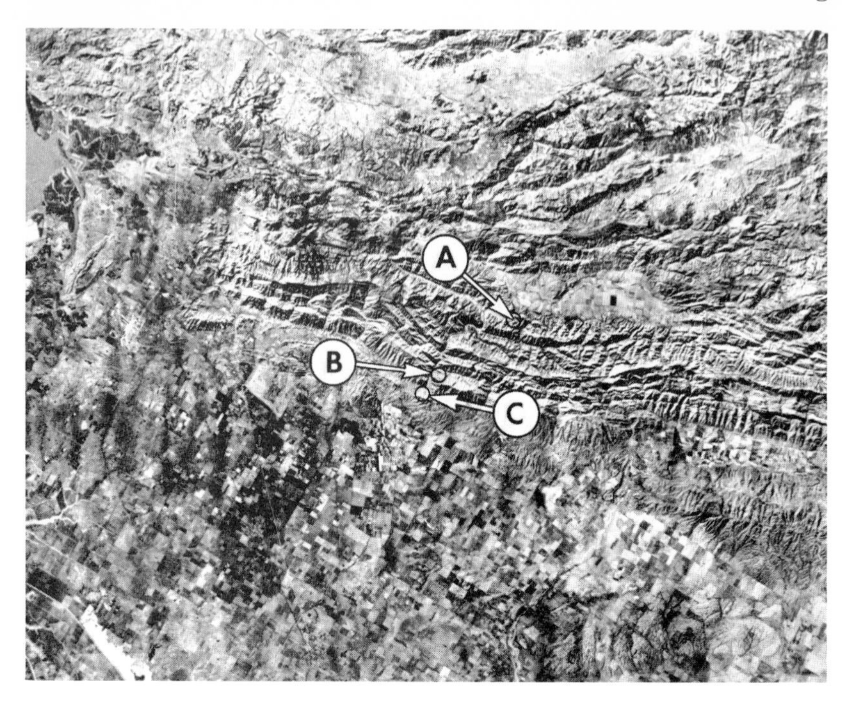

FIGURE 18 A simulated space photograph of a mountainous area. The presence of dark-toned woody vegetation and stream channels indicates moderate rainfall, probably 30–50 inches a year. Nearby are large fields in rectangular pattern, without deeply incised drainage; these fields must be moderately flat and, judging from their individual sizes, must presently be planted to dry-land crops. This combination of conditions suggests that the soil is fertile and that food production would be greatly increased if water were impounded in a reservoir and used to irrigate the fields. Further study of the photograph reveals that a large stream is constricted in a deep canyon at point *A*. Here a dam could be built, converting the mountain valley immediately upstream into a large reservoir. This valley does not appear to be densely populated, so the property probably could be acquired at reasonable cost. The watershed draining into this mountain valley is large enough to supply ample water to the proposed reservoir. Since the canyon at point *A* is deep, a dam built there probably would permit a waterfall sufficient to generate great amounts of hydroelectric power. There appears to be a rapid fall in the stream from the proposed dam site to a nearby point, *B*. From this point, water might be fed into a second power plant at *C*. Closer study, however, shows that this portion of the plan would not be feasible because the water conduits would have to traverse many deep canyons between points *A* and *B*. The top half of the photo shows alternating mountains and valleys, and the topography and culture are seen to be unfavorable for dam construction, crop irrigation, and hydroelectric development.

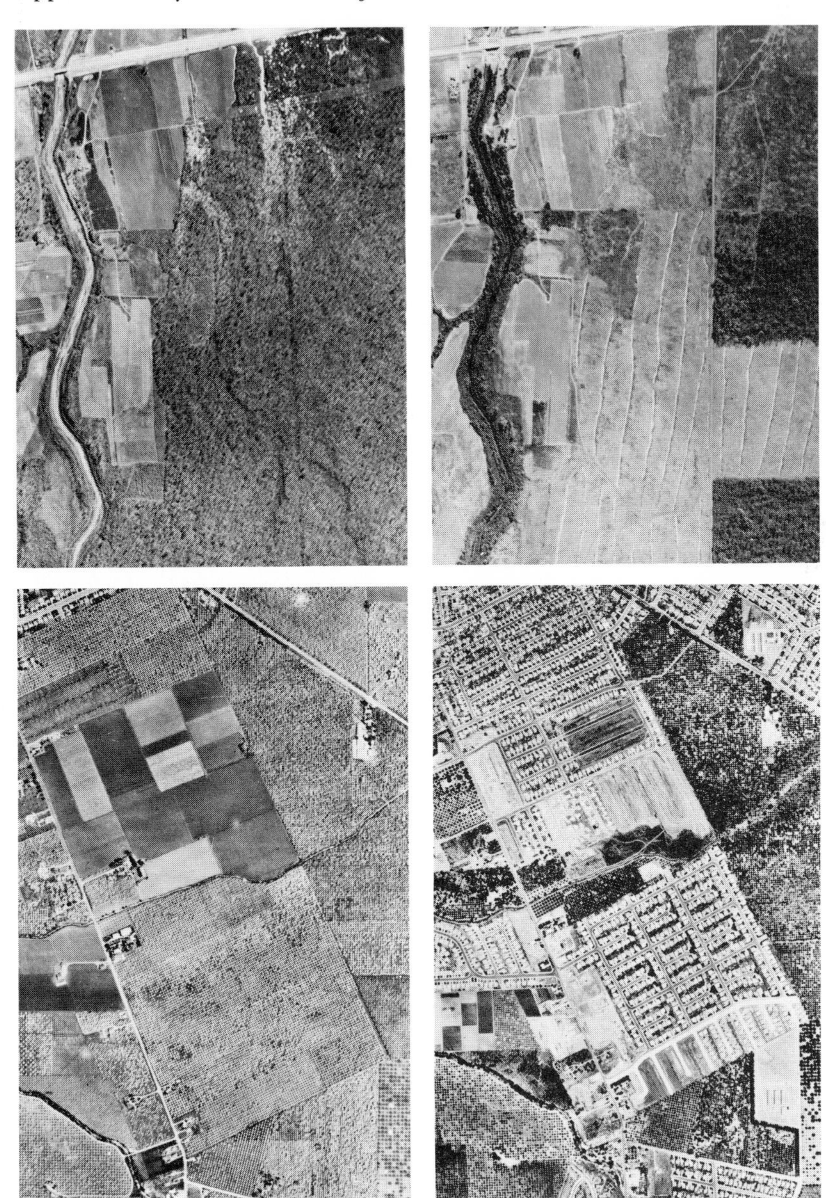

FIGURE 19 Sequential aerial photos such as these often are taken at roughly 5-year intervals to permit the agriculturist or forester to determine the extent to which forestland is being converted to agricultural land (*top pair*) and agricultural land to urban development (*bottom pair*). (*Courtesy U.S. Department of Agriculture.*)

FIGURE 20 Aerial photograph stratified and marked after photo interpretation. Numerator: N = nontimber site; O = old; Y = young; first digit = density of sawlog-size timber; second digit = density of all timber, regardless of size (digits represent percentage brackets). Denominator: letter codes for tree types: C = conifers; H = hardwoods; S = shrubs; Hy = hardwoods (young); R = rocks. (*Courtesy U.S. Forest Service.*)

been purposely spaced so as to receive full sunlight. Consequently, these plants are uniformly exposed to the aerial view and exhibit a remarkably uniform appearance on aerial photographs. In contrast, many plants that a forester wishes to identify as to species or type from a study of aerial photos, are in the forest "understory"; since they are obscured to varying degrees by overtopping trees, their photo appearances also are quite variable. This makes species identification difficult.

Topography, stand density, elevational range, and the size and configuration of dominant trees in the stand are more easily interpreted from aerial photos than are species. A forester can often establish a high correlation between these factors and species composition. Nevertheless, a great deal of confirmatory field checking is required.

Figures 20 and 21, and Plates 13 and 14 show various approaches to the identification of tree species and the classification of timber stands by means of aerial photography.

Determining Tree Vigor and Agents Responsible for Vigor Loss

The photo-interpretation approach used in trying to obtain information about forest tree vigor is almost exactly the same as for farm crops. The trees most commonly attacked by insects and pathogens in an uneven-aged timber stand are the dominant, over-mature

FIGURE 21 Large-scale vertical stereogram taken from a helicopter. Near left center, note white dot (a small gas-filled balloon slightly above treetop level to facilitate location of sample plots). The most important tree species (fir, cedar, hemlock) are identifiable. Tree heights can often be determined more accurately from stereoscopic parallax measurements than they can from conventional on-the-ground measurements. By image comparison, the photo interpreter can extrapolate information he has obtained from large-scale photography such as this to similar adjacent areas for which only small-scale photography is available. (*Courtesy E. H. Lyons and British Columbia Forest Service.*)

ones, as shown by the bottom pair of photos in Plate 13. Furthermore, the earliest symptoms of these attacks are quite commonly registered in the top one fourth to one third of the tree crown. Usually, the denser the stand, the greater its value, the greater the forester's interest in protecting it, the greater the opportunity for rapid spread of insects and pathogens, and the greater the difficulty of detecting from the ground the presence and rate of spread of insects and pathogens, many of which strike primarily in the tops of the crowns of the dominant trees. Although the tops of dominant tree crowns are difficult to see from the ground, they are the most conspicuous features when the stand is viewed from the air, or on an aerial photograph. Furthermore, in timber stands, just as in farm crops, the earliest symptoms of attack may be in the form of a loss of near-infrared reflectance, which the human eye cannot see. For all these reasons, aerial photos, particularly when film, filter, scale, and camera inclination are specified, may be superior to ground survey. As with most of the other uses of aerial photos discussed in this chapter, it would be a mistake, however, to insist that aerial photos replace ground-survey methods. Realistically, one should use photos to the maximum possible extent because of the speed and economy (and sometimes additional accuracy) they offer. However, one should always verify and augment the photo data with direct on-the-ground observation.

Species identification and vigor estimation of trees, as determined from aerial photographs, are well illustrated from Croxton's study[9] and by Plate 14. Here, color photography has been used to differentiate various hardwood species and to detect the incidence of dieback disease on ash trees. The following key explains the numbers on the stereograms comprising Plate 14.

(1) Eastern white pine, *Pinus strobus* L. Silvery blue-green. Massive triangular branches horizontal in the lower crown, ascending in the upper crown. Very young trees have a broadly conical crown apex, but the tops of older trees are rounded (broadly oval). In mature and over-mature stands, white pine is considerably taller than associated hardwoods.

(2) Eastern hemlock, *Tsuga canadensis* (L.) Carr. Distinctive silvery olive-drab. Apex obtuse-conical. Texture feathery.

(3) Red maple, *Acer rubrum* L. Healthy trees decidely yellow-green. Often has yellow spots at its twig tips. Has distinctive fine columnar branches that emerge and ascend from general crown level. Entire crown therefore has a pincushion or porcupine appearance. Found on wet sites.

(4) Sugar maple, *Acer saccharum* Marsh. Healthy foliage yellow-green, but twigs end in a very small rounded clump of foliage often with a dark spot at

the center. Crown is extremely dense, periphery is smooth. Requires somewhat better drainage than red maple.

(5) White Oak, *Quercus alba* L. Foliage in many medium-sized clumps, each containing several very tiny rosette clumps with dark centers. Foliage glossy dark blue-green. Found on upper slopes and ridgetops.

(6) Northern red oak, *Quercus rubra* L. Same as white oak except that color is green, not blue-green. Dominant or emergent species wherever found.

(7) Bigtooth aspen, *Populus grandidentata* Michx. Foliage blue-green. Light-colored bark often shows through the thin crowns. Foliage arranged in tip-clusters much larger than those on ash with dieback disease. Color even blue-green, unlike the pale sickly yellow-green of the diseased ash and the dark green of the healthy ash. Twig tips unlike those of white ash.

(8) Healthy American elm, *Ulmus americana* L. Dark-green foliage similar to white ash but crown is finer textured. Upper side branches not vertically oriented. Branch ends that protrude beyond the crown margin are lanceolate sprays. Light spots common at twig tips. Crown may be tufted, but the tufts are fine-textured, flat sprays rather than columnar, and not borne erect. Usually grows on wet sites.

(9) American elm recently killed by Dutch elm disease. Dichotomous branching habit commonly discernible. Branches horizontal or drooping, crooked, and often end in a group of twigs that all curve abruptly in the same direction. Upper crown looks like a puff of dark smoke. (The glossy yellow-green foliage on some of the tree boles is that of poison ivy vines, *Rhus radicans* L.)

(10) Healthy white ash, *Fraxinus americana* L. Dark green. Twigs upswept, topmost ones widely separated (intertwig openings appear as dark shadows), adjacent upper ones at varying heights give ragged appearance to crown surface. Twig tips vertically oriented, with pale green spot, often have a tiny four-pointed star shape (large opposite compound leaves). Found on moist to wet sites.

(11) White ash with a severe amount of dieback. Sickly yellow-green. Foliage sparse with leaves clumped at tips of twigs. Leaves dwarfed. Often a large portion of the crown is entirely dead. Dieback sometimes stimulates premature leaf fall, and foliage may have fall colors also.

(12) White ash recently killed by dieback. Branches and twigs strongly ascending. Bark pink. Twigs very coarse. Twig tips straight or slightly arcuate but never sharply curved at tips. Branches seem to radiate in straight lines from a central point. Occasional opposite branching.

(13) Paper birch, *Betula papyrifera* Marsh. Trees alive but seasonally browned by birch leaf miner, *Fenusa pusilla* Lep. Foliage very fine textured. Chalk-white branches often show through foliage. Many leaning trees, showing almost the entire white bole. Found usually on wet sites, often in clumps.

(14) Black willow, *Salix nigra* Marsh. Distinctive silver color. Foliage dense, fine textured and arranged in large separate masses. Found on wet sites.

(15) Black birch, *Betula lenta* L. Like American elm except lacks protruding tufts and has a narrow oval apex. Grows upslope from elm—less tolerant of poor drainage. May be pale green.

In California, aerial photographs on panchromatic film with a minus blue filter (No. 25A) on a scale of 1 : 20,000 were used in several phases of blister-rust control work.[8] A technique was worked out in which aerial photographs were marked to show sugar pine stands suitable for protection against white pine blister rust.

Areas where root rot was killing Douglas fir trees appeared as small openings or breaks in the forest canopy on panchromatic photographs. The openings could not be attributed to root rot alone. Where there were no openings in the canopy, however, no serious root rot was present, and the absence of openings alleviated the need for ground inspection.

Color photography was tested for oak wilt detection in eastern Tennessee and was found to be too expensive in relation to conventional aerial sketch mapping. However, nonphotographic aerial methods sometimes can be successfully employed. The sketch-map technique was used to detect and spot pole-blight infections of western white pine. More recently Heller and Bean[10] have made very effective use from the air of a device known as the "operation recorder" when recording the incidence and the severity of insect infestations on coniferous timber stands.

Estimating Timber Stand Acreage

Attempts to measure acreages from aerial photos of rolling to mountainous terrain are subject to error. However, the error is remediable. These are the areas where timber stands are most commonly found. The magnitude of the error introduced by this elevational factor can best be indicated by a situation commonly encountered in the Sierra Nevada mountains of California. In this area, timber species *A* is confined largely to canyon bottoms at elevations averaging 5,000 feet. Species *B*, having characteristics vastly different from those of species *A*, and hence vastly different merchantable values, is confined largely to ridges and plateaus at elevations averaging 6,000 feet. Let us assume that vertical aerial photography is taken from an altitude of 15,000 feet above sea level of a typical portion of this area wherein the lowlands have an average elevation of 5,000 feet and the highlands an average elevation of 6,000 feet. In this instance the ratio of the *linear* photo scale in the highlands to that in the lowlands is 10 : 9, but the ratio of the *area* photo scale is 100 : 81. Because of differences in the timber species composition and merchantable values at the two elevations, this discrepancy in area determination is cumulative rather than compensating. Since the error is on the order of 20 percent, it can be serious if no attempt is made to correct for it.

The remedy is first to delineate the timber stands on the aerial photos, and then to map the stands at uniform scale from the photos with the aid of suitable plotting equipment. The acreage in each stand is then measured on the map rather than on the photographs.

Estimating Timber Stand Volumes

There are several ingenious approaches to estimating timber stand volumes on aerial photographs.

1. Photogrammetrically estimate the volumes of the individual trees comprising the stand. The ground observer, in estimating a tree's volume, commonly measures its stem diameter 4½ feet above the ground level and the merchantable height of the stem. It is possible to measure accurately on aerial photos both the apparent diameter of a tree's crown (which is closely related to the diameter of its stem 4½ feet above the ground level) and the apparent height of the tree's crown (which is closely related to the merchantable height of the tree's stem). Once such measurements have been made on a particular tree, a previously compiled aerial-photo tree-volume table can be consulted that will give the volume in cubic feet of wood, or in board feet of lumber, that is obtainable from a tree with the given dimensions.
2. Photogrammetrically estimate the volume per unit area of the timber stand itself. In many timber stands, the volume per unit area is closely associated with (1) the average apparent height of the dominant and codominant trees in the stand (roughly, the tallest 10 to 20 percent of the trees) as measured on aerial photos, and (2) the proportion of the total ground area within the timber stand boundary that is obscured by the crowns of the trees, also as estimated from the photos. Once the photo measurements in a particular stand have been made, a previously prepared photo-stand volume table can be consulted that gives the volume in cubic feet of wood and in board feet of lumber obtainable per acre from a stand with the given dimensions.
3. Ocularly compare the photographic appearance of an individual tree or stand with photographic examples of similar trees or stands for which the volumes have been determined by on-the-ground measurements. This method is comparable to the last of the methods for estimating crop yields discussed earlier in this chapter.
4. Delineate on aerial photos the boundaries of timber stands within

which conditions indicative of volume appear to be homogeneous. Once the boundaries have been delineated, several representative portions of each timber stand class are visited on the ground, and the average volume by species per unit area within the class is thus determined by conventional "timber-cruising" methods. In one vast forested area, comprising approximately 16 million acres, the value of this method has recently been demonstrated. With conventional aerial photography, the various kinds of timber stands could be so accurately delineated (on the basis of age or size class, stand density, and broad species composition, all of which are closely related to timber volume) that only about 700 acres (actually three 1/5-acre sample plots at each of 1,125 locations) needed to be cruised on the ground to permit the over-all volume of the 16 million acres to be estimated to within 5 percent.

5. Delineate on aerial photos the spots or strips within which conventional ground-cruising methods should be employed. Just as method 4 above is most commonly used on very large forest properties (comprising perhaps millions of acres), so this method is most commonly used on very small forest properties (comprising only a few hundred or a few thousand acres). In this method, advantage is taken of the fact that aerial photos permit one to see the entire forest property at a glance, thus facilitating selection of representative areas for ground cruising. Furthermore, when later taken into the field, these photos prove invaluable in guiding the field crews to the exact spots where representative plots or strips are to be cruised.

Other Current Applications

Photogrammetry and photo interpretation can be useful in combating forest fires under conditions where it is of the greatest importance to know accurately and quickly the conditions of vegetation and topography affecting fire behavior. Given this information, an intelligent plan for controlling the fire can be quickly developed. As shown in Figure 22, thermal-infrared imagery reveals the exact perimeter of a forest fire even when a thick, dense cloud of smoke lies between the fire and the sensor.[11]

A second application is in forest engineering projects, where the objective is to lay out an efficient system of logging roads, landing sites, ski trails, or recreational facilities (Figure 23).

Still another application is in multiple-resource management of forested areas, where by careful planning it may be possible to conduct logging operations without greatly impairing the usefulness of the

FIGURE 22 For delineating a forest fire's perimeter, the superiority of thermal-infrared heat-mapping imagery (bottom photo) over conventional panchromatic photography is clearly shown here. Despite a layer of smoke 1,500 ft thick, the infrared image shows the outline and internal details of the area engulfed by the fire, as well as some detail of the roads in the area; furthermore, hot spots remaining in the area that was burned on the previous day are seen at the edge of the heat-mapping image, even though the entire area is obscured by smoke on the panchromatic photo. (*Courtesy Stan Hirsch, U.S. Forest Service, Missoula, Montana.*)

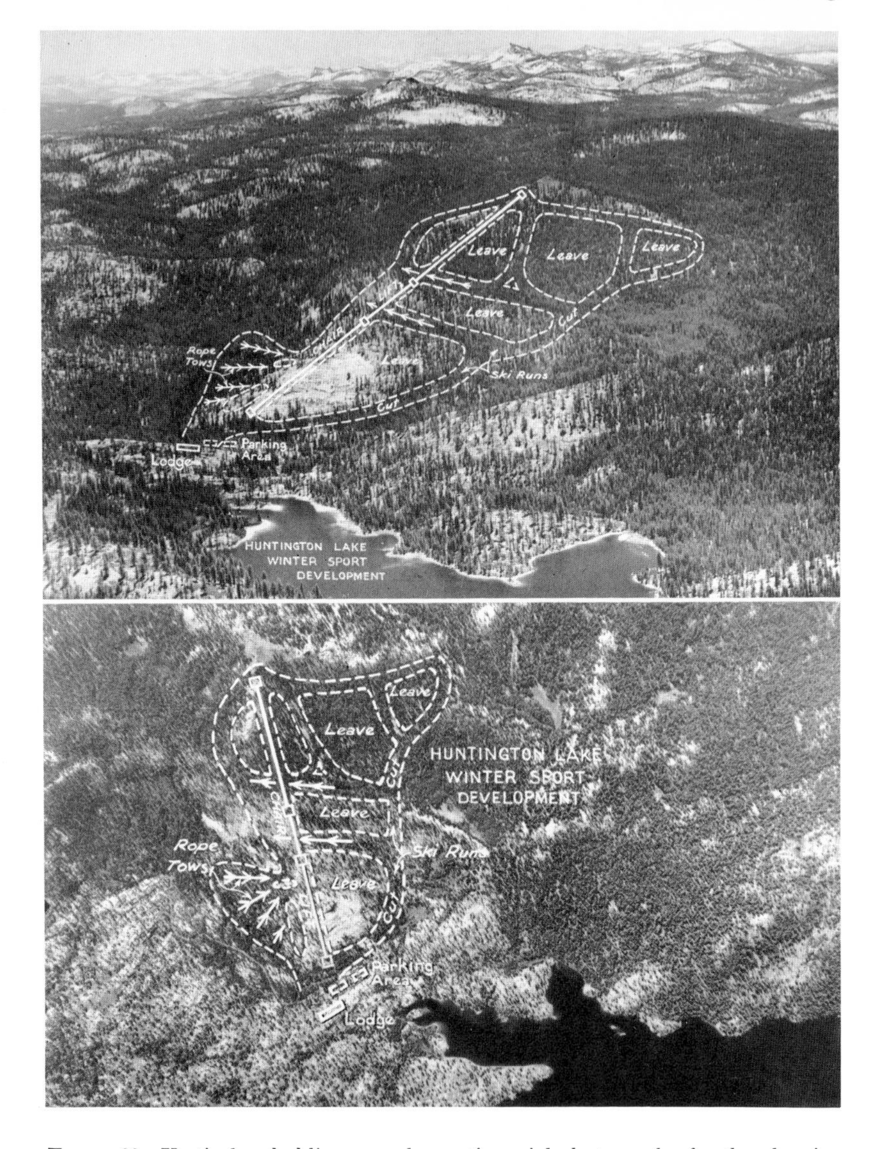

FIGURE 23 Vertical and oblique panchromatic aerial photography for the planning of recreational improvements in wildland areas (Huntington Lake, California). Note that the proposed siting of buildings as well as the paths where timber must be cut have been clearly indicated. The recreationist finds the oblique photo more interpretable and appealing, but the engineer or forester who must delineate the "cut-and-leave" boundaries, tree-by-tree, finds it far easier to position himself in the field through use of the vertical photograph. (*Courtesy U.S. Forest Service.*)

area as a watershed, fish and game sanctuary, livestock browsing area, or recreational site. In making decisions of this complexity, the forest manager often wishes he had the opportunity of hovering over the forest, so that he could look directly down on it, make a few pertinent measurements, and in general increase his appreciation of "how much" of "what" there is "where" in the forest. If he is skilled in the use of aerial photos, he has precisely this opportunity, in the comfort of his office, with precision measuring equipment, and without the time and fatigue limitations that ordinarily would be imposed on him if he were to attempt to make the same observations from an aircraft. Furthermore, these many pertinent observations are accurately recorded and positioned on the photographs. This permanent record is far superior to the alternative of relying on one's memory after having merely flown over the area to make a direct visual appraisal of it. The use of thermograms (Figure 24) and radar images (Figure 25) in the inventory of certain wildland resources are also shown.

Livestock Inventories

Among the inventory data desired by livestock managers are counts in each field, by kind of animal (cattle, sheep), use (dairy cattle, beef cattle), breed, sex, age, and vigor. In addition, they need an estimate of each factor affecting the animal-carrying capacity of an area, including the amount, palatability, accessibility, and nutritive value of each species of forage; the location and state of repair of corrals and livestock fences; and the location of stock-poisoning plants, noxious weeds, springs, salt grounds, watering places, rodent concentrations, highly erodible sites, and areas in need of reseeding.

The Vidya Study

Preliminary aerial photographic tests of livestock detectability were made by the Vidya Division of Itek Corporation in June 1963 over pastureland near Milpitas, California, at flight altitudes of 5,000, 10,000, 15,000 and 20,000 feet, with both high and low sun angles, using a high-resolution panchromatic film and a light red filter. Interpretation of this photography indicated that both cattle and sheep could be detected in aerial panoramic photographs at scales as small as 1 : 20,000. Reliable age and sex distinctions were not possible at this scale, however, and in the words of the Vidya report, "some cattle

FIGURE 24 Panchromatic photograph (0.4–0.7-μ wavelength region, minus blue filter, Wratten No. 12) (left) and thermal-infrared image (8–14-μ wavelength region) of a portion of the NASA Forestry Test Site near Bucks Lake, Sierra Nevada Mountains, California. Although the panchromatic photo has higher resolution and is far superior for estimating timber-stand densities, the thermal-infrared image clearly differentiates the moist from the dry portions of the meadow and highlights watercourses and the confluence of

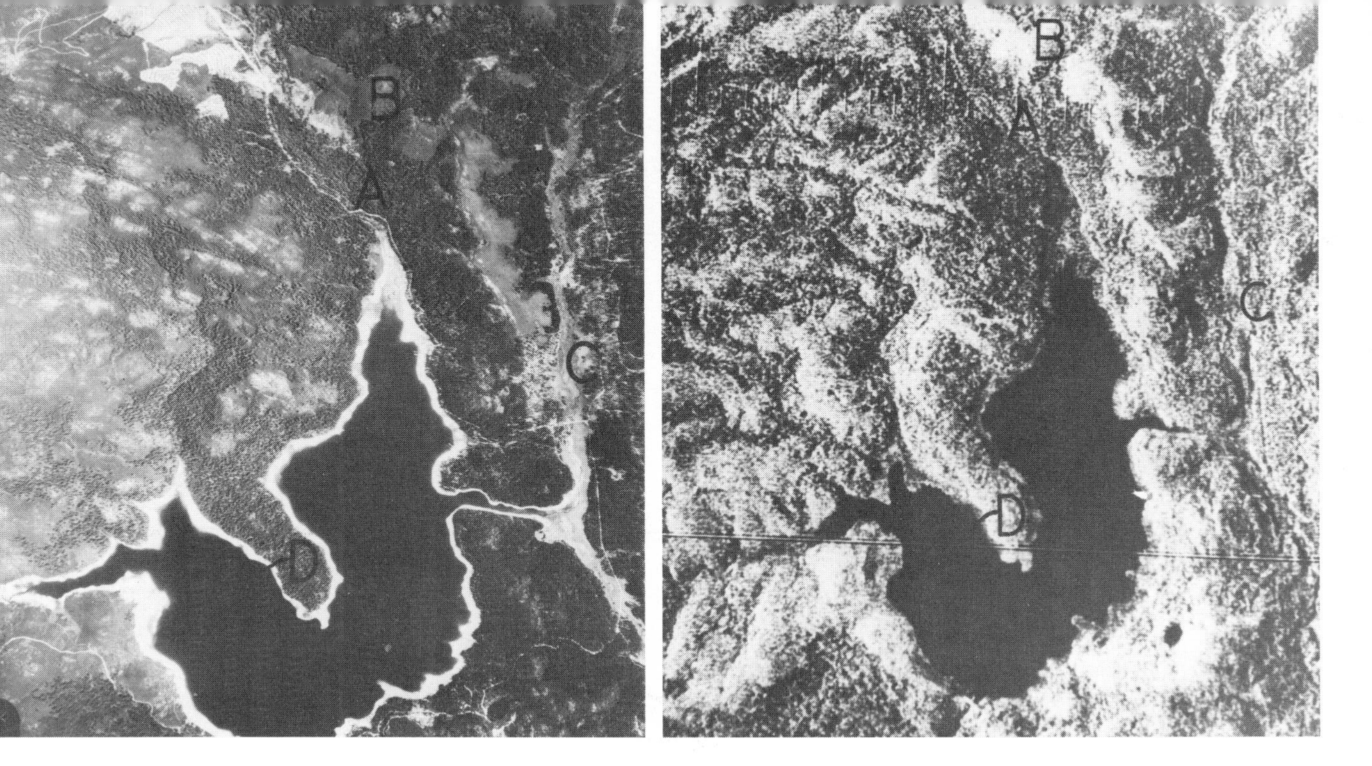

FIGURE 25 Panchromatic photo (0.5–0.7 μ, Wratten No. 12) (left) and radar image (1.0–3.0 cm) of a portion of the Bucks Lake Test Site. (*A*, timber; *B*, brush; *C*, meadow; *D*, water.) The panchromatic photo is superior for estimating vegetation density and identifying certain species, whereas the radar image is superior for discerning other vegetation types and for delineating drainage nets. (*Radar image courtesy Westinghouse Corp.*)

probably were overlooked." Because accurate on-the-ground cattle counts at the time of photography were lacking, no more-definite statements could be made.

Simulated operational flights were then made by Vidya, covering areas in Utah, Colorado, and Wyoming that the Department of Agriculture considered representative for a feasibility study on the making of livestock inventories. In these flights, 27 passes were made over 11 flight lines, mostly at an altitude of 12,000 feet above the terrain. The photographic flight lines for this 1 : 22,000 scale coverage totaled 190 linear miles and an area of nearly 700 square miles. Most of this photography was flown between the hours of 8:00 A.M. and 10:00 A.M. in order to combine the advantages of clear atmosphere, optimum light angle for interpretable shadows of livestock, and visibility of animals in the open during the cool of the day. (Animals seek shade in the middle of the day and thus may be hidden from view on aerial photos taken at that time.)

By means of Vidya's large-screen rear-projection equipment, negatives obtained from this photography were projected for viewing and interpretation. Livestock counts were made from the projected images in several hundred areas, including dairy feedlots and pastures, family-type barnyards, and range areas, on a total of 376 exposures. Where possible, the animals counted were tabulated by type, breed, sex, age, and environment.

Efforts were next directed toward checking the livestock counts against those made on the ground by Department of Agriculture field personnel and discovering the causes of discrepancies. Although the correlations between photo interpretation and ground sampling were considered satisfactory for a basic judgment of feasibility, Vidya concluded that denser and more-precise ground sampling would be necessary to make a detailed assessment of reliability, and particularly to establish a valid statistical confidence level.

Despite these limitations in the Vidya study, the authors of the report[12] concluded: (1) HyAc panoramic photography at scales of 1 : 7,000 to 1 : 8,000 is adequate for a livestock survey program. (2) Photographic passes made at these scales have been successfully used for detection and identification of animals by type and often by use class and breed. (3) The best black-and-white film–filter combination for use in making livestock inventories appears to be one that combines a high-resolution panchromatic film with a filter that eliminates blue, blue-green, and red light while transmitting green, yellow, and orange light. (4) Correct head counts can often be made on photography at

1 : 12,000 or even smaller scale. (5) Within the total head count, however, distinctions of age, breed, and particularly sex, are not consistently reliable at 1 : 12,000. (6) The primary photo-image characteristics that are useful in making livestock inventories are shape, size, color or tone, shadow, and site. The potential usefulness of these characteristics is discussed in detail in the Vidya report.

The encouraging but somewhat tentative conclusions drawn from the Vidya studies indicated that additional research might well be performed on the usefulness of aerial photography in making livestock inventories. Such studies should be well fortified with ground truth,* obtained, whenever possible, at the instant of photography. It seemed probable that the completeness and accuracy of livestock interpretation made on small-scale aerial photos might be improved if some basic studies were performed with large-scale simulated aerial photos. Such photos might be taken under carefully controlled conditions, from the top of a tower, for example, with representative livestock on display near the base of the tower. The near-vertical photos thus obtained would permit the photo interpreter to make a realistic and detailed analysis of each animal image and of the shadow cast by the animal. Many photos of this livestock array could be taken from the tower quite economically, at various sun angles, with various film–filter combinations, and with varying but precisely known amounts of stereoscopic parallax. Furthermore, the animals comprising this array could be observed at various stances representative of those encountered on operational aerial photos. It seemed likely that a study of this photography and of actual aerial photography obtained of the same target array at the same time would indicate the most suitable film–filter combination and the best time of day for photographing livestock. The aerial and tower photography would also contain valuable examples for the construction of photo-interpretation keys to livestock.

Once this work had been performed, simulated operational photography of representative range and pasture lands could be flown to optimum specifications. Then, through use of the photo-interpretation keys, livestock inventories could be attempted on this photography under truly optimum conditions. It seemed necessary to strive for such conditions because the Vidya studies indicated that interpretation of aerial photographs of livestock is a difficult task.

*The term "ground truth" pertains to information obtained by direct on-the-ground observations in any given geographical area. Its purpose is to establish the accuracy of image analyses.

There is a far better prospect of identifying each animal as to type, breed, sex, age, and vigor if both the vertical and the horizontal views are considered than if only one of these views is considered. The photo interpreter who realizes this important fact will take pains to interpret both the animal and its shadow whenever possible, as he attempts to make livestock inventories from aerial photographs.

To activate the reasoning above, photos were taken from a water tower both at midday and in the late afternoon to determine the optimum elevation of the sun above the horizon in relation to livestock-shadow interpretations. An elevation of 10° to 30° above the horizon provided the most interpretable imagery and also corresponded to the cooler times of day when the animals would not be seeking shade.

When the *simulated* aerial photos of the target array were being obtained from the water tower, a limited number of *actual* aerial photos were being obtained of the same array from a Cessna 180 aircraft. Flight altitudes for the aerial photos were 300, 600, 1,200, and 2,400 ft, respectively. The Zeiss aerial camera used had a focal length of 6 in., a negative size of 9 × 9 in., and an average resolution over the entire field of approximately 40 line pairs/mm, as compared with only about 20 line pairs/mm for most aerial cameras. The largest scale of photography thus obtained was roughly the same as that obtained from the water tower; hence, the simulated and actual aerial photos could be directly compared. When this was done, it became apparent that an aerial photo interpreter can draw valid conclusions from the economical tower photography (as to the relative merits of various film–filter combinations and the photo interpretability of livestock, for example), thus eliminating the need for flying large amounts of costly aerial photography when conducting the research.

California Studies

The photos comprising Figure 26 were taken on the Davis Campus of the University of California and are representative of the more than 200 vertical photographs taken from the catwalk of a water tower there.[13] They illustrate how livestock of various kinds, breeds, sizes, and sexes appear in the near-vertical view when photographed from an altitude of 150 ft with a camera having a focal length of 6 in. All these simulated aerial photos were taken with panchromatic film and a Wratten No. 12 (minus blue) filter. However, several other films were tested, including orthochromatic, infrared, negative color, Ektachrome, and Infrared Ektachrome, each in conjunction with various filters.

FIGURE 26 Possible uses of aerial photos in the inventory of livestock are indicated here. These photos were taken from the catwalk of a water tower, 150 feet above the livestock target array. Being of large scale, these photos highlight image-recognition characteristics of the major livestock that might lend themselves to an aerial photo inventory. The photos have been mounted in 6 vertically oriented pairs, with 3 pairs in the top half of the figure and 3 pairs in the bottom half. The animals illustrated are as follows: *top half, left pair:* Hereford bull and steer; *middle pair:* quarterhorse and thoroughbred; *right pair:* mules; *bottom half, left pair:* colts; *middle pair:* calves and goats; *right pair:* pigs and sheep. *A:* shadows cast perpendicular to sun. *B:* shadows cast parallel to sun. *C:* shadows insignificant because animals are lying down.

The photos of Figure 26 have been mounted so that shadows cast by the animals fall toward the bottom of the page, i.e., toward the observer. This has been done in accordance with accepted photo-interpretation procedure to avoid a pseudoscopic (false stereo) effect. Nevertheless, the reader may find it informative to rotate the figure 180° and then note the very close similarity between the shadow of each animal and the silhouette it would exhibit in the familiar ground view.

Actually, through the use of two identical cameras for each film–filter combination, two photos were taken simultaneously from the water tower in order to eliminate troublesome stereoscopic effects that otherwise would have been introduced by movements of the animals of the target array during the interval between exposures. The tower tests were followed by tests made with aerial photos of livestock on representative range and pasture. The example shown in Figure 27 is representative of the photography that has recently been obtained in California. In this instance, as in all others, a special effort was made to obtain simultaneous terrestrial photography in order to provide irrefutable ground truth against which to check the interpretation of the aerial photos. On the original of Figure 27*A*, the man taking the on-the-ground photos could be discerned standing on top of the automobile. From that point he took a panoramic series of five photographs of sheep in the nearby field, in rapid succession while the photographic aircraft was following the prescribed flight line overhead. Two of the five are shown (Figure 27*B* and *C*). Although an animal-by-animal tie-in can be made between the aerial photo and the corresponding ground photos, careful study shows that a few of the animals are not in identical positions on the aerial and terrestrial photos; this is explained by the fact that the herd is practically never static.

In some experiments, two-way radio communication between the aerial photographer and the ground photographer has been maintained so that the man on the ground would know exactly when to take a photo that would tie perfectly to a corresponding aerial photo. This degree of refinement may seem totally unnecessary in research of this type, but it has been shown that, without careful on-the-ground checking at the instant of photography, discrepancies might be attributed to inability to detect animals consistently rather than to the true reason (their rapid movement from one location to another, for example).

The following conclusions can be drawn from the California work:

1. Enumeration of livestock on photos at a scale of 1 : 6,000 is quite feasible.

FIGURE 27 *Top:* Panchromatic photograph (minus blue filter, Wratten No. 12) of sheep on pasture near Hopland, California. Scale 1 : 2,000. Low sun angle. Total number of sheep in the five sectors = 67. *Bottom:* Simultaneous terrestrial photos of Sectors 1 and 3.

2. In representative fields, e.g. Figure 27, errors of greater than 5 percent in total livestock count are rarely made by the photo interpreter.

3. In areas having a denser cover of tall vegetation than is shown in Figure 27, the accuracy falls off very rapidly to a point where less than 10 percent of the animals can be detected on aerial photos.

4. Many errors in interpreting small-scale photographs can be

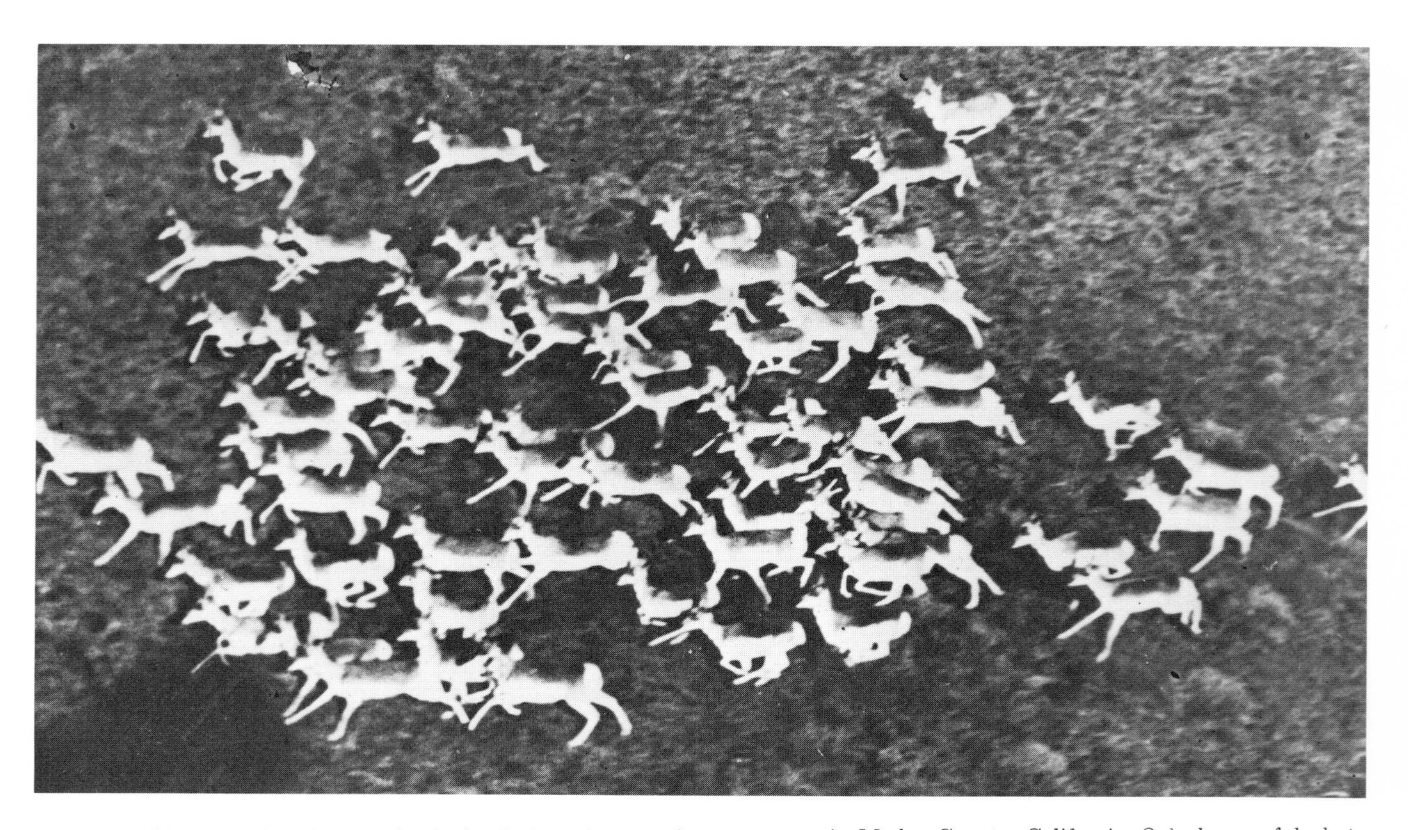

FIGURE 28 Oblique aerial photograph of a herd of antelope on the open range in Modoc County, California. Only by careful photo interpretation can all 73 antelope be detected, but such a method is vastly superior to on-the-ground inventory methods. (*Courtesy U.S. Fish and Wildlife Service.*)

avoided if the interpreter first studies large-scale photographs (e.g., Figure 26).

5. Photos taken at low sun angle usually are preferable because of greater shadow detail and the tendency of animals to be in the open during the cool of the day.

6. The best time of year for livestock inventory is late spring. Livestock data tend to be most interpretable at that time of year because: (a) the weather is cool enough that cattle are not so likely to seek shade; (b) most of the grass is green and contrasts better with sheep and other light-toned animals; (c) sheep have only recently been shorn and hence are very light in tone; (d) trees may not be in full leaf and hence may not obscure livestock as much as in midsummer.

Wildlife Inventories

Many of the conclusions just cited with reference to livestock are equally applicable when the objective is to inventory wildlife on the open range. Examples are antelope and deer in dry plateau areas (Figure 27) and caribou in snow-covered clearings (Figure 28). The objective when inventorying wild animals frequently goes beyond that of making a count of the total animals present. Often it is also used to obtain the ratio of juveniles to adults and, among the adults, the ratio of males to females. In photos of caribou, such as that shown in Figure 29, juveniles can be distinguished from adults, but since both the male and female adults have horns, it rarely is possible to differentiate them.

Possibilities for inventorying seals and salmon by means of aerial photography are illustrated in Figures 30 and 31.

Forage Inventories

Forage inventories usually seek to determine the volume and species composition of herbage on rangelands in an effort to estimate animal-carrying capacities (the number of animals of any given type that can be grazed on the range for a given period of time). Important differences in range herbage can be detected on small-scale aerial photographs, mainly on the basis of differences in photographic tone or color (see Figure 32).

The accuracy of such a classification of rangeland conditions depends on the scale of the aerial photography, the film–filter combina-

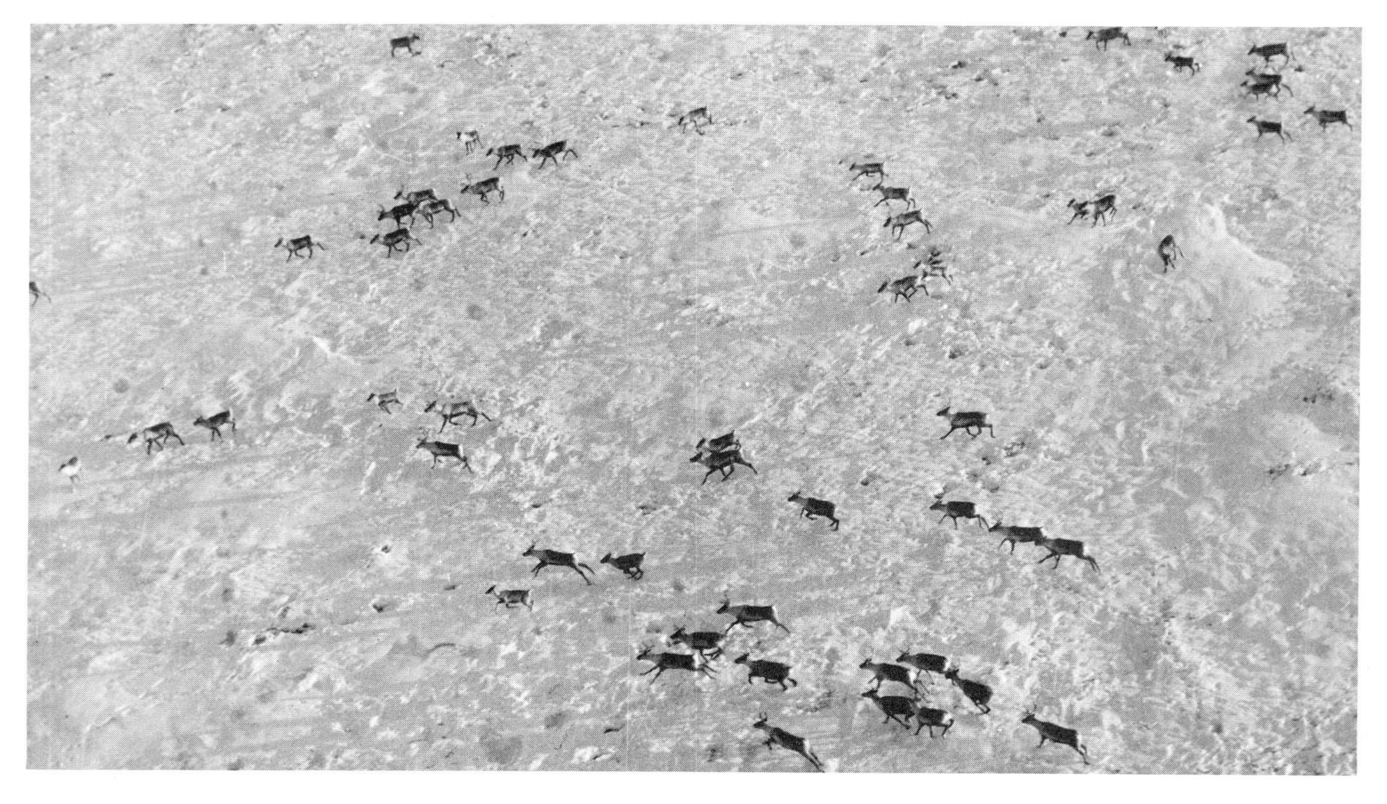

FIGURE 29 Oblique aerial photo of Stone's caribou in the Lake Clarke area, Alaska Peninsula, showing the feasibility of inventorying these animals against a background of snow. (*Courtesy U.S. Fish and Wildlife Service.*)

FIGURE 30 Aerial oblique photo showing harbor seals resting on a small island formed at low tide in the shallow part of the Bering Sea. (*Courtesy U.S. Fish and Wildlife Service.*)

tion employed, and the seasonal stage of development of the forage. Viewing the aerial photographs as stereo pairs makes differences even more obvious.

Rangeland conditions in several areas of Contra Costa County, California (Figure 33), were analyzed after they had been classified by interpretation of aerial photos. Ground checks revealed three classes of per-acre carrying capacity, the first nearly three times greater than the second, and eight times greater than the third. These differences

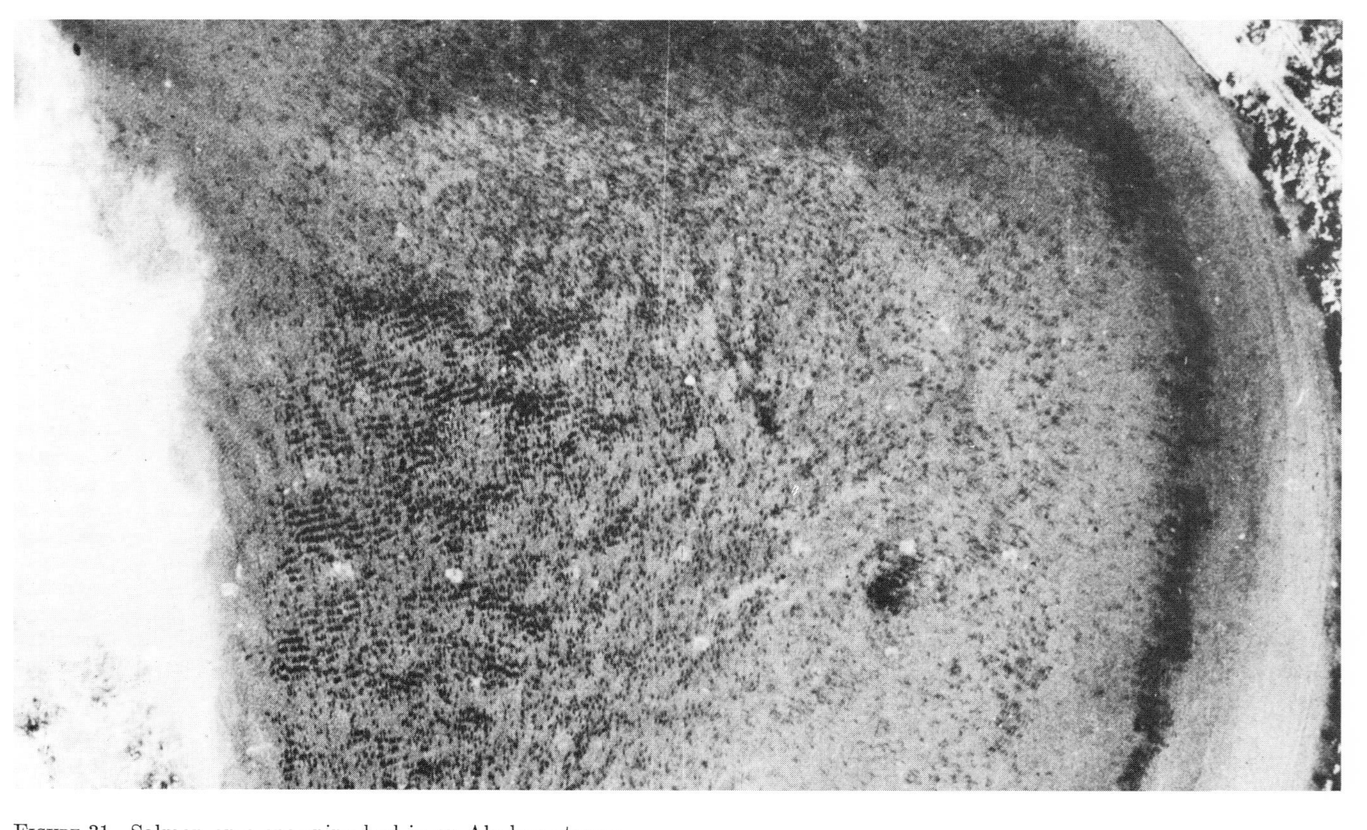

FIGURE 31 Salmon on a spawning bed in an Alaskan stream.

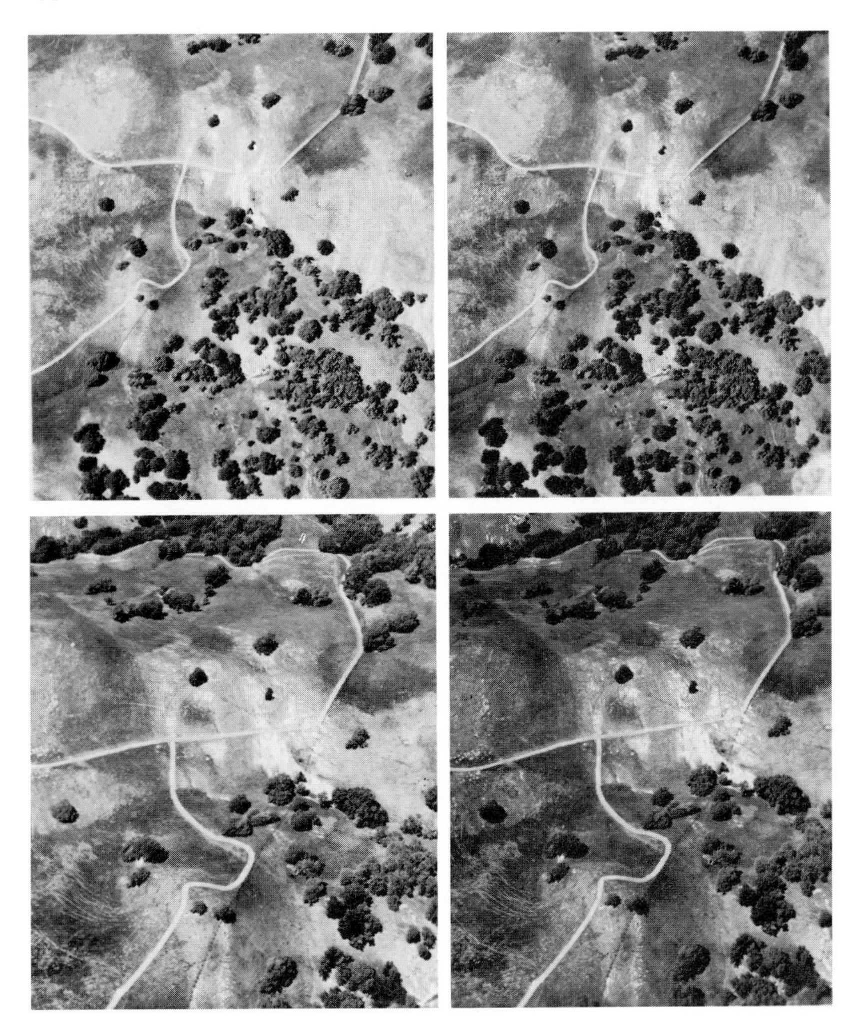

FIGURE 32 Two panchromatic aerial stereograms of rangelands in Contra Costa County, California, as they appear in mid-May each year. *Top* = vertical; *bottom* = oblique. Very light-toned areas are rock outcroppings and bare ground. In the darkest toned areas that are grass covered, the forage is still green and will have nearly three times greater animal-carrying capacity per acre than adjacent grass-covered areas that are only light gray in tone.

were consistently significant at the 95 percent level of probability. Total acreage within each of the three classifications was also readily determined photogrammetrically.

Photographic specifications have been developed for range inventory. Both aerial and ground photos of representative areas were taken on

FIGURE 33 Photo interpretation and associated fieldwork have enabled the range surveyor to stratify this area into essentially homogeneous units. Lines with crosses represent fences; lines without crosses represent type boundaries. Within each stratum a symbol gives three types of information: (1) the range type (e.g., 1 signifies grassland other than meadows, 6 signifies coniferous timber land with an understory of forage plants; (2) the most abundant species (e.g. "Far" signifies fescue grass, *Festuca arizonica,* PP signifies ponderosa pine); (3) the vegetation density (e.g., .25 signifies that 25 percent of the ground surface is covered with vegetation). (*Courtesy U.S. Forest Service.*)

specific dates during each of the four seasons, at four scales from 1 : 2,000 to 1 : 20,000. The following film–filter combinations were used:

Film	Filter Number	Filter Description
Panchromatic	25A	Light red
Infrared	89B	Dark red
Aero Ektachrome (color)	HF-2	Haze-cutting
Camouflage detection (color)	15	Orange

These trials showed that the optimum time for aerial photography of rangelands in most of the California foothill country is in late spring; the next best time is one month after the first soaking rains in fall. At these periods, developing vegetation provides the greatest differentiation. The optimum scale for aerial photography, when cost must be balanced against useful information, is about 1 : 5,000, although spot coverage of a few representative areas at a scale of 1 : 2,000 also is desirable.

Of the four film–filter combinations tested, Aerial Ektachrome film with a haze-cutting filter gave the best results, and panchromatic film

FIGURE 34 Portion of NASA Range Resources Test Site, Harvey Valley, California. *A*, Aerial oblique photo, infrared film, dark red filter (No. 89B). Light-toned bushes are big sage, and dark-toned bushes are bitter brush (identified in field check), the two most important forage species for both cattle and deer (preference varying with type of animal and time of year). Only this film–filter combination adequately differentiates these species. *B*, Same area, vertical view, panchromatic film, minus blue filter (No. 12). *C*, Same area, vertical view, infrared film, dark red filter (No. 89B). Arrow indicates camera station from which *A* was taken. Compare *B* and *C* with Plate 15. (*Courtesy David Carneggie, University of California.*)

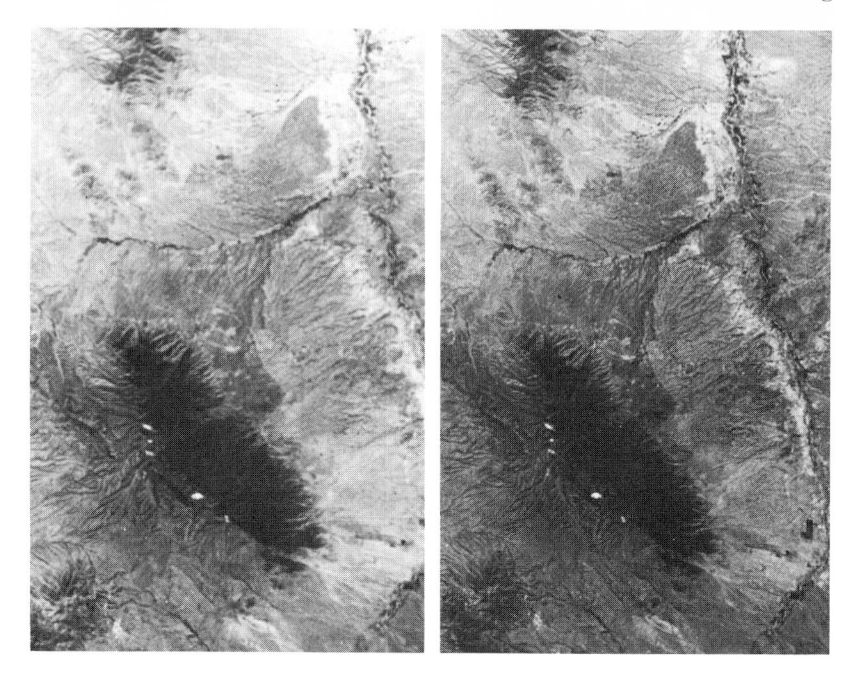

FIGURE 35 Stereogram made by space photography (Gemini IV). Compare with Plates 1 and 15, which show the same area.

with a light red filter was almost as good. Photographs on these films should not be taken on hazy days.

Figures 34 and 35 and Plates 15 and 16 show additional applications of remote sensing in the inventory of rangeland resources. See also Figure 5, Chapter 1; and Plate 1.

CONCLUDING NOTE

The very great promise of photogrammetry and photo interpretation in agriculture, forestry, and related fields, of which but a glimpse is contained in this chapter, should not be discredited on the strength of some early, inevitable failures that are on record. Such failures usually happened for one of two reasons: (1) enthusiastic writers have failed to stress sufficiently the limitations of photographic methods; and (2) readers, when attempting to apply what they have read on some new method, often lack the training, experience, and equipment necessary to insure even limited success.

The authors recommend the *Manual of Photographic Interpretation*[14] as an excellent basic reference work for the potential user of photography as a means of acquiring agricultural information. Within the limits and with the precautions noted therein, and with the discussions and illustrations of this chapter as examples, the value of photography as a means of remote sensing immediately becomes apparent.

Although nonphotographic sensing is less well developed, its own great potential should not be underestimated.

References

1. Steiner, D., and H. Mauer. 1967. Development of a qualitative semi-automatic system for the photo identification of terrain cover types. *In* Proceedings of Second CIS/ICAS Symposium on Air Photo Interpretation. Ottawa, Canada.
2. Simonett, D. 1966. Application of color-combined multiple polarization radar images to geoscience problems. Computer Applications in the Earth Sciences, A Colloquium. Computer Contributions #7, Kansas Geological Survey, University of Kansas, Lawrence. p. 19–23.
3. Brenchley, G. H., and C. V. Dadd. 1962. Potato blight recording by aerial photography. NAAS Quart. Rev., London, England. 57:21.
4. Manzer, F. E., and G. R. Cooper. 1967. Aerial photographic methods for potato disease detection. Maine Agr. Exp. Sta. Bull. 646.
5. Colwell, R. N. 1956. Determining the prevalence of certain cereal crop diseases by means of aerial photography. Hilgardia 26(5).
6. Walker, W. R. 1963. VELA uniform project, Semi-Annual Tech. Rep. No. 4. Itek Corp., Vidya Div., Palo Alto, Calif.
7. Colwell, R. N. 1968. Determining the usefulness of space photography for natural resource inventory. *In* Proc. 5th Symp. Remote Sensing. Univ. of Michigan.
8. Harris, T. H. 1951. Uses of aerial photographs in control of forest diseases. J. For. 49(9):630–631.
9. Croxton, Ralph J. 1966. Detection and classification of ash dieback on large-scale color aerial photographs. Pacific SW For. Range Exp. Sta., U.S. For. Service Res. Paper PSW-35. Berkeley, Calif. 13 p. (illus.).
10. Heller, R. C., and J. L. Bean. 1952. Aerial surveying methods for detecting forest insect outbreaks. U.S. Dep. Agr., BEPQ Div. of Forest Insects. (mimeo.)
11. Hirsch, S. N. 1962. Applications of remote sensing to forest fire detection and suppression. *In* Proc. 2nd Symp. Remote Sensing of Environment. Univ. of Michigan, Inst. Sci. Tech.
12. Strandberg, C., and C. Lukerman. 1963. Development of livestock and crop survey techniques—Phase I. Vidya Report No. 118. Itek Corp., Palo Alto, Calif., December 13.
13. Colwell, R. N. 1964. Uses of aerial photography for livestock inventories. *In* Proc. Ann. Meeting, Agr. Res. Inst. Nat. Res. Counc.
14. American Society of Photogrammetry. 1960. Manual of photographic interpretation. George Banta Co., Inc., Menasha, Wis.

5

Physical and Physiological Properties of Plants

David M. Gates Missouri Botanical Garden and Washington University, St. Louis, Missouri

Introduction

The appearance of plants and of vegetated surfaces to multispectral sensors or to the human eye depends on their interaction with radiation. A plant or vegetated surface may be viewed actively by reflected sunlight and skylight or passively by the emission of thermal radiation from the plants.

The precise spectral quality and intensity of plant reflectance and emittance depends on leaf geometry, morphology, physiology, chemistry, soil site, and climate. It is the purpose here to discuss those physical and physiological properties of plants that are significant for multispectral sensing of vegetation. Some description is given of the appearance of vegetation and soils.

Plant Reflectance, Transmittance, and Absorptance

A plant leaf reflects and transmits incident radiation in a manner that is uniquely characteristic of pigmented cells containing water solutions. Typical spectral reflectance, transmittance, and absorptance curves for plants are shown in Figures 1 and 2. The striking features of leaf spectra are the high absorptance in the ultraviolet and the blue, the reduced

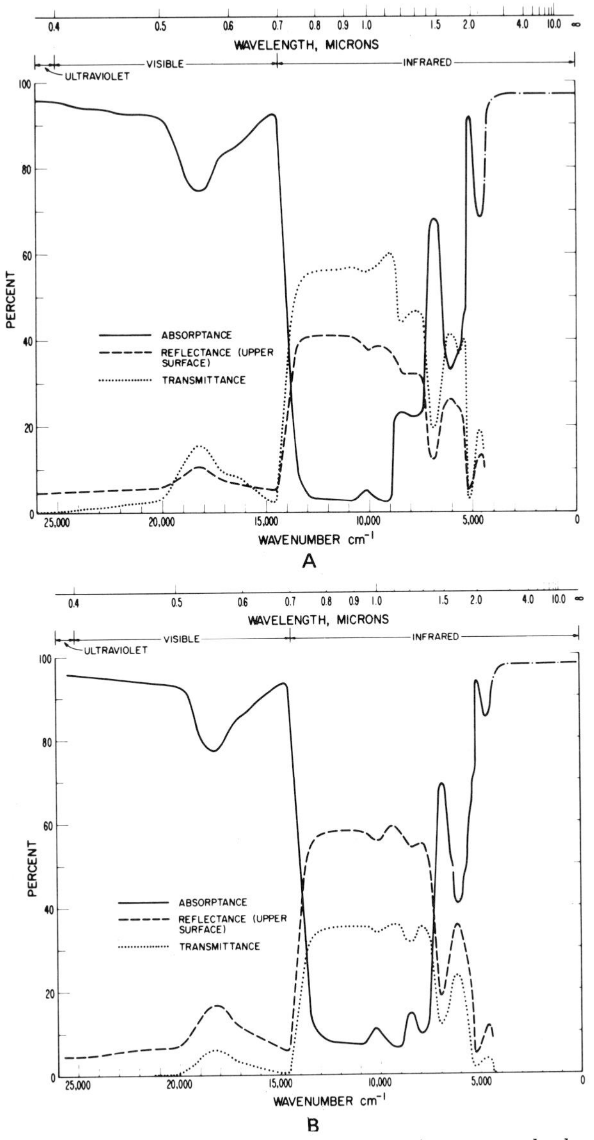

FIGURE 1 Spectral reflectance, transmittance, and absorptance for (*A*) *Populus deltoides,* and (*B*) *Nerium oleander.*

absorptance in the green, the high absorptance in the red, the very low absorptance and high reflectance and transmittance in the near infrared (0.7–1.5 μ), and the very high absorptance in the far infrared.

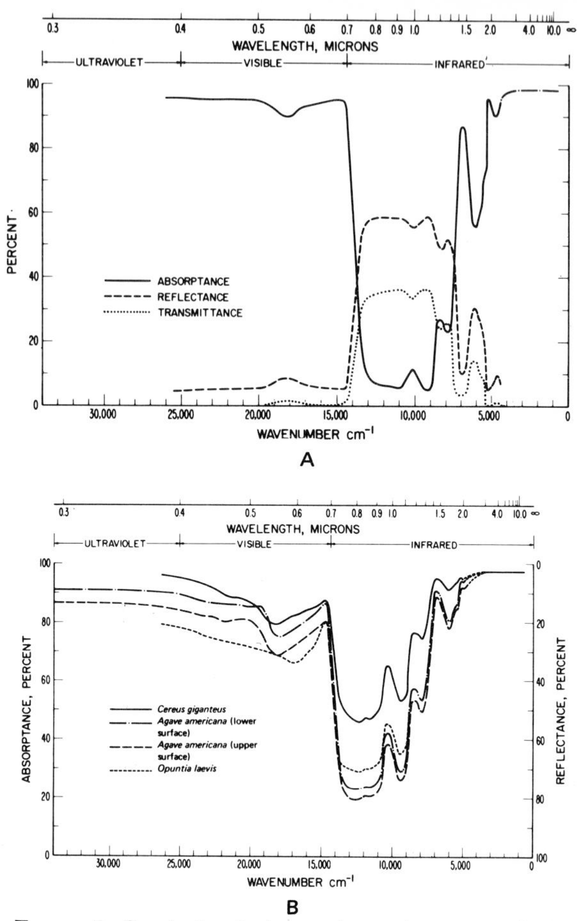

FIGURE 2 Spectral reflectance, transmittance, and absorptance for (*A*) *Raphiolepis ovata,* and (*B*) three succulent species.

The very abrupt increase in reflectance near 0.7 μ and the fairly abrupt decrease near 1.5 μ are present for all mature, healthy green leaves. The peaks in the absorptance curve at infrared wavelengths of 0.9, 1.1, 1.4, and 1.9 μ are liquid-water absorption bands. Liquid water in a leaf is largely the cause of the very strong absorption throughout much of the far infrared. Figure 3 illustrates possible paths for light interacting with a leaf. A small amount of light is reflected from the leaf cuticle; much is transmitted into the spongy mesophyll, where the rays have frequent encounters with cell walls and are critically reflected if the angles of incidence become sufficiently large. The change in the index of refraction encountered at the cell walls causes Fresnel reflection. The multiple reflection is essentially a "random walk" as the rays frequently change direction within the leaf. Because of the numerous cell walls, nearly as many rays are reflected back toward the source as are transmitted through the leaf. Generally, the transmittance is greater than the reflectance for thin leaves, but with thicker leaves the transmittance is substantially less than the reflectance. The very thick

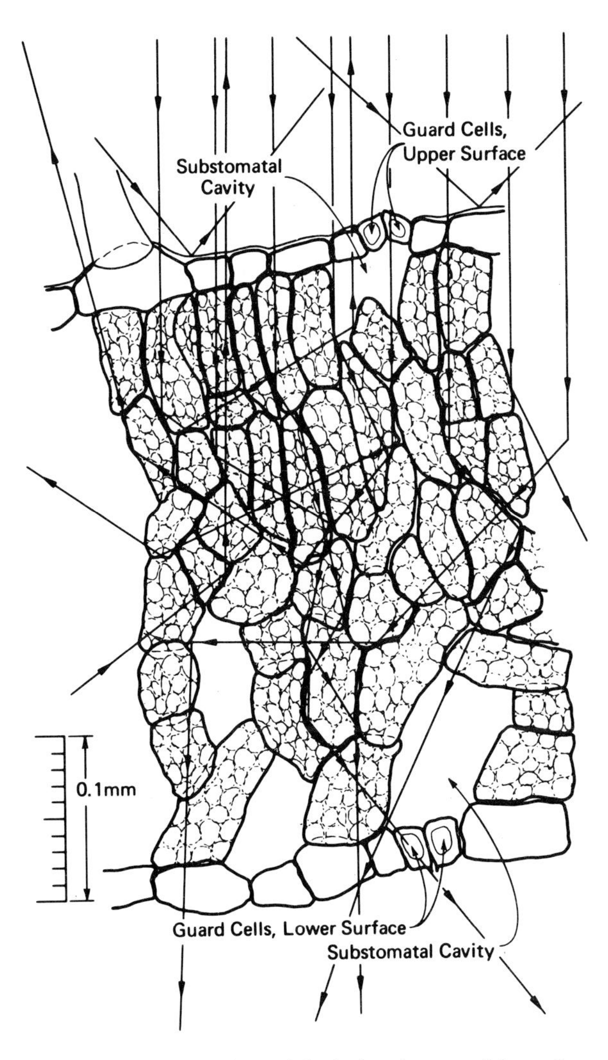

FIGURE 3 Cross section of leaf showing possible pathways for light.

dark leaves of xeric plants and the fleshy stems of cacti transmit no radiation and reflect very strongly in the near infrared. A considerable amount of reflectance occurs at the outermost layer of thick waxy cuticle covering many succulent plants.

In Figure 4 the principal absorption bands of some plant pigments and of liquid water are shown. Although the absorption coefficients in the red are substantially less than in the blue, scattering within the mesophyll cells produces nearly complete absorption over a narrow wavelength region in the red. In addition, scattering also closes the transmission gap in the green and broadens the absorption. The reflection in the green is generally relatively low (10 to 20 percent), but

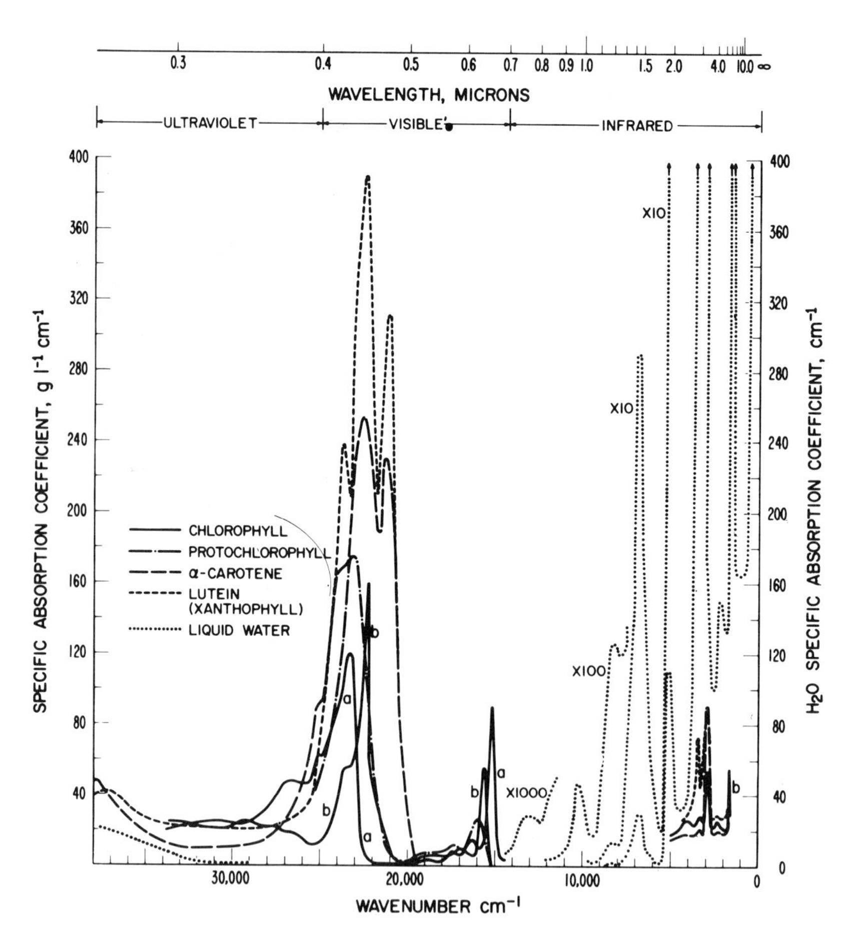

FIGURE 4 Spectral absorption of some plant pigments and liquid water.

because of the enormous sensitivity of the human eye, in the green, the eye sees light of this wavelength reflected with great contrast.

The greatest region of little absorption in the entire transmission spectrum of leaves is between the red bands caused by electron transitions in pigments and the infrared bands of liquid water, which do not become very strong until 1.9 μ. The pigment and liquid-water absorption bands are physically very different. The pigment absorptions are caused by electron transitions within the pigment molecular complexes. The liquid-water absorptions are caused by transitions of the vibrational and rotational states of the water molecules. Electron transitions require substantially higher energies than the vibrational-rotational transitions. Therefore, the electronic absorption bands are in the ultraviolet and visible regions, and the vibrational-rotational absorption bands are in the long-wave infrared. For water, the fundamental vibrational band with the highest frequency is at 2.66 μ. Those bands at 1.9, 1.4, 1.1, and 0.9 μ are all overtone and combination bands that have lower transition probabilities and are therefore successively weaker bands than the fundamentals. The other two fundamental bands of water are at 2.74 and 6.3 μ. That plant leaves reflect and transmit strongly in the near infrared is a fortuitous result of the separation of the physical processes governing high-frequency and low-frequency absorptions. However, the low absorptance of the near infrared of many plant leaves materially reduces the amount of incident solar energy absorbed by the leaf. As a consequence, the temperatures of sunlit leaves are as much as 10° C lower than they would be if they strongly absorbed in the near infrared, and the plant pigments are thus saved from denaturation during warm days.

The spectral reflectance of most chlorophyll-containing surfaces is similar. Examples of the spectral reflectances of several plant groups are shown in Figure 5. The spectral reflectance of the bracket fungus is very different from that of leaves, and so is the reflectance of barks shown in Figure 6. A deciduous forest in the winter, presenting a surface of bark and a ground cover of dried leaves to a spectral scanner, will give a rather neutral signal as a function of the wavelength; this is particularly evident when compared with the sharp differences of reflectance with wavelength for chlorophyll-containing leaves.

The spectral reflectance of leaves undergoes strong changes both early and late in the growing season. Figure 7 gives an example of such changes. With the juvenile leaf, the blue and green reflectance is low, the yellow and red relatively strong, and the near-infrared very strong. The reason is that the juvenile leaf (white oak, *Quercus alba*) has a dense covering of pubescence comprised of purplish hairs. As the leaf

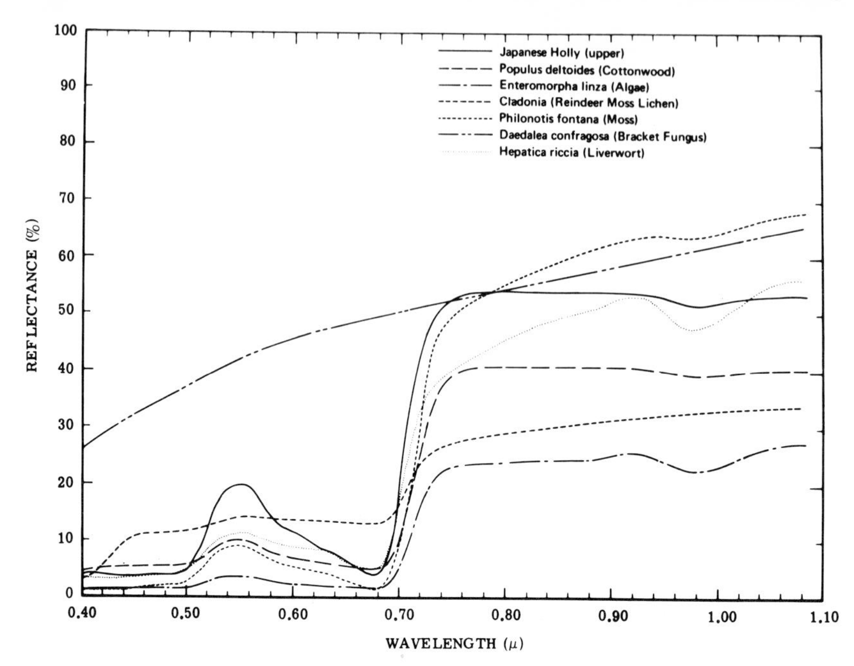

FIGURE 5 Spectral reflectance of seven species of plants representing diverse groups.

grows and expands quickly, the hairs spread out, exposing the underlying leaf to view. The near-infrared reflectance drops, and some absorption (Figure 7*A*) in the red begins due to the presence of chlorophyll. The stronger green absorption caused by the dense pubescence is now reduced, and the result is an increase in green reflectance. A surface pubescence or bloom will affect the spectral reflectance not only at visible wavelengths but also in the near infrared. On April 22 the leaf exhibited a yellowish appearance. The near-infrared reflectance continued to drop as the leaf expanded, unfolding and spreading out the hairs. By May 5 the chlorophyll-absorption bands in the blue and red were strongly in evidence, and the near-infrared reflectance reached a minimum value. By May 11 the chlorophyll content of the leaf had increased, causing further increase in blue and red absorption, which overlapped into the green causing a reduction in green reflectance and a visible darkening of the leaf to the eye. The near-infrared reflectance increased substantially, apparently as the intercellular air spaces developed in the mesophyll, thus increasing once again the number of reflecting surfaces with abrupt changes in index of refraction. The leaf chlorophyll content increased very little between May 11 and 18 because the green reflectance dropped only slightly. The near-infrared reflectance increased significantly as the mesophyll developed further and the intercellular air spaces increased. During the period May 11 to 18 very little enlargement of leaf area occurred, but presumably an expansion of the mesophyll thickness continued.

From May 18 on, throughout the growing season, the spectral reflectance characteristics remained nearly invariant. Once the reflectance, absorptance, and transmittance are known for a mature leaf they remain fixed until some external factor changes the leaf chemistry. Such a factor may be a change in soil chemistry or climate or a change caused by a plant pathogen. It is evident that several very different processes may affect the spectral reflectance. Among these are (1) a change in pigmentation, (2) a change in mesophyll cell structure, (3) a change in water content, and (4) a change in the surface coat of the leaf (pubescence, bloom, or a mold).

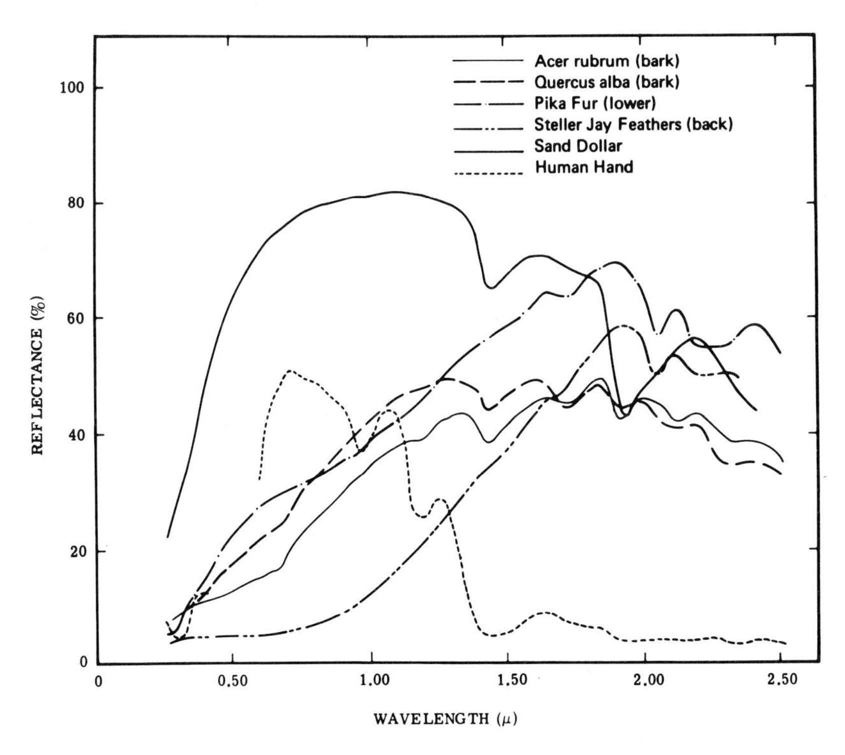

FIGURE 6 Spectral reflectance of bark from two species of trees and parts of several animals.

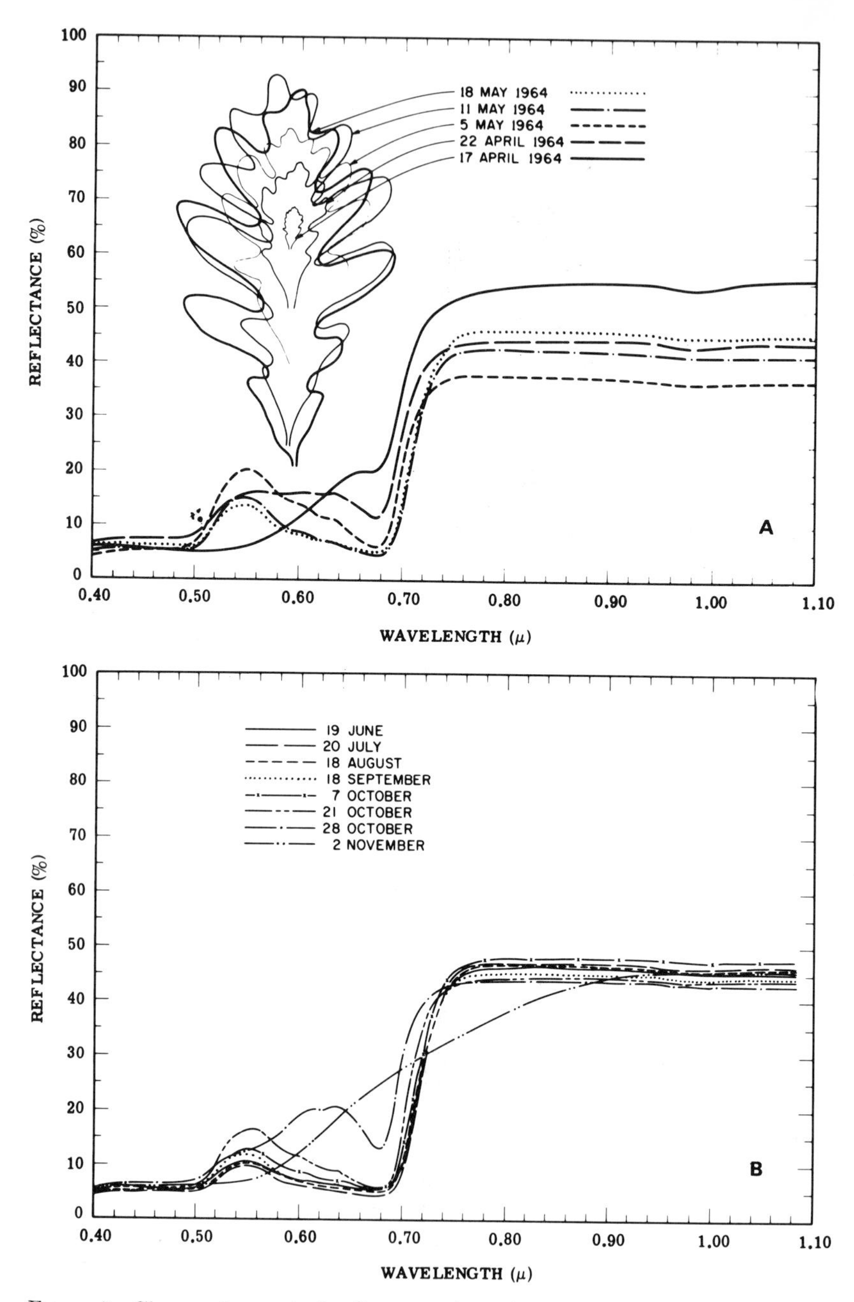

FIGURE 7 Changes in spectral reflectance throughout growing season in *Quercus alba: A,* April–May; *B,* June–November.

By far the most sensitive of these changes are those due to pigmentation and to surface effects. Relatively slight changes in pigmentation will appear in the quality of the visible spectral reflectance. Pigmentation changes, such as leaf chlorosis, can occur because of abnormal soil chemistry.

The spectral reflectance of the near infrared is largely the result of interaction of the incident radiation with the mesophyll structure. Once the leaf mesophyll is fully developed it would appear unlikely that further changes will be caused by small external causes. Certainly, very large variations in such external factors as water availability will produce changes in the mesophyll water–air relationship. Perhaps in such instances substantial differences in the near-infrared reflectance will occur.

A change in the total water content of the leaf will cause a change in reflectance, primarily beyond 1.5 μ. However, only large changes in water content, such as drying up, will produce significant changes in the infrared reflectance beyond 1.5 μ. A completely dry leaf shows diminished reflectance in the near infrared, compared with that of a normal leaf, and an increased reflectance at the long wavelengths because of a reduction of water absorption.

The changes in spectral reflectance that occur at the end of the growing season are seen in Figure 7*B*. The spectral reflectance remains stable until very close to the time of leaf senescence, when pigment change from chlorophyll to anthocyanins occurs very rapidly. This change is seen first in the curve dated October 28, when increased reflectance occurred at all visible wavelengths as chlorophyll breakdown began. Two days later, the loss of chlorophyll was greater, the anthocyanin pigment absorption in the blue was strong, and the yellow reflectance was very much enhanced. The further destruction of pigments continued, and by November 2 the leaf had dried and the blue and green absorption had increased. The drying of the leaf produced reduced reflectance at 0.8 μ, but the infrared reflectance at 1.0 μ remained near its midseason value. The reflectance in the far infrared probably increased somewhat because of a drastic reduction in the absorption by water.

Directional Reflectance

Recently Coulson[1] has shown interesting directional monochromatic reflectance data for a short grass turf, soil, and sand, as a function of the angle of incidence and reflectance (Figure 8). It is noteworthy

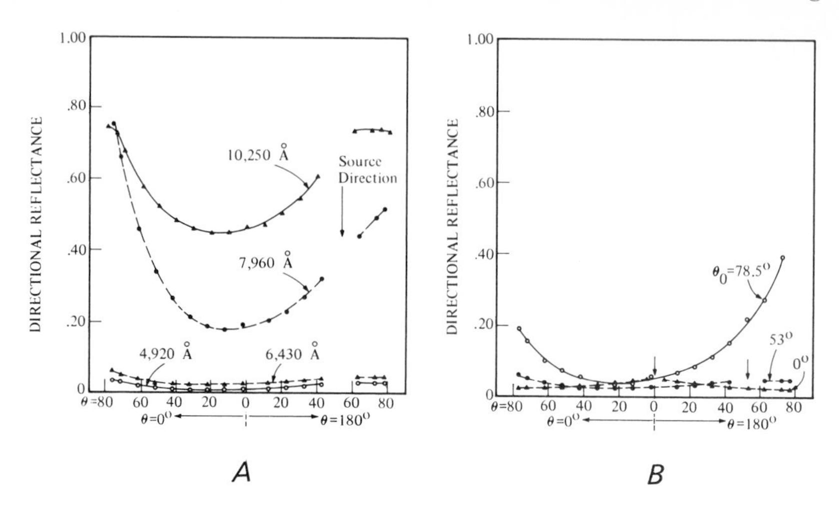

FIGURE 8 Directional reflectance of green grass turf. θ = angle of incidence; ϕ = reflection angle from nadir. *A,* At four different wavelengths. Principal plane, with $\theta_0 = 53°$. *B,* At three different angles of incidence. Principal plane; $\lambda = 6{,}430$ Å. (*Adapted from Coulson.*[1])

that for those wavelengths shorter than 0.7 μ, where strong pigment absorption exists, the change in reflectance with angle of reflectance is very slight except when the angle of incidence $\theta_0 = 78.5°$. For these wavelengths in the visible the spectral reflectance is always less than that for soils and sand, for which there is strong angular dependence. However, for wavelengths of 7,960 Å and 10,250 Å (0.796 and 1.025 μ), the reflectance for short grass turf has a strong angular dependence that is stronger than for black loam soil and not as strong as for desert sand. When substantial reflection occurs, as it does at infrared wavelengths, the orientation of the blades of grass becomes important. Note in Figure 8 how the directional reflectance drops to a minimum when the turf is viewed in the nadir direction ($\theta = 0°$) for all angles of incidence except the vertical.

POLARIZATION PROPERTIES

Nonpolarized light striking a surface may reflect with or without some polarization of the beam. However, as a property to aid in interpretation of surface features, polarization as well as reflection is useful.*

*Additional discussions of polarization are included in Chapter 3 and Chapter 9.

The degree of polarization is defined as follows:

$$P = (I_{max} - I_{min})/(I_{max} + I_{min}), \tag{1}$$

where I_{max} and I_{min} are the maximum and minimum intensities of radiation transmitted by an analyzer that is rotated about an axis parallel to the direction of propagation of the radiation. The polarization of the spectral directional reflectances of short grass turf, sand, and soils are given by Coulson;[1] see Figure 9. The angle for maximum polarization of light reflected by short grass turf is smaller than that from soils and sand. Whether this is truly significant is difficult to say without additional investigation. Materials with a high reflectance, such as metals, do not polarize light well. Dielectrics are generally the best polarizers. For those wavelengths that are strongly absorbed by the plant leaf, the degree of polarization is greatest; and for the near-infrared wavelengths, which are well reflected, the degree of polarization is the least. It is difficult to know whether the higher polarization of the reflected radiation, which is associated with absorption by pigments, is caused by an alignment of the chloroplasts or whether it is just a property of highly absorbing dielectrics. Notice in Figure 9 that negative values of polarization are shown. This comes about by the convention of

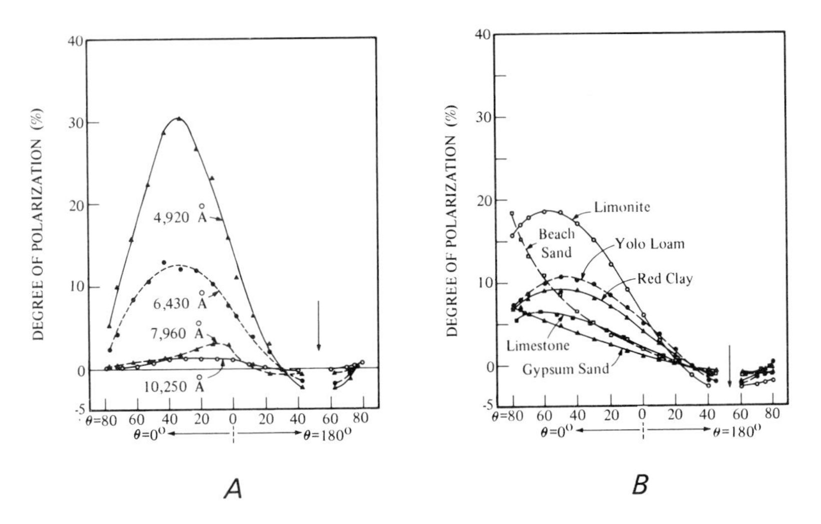

FIGURE 9 Degree of polarization of radiation. θ = angle of incidence; ϕ = reflection angle from nadir. *A*, Green grass turf, four different wavelengths. Principal plane, with $\theta_0 = 53°$. *B*, Various types of mineral surfaces. Principal plane, $\lambda = 4{,}920$ Å; $\theta_0 = 53°$. (*Adapted from Coulson.*[1])

always taking I_{max} as the intensity component normal to the principal plane and I_{min} as the component parallel to the plane.

Soil and Plants

A soil surface may have complete plant cover, no plant cover, or some fraction of plant cover. Let f be the fraction of soil surface that is sunlit. The downward stream of direct sunlight and skylight is $(S + s)$. Let the reflectivity of vegetation be r_l and of the soil, r_g. Then the energy reflected by soil and plants is written to a first approximation as:

$$(g + l)^{Q_{refl}} = [(1 - f)r_l + fr_g](S + s), \tag{2}$$

where the very small amount of long-wave thermal radiation from the atmosphere that is reflected by the plants or soil and the small amount of light transmitted through the leaves and reflected from the underlying soil to return upward through the spaces between the plants are neglected, and where Q = thermal energy. It is interesting to compare the reflectance of an area showing partly bare soil and partly vegetation with an area of bare soil only. Dividing Equation 2 by $r_g(S + s)$:

$$f + (l - f)\frac{r_l}{r_g}. \tag{3}$$

It is clear from Equation 3 that, when f is small, the ratio r_l/r_g becomes very significant, especially when r_l/r_g is large. In the visible, soils usually have a higher reflectance than plants, and a surface of mixed bare soil and vegetation will reflect less than the bare soil. However, plants reflect somewhat better than soils in the near infrared, and $r_l/r_g > 1$.*

Solar Radiation and Plants

The spectral quality of light reflected or absorbed by vegetation will depend not only on the spectral reflectance and transmittance of the plant but also on the spectral character of the incident radiation. Sunlight traversing the earth's atmosphere to reach the surface is strongly

*For additional material on this topic, see Chapter 6.

absorbed by atmospheric gases and scattered by air molecules, aerosols, and dust (see also Chapter 3).

The spectral distribution of direct sunlight at the surface of the earth on a clear day is shown in Figure 10. At the ultraviolet end of the spectrum it is strongly terminated by ozone absorption, and at the infrared end it is more gradually reduced by water vapor and carbon dioxide absorption. The absorption bands are clearly seen in the spectral distribution. The distribution shown in Figure 10 is given in wavenumbers (reciprocal of the wavelength), which are proportional to frequency. (The advantage of this is that the infrared wavelengths can be compressed into a limited size of graph. On a direct wavelength plot, the infrared portion of the scale extends indefinitely without a reasonable limit, and the ultraviolet end becomes seriously compressed.) On a wavenumber scale, the spectral intensity of direct sunlight exhibits a broad peak between 0.7 and 1.0 μ.

The distribution of skylight on a clear day is relatively rich in ultraviolet and blue light, and, of course, for this reason the human eye sees the sky as blue. The distribution of cloudlight on an overcast day is quite different from the distribution of direct sunlight or of

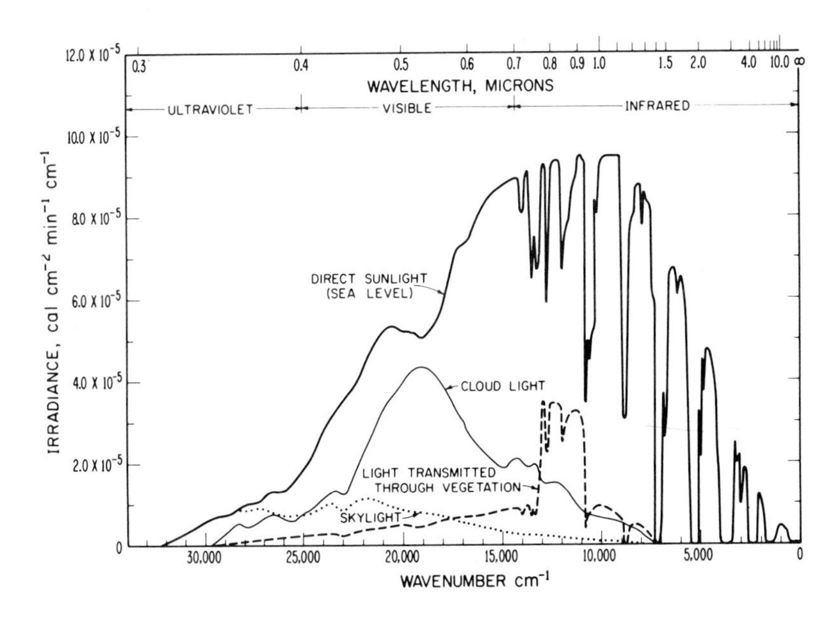

FIGURE 10 Spectral distribution of direct sunlight, cloud light, skylight, and light diffusing through forest vegetation.

skylight, as is seen in Figure 10. The interior of a forest or a crop, illuminated primarily by light diffusing through the canopy, is relatively high in infrared radiation, as shown in Figure 10. It is obvious that the spectral quality of the light absorbed or reflected by a plant depends strongly upon the quality of the incident radiation. An example of the quality of reflected radiation from a poplar leaf (*Populus deltoides*) illuminated by direct sunlight or cloudlight is shown in Figure 11. The spectral signature of the reflected light is very different for the two modes of illumination. A single value of the reflectance or absorptance of a leaf cannot be stated unless one specifies in detail the spectral character of the incident radiation. The total area beneath the reflected sunlight curve of Figure 11 divided by the total area beneath the direct sunlight curve will give the average reflectance under that condition of illumination. For this reason, when a single value of leaf reflectance is given in the literature, unless otherwise specified, it must be assumed to refer to average sunlight conditions at the surface of the earth; but it gives rise to considerable uncertainty because the spectral distribution of the incident radiation can only be conjectured.

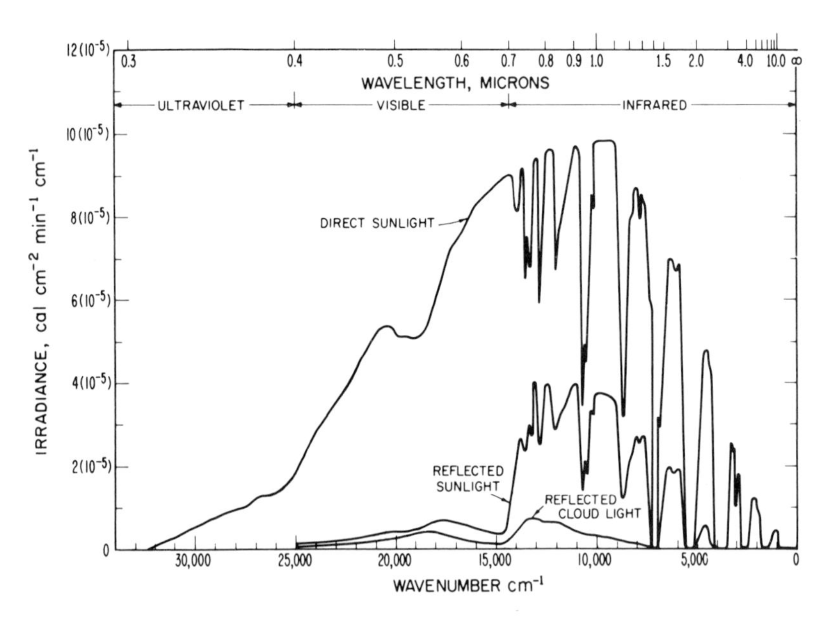

FIGURE 11 Quality of reflected radiation from a leaf of *Populus deltoides* illuminated with direct sunlight, reflected sunlight, and reflected cloud light.

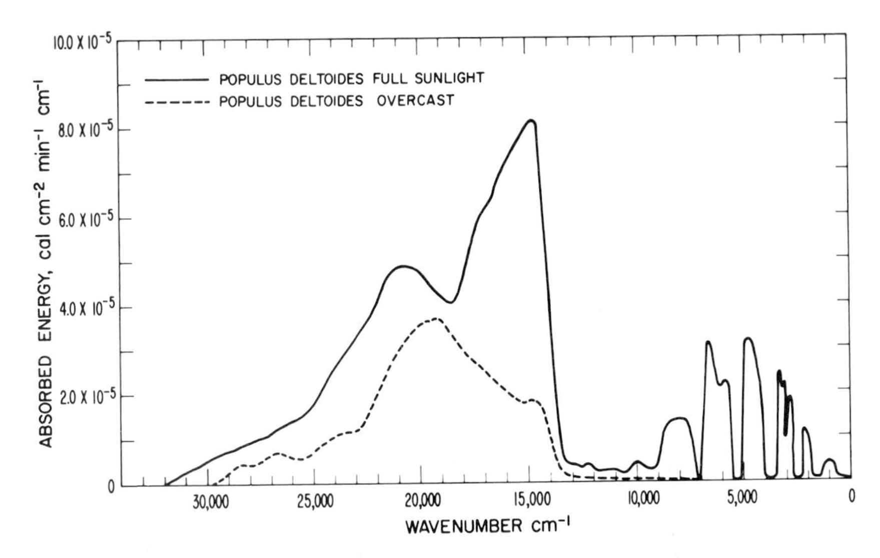

FIGURE 12 Spectral distribution of absorbed sunlight and cloud light for a leaf of *Populus deltoides*.

The spectral distribution of absorbed sunlight and cloudlight is shown in Figure 12 for a leaf of *Populus deltoides*. The same conditions apply to a single absorptance value as to a single reflectance value. The spectral quality of sunlight absorbed by leaves deep within a plant canopy is very different from that of leaves directly exposed to sunlight. To demonstrate the wide variations that exist in the absorptances, reflectances, and transmittances of plants, mean values are given in Table 1 for plants exposed to the incident solar energy of 1.20 cal cm^{-2} min^{-1} and to cloudlight of 0.38 cal cm^{-2} min^{-1} for the spectral distributions shown in Figure 10.

A plant leaf absorbs the visible wavelengths effectively and utilizes the energy content of these frequencies for photosynthesis. A plant leaf reflects and transmits the near infrared well and absorbs relatively little of the energy between 0.7 and 1.5 μ, where the incident sunlight contains the great bulk of its energy. The leaf does not use absorbed energy of these frequencies or quanta for photochemical purposes but converts it into heat within the leaf.

The fact that a leaf absorbs the near infrared very inefficiently is fortunate because as a result leaf temperatures of fully sunlit leaves remain substantially cooler than they would if they were "black" at those frequencies. At wavelengths beyond 2.0 μ, leaves behave almost as blackbodies, the absorptivity and therefore the emissivity being about

TABLE 1 Absorbed Energy and Mean Absorptance Values[a]

Plant	Sunny		Cloudy	
	Energy Absorbed (cal cm^{-2} min^{-1})	Mean Absorptance	Energy Absorbed (cal cm^{-2} min^{-1})	Mean Absorptance
Populus deltoides	0.72	0.60	0.26	0.70
Mimulus cardinalis (Los Trancos)	0.68	0.57	0.25	0.66
Mimulus cardinalis (Priest's Grade)	0.62	0.51	0.22	0.57
Nerium oleander	0.76	0.64	0.28	0.74
Raphiolepis ovata	0.81	0.68	0.30	0.79
Cereus giganteus	0.88	0.74	0.28	0.75
Agave americana	0.88	0.73	0.27	0.72
Agave lecheguilla	0.76	0.63	0.28	0.74
Opuntia gasseliana	0.84	0.70	0.25	0.67
Opuntia laevis	0.74	0.62	0.26	0.69
Opuntia aciculata	0.98	0.82	0.29	0.76
Mamillaria lasiacantha	0.71	0.59	0.25	0.66
Larrea divaricata	1.00	0.83	0.30	0.81
Ilex cornuta	0.69	0.57	0.26	0.69
Agropyron repens	0.72	0.60	0.28	0.73
Bambusa sp.	0.72	0.60	0.27	0.72
Thuja occidentalis	1.06	0.88	0.33	0.88
Pinus strobus	1.07	0.89	0.33	0.88

[a] Plants were exposed to incident solar energy of 1.20 cal cm^{-2} min^{-1} and to cloud light of 0.38 cal cm^{-2} min^{-1}. Spectral distributions are shown in Figure 10.

0.97. This was shown to be true from the reflectivity observations by Gates and Tantraporn.[2] This is again fortunate for the leaf because of the energy exchange. Since the intensity of solar radiation is very low at wavelengths greater than 2.0 μ, the amount of energy absorbed is small. However, the high emissivity of the leaf permits it to radiate efficiently in the far infrared according to the fourth power of its absolute temperature. Hence a leaf is a good absorber at visible frequencies where it utilizes the energy photochemically, is a poor absorber in the near infrared where the bulk of the incident sunlight exists, and is a good emitter in the far infrared where it effectively reradiates. This beautiful interplay of the absorptance and emittance characteristics is not an adaptive or evolutionary consequence but is purely fortuitous. The spectral reflectance of aquatic plants is of the same form as that of terrestrial plants, thus indicating no adaptive differences.

Thermal Radiation from Plants

A plant or a stand of plants radiates infrared energy according to the fourth power of the absolute temperature. The amount of energy flow between a plant and its environment affects the plant temperature. A plant may be warmer than the air temperature if it receives sufficient energy, such as irradiation by sunlight, and it may be cooler than the air if it loses energy rapidly by radiation. Three basic processes contribute to the energy budget of a plant and affect the plant temperature: solar and thermal radiation, free and forced convection, and transpiration. The contribution of metabolic processes to the energy budget and plant temperature is negligible. Except for short-order transient changes, a plant is always in energy equilibrium with its environment: it is losing energy as rapidly as it is gaining energy. The energy relationships of a plant can be expressed in the following form:

$$\frac{\alpha_1(S+s)+\alpha_2 r(S+s)}{2}+\frac{\alpha_3(R_a+R_g)}{2}=\epsilon\sigma T_l^4 \pm C \pm LE, \quad (4)$$

where S = direct solar radiation
s = scattered skylight
r = reflectance of ground or underlying surface
R_a = thermal radiation from atmosphere
R_g = thermal radiation from ground
α_1 = absorptance of direct solar and sky radiation
α_2 = absorptance of reflected solar and sky radiation
α_3 = absorptance of thermal radiation
ϵ = emittance of thermal radiation = α_3
σ = Stefan-Boltzmann radiation constant
T_l = leaf temperature in °K
C = energy lost or gained by convection
L = latent heat of vaporization
E = transpiration rate

The exchange of energy between the leaf and the air by convection is written as follows for free convection:

$$C = 5.8 \times 10^{-3} (\Delta T/D)^{1/4} \Delta T, \quad (5)$$

where $\Delta T = T_l - T_a$ = difference of leaf from air temperature in °C
D = width of leaf or characteristic dimension in cm
C = energy lost or gained by convection in cal cm^{-2} min^{-1}

and for forced convection:

$$C = 6.0 \times 10^{-3} (V/D)^{1/2} \Delta T, \tag{6}$$

where V = windspeed in cm sec^{-1}
D = width of leaf or characteristic dimension in cm

The transpiration rate of water from a leaf according to the law of diffusion may be written

$$E = \frac{{}_sd_l(T_l) - d_a(T_a)}{R} = \frac{{}_sd_l(T_l) - \mathrm{rh}\,{}_sd_a(T_a)}{R}, \tag{7}$$

where ${}_sd_l(T_l)$ = density in g cm^{-3} of water vapor at the mesophyll cell walls at the leaf temperature T_l
$d_a(T_a)$ = density in g cm^{-3} of water vapor of the air at the air temperature T_a
${}_sd_a(T_a)$ = density in g cm^{-3} of water vapor of the air at saturation at the air temperature T_a
R = resistance of the diffusion pathway through the substomatal cavity, the stoma, and the boundary layer in sec cm^{-1}
rh = relative humidity

A complete expression of the relation of climate or environment to plant temperature and therefore to the emission of radiation from a plant is given by Equations 4 and 7 in combination with either 5 or 6. The plant automatically "solves" these equations simultaneously and adjusts its temperature so that it loses energy as rapidly as it gains energy. Given the values of the environmental parameters S, s, r, R_a, R_g, rh, and T_a, it is possible to solve the equations for the leaf temperature T_l. The characteristics of the particular plant enter the equations through the leaf dimension, expressed by D, and the diffusion resistance to water vapor, R, which depends on leaf morphology. The equations can be either programmed on a computer or solved graphically. A graphical solution of the energy equation is illustrated in Figure 13 for a leaf of width 5 cm. A graphical solution of the diffusion equation for transpiration is given in Figure 14.

The diffusion resistance of leaves will vary from values as low as 3.0 sec cm^{-1} to values as high as 400 sec cm^{-1}. For the following calculation, a resistance of 3.0 sec cm^{-1} is used. Leaf width is highly variable also, but a reasonable value is 5 cm. This is the width for which the convec-

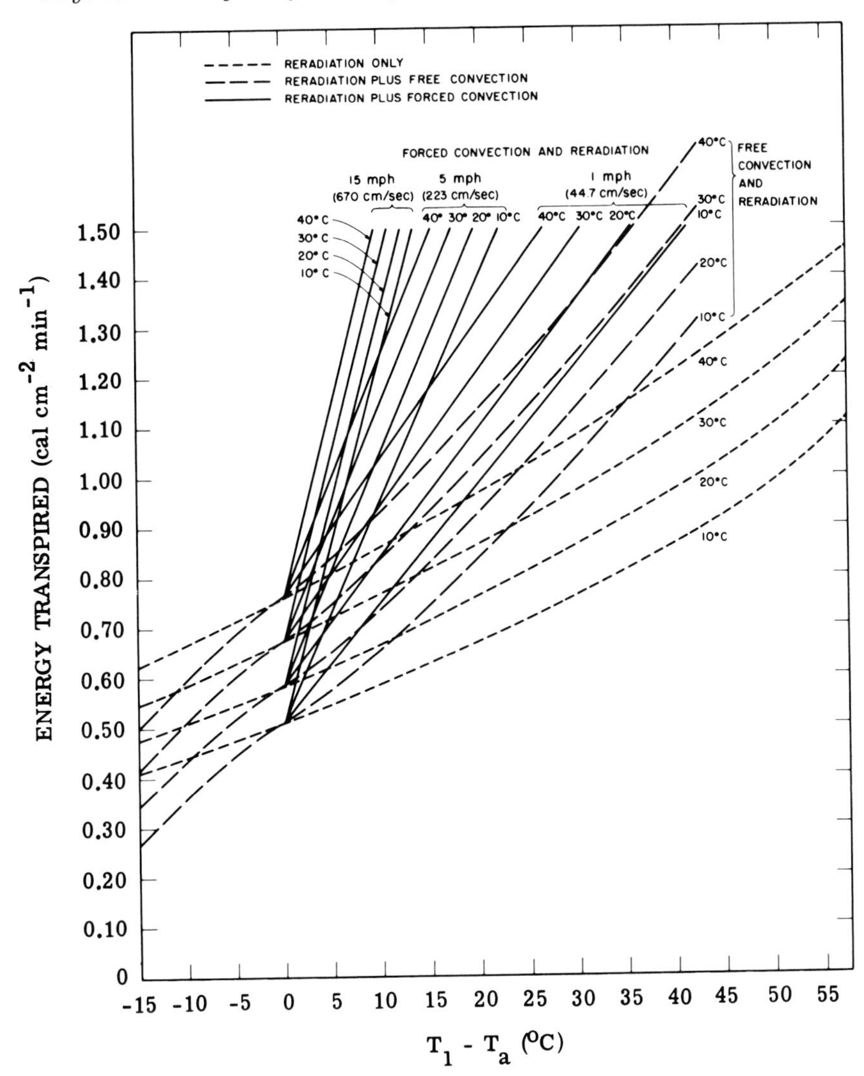

FIGURE 13 Graphical solution of energy equation for a leaf 5 cm wide.

tion component in Figure 13 was derived. Let the following values be considered as an example.

$$S = 1.20 \text{ cal cm}^{-2} \text{ min}^{-1}$$
$$s = 0.20 \text{ cal cm}^{-2} \text{ min}^{-1}$$
$$R_a = 0.50 \text{ cal cm}^{-2} \text{ min}^{-1}$$

$$R_g = 0.70 \text{ cal cm}^{-2} \text{ min}^{-1}$$
$$r = 0.15$$
$$T_a = 30° \text{ C}$$
$$\text{rh} = 50 \text{ percent}$$
$$\text{Air Velocity} = 0$$
$$\alpha_1 = \epsilon = 0.60$$
$$\alpha_3 = \alpha_2 = 0.97$$

Hence,

$$\frac{1}{2} [\alpha_1(S + s) + \alpha_2 r(S + s) + \alpha_3(R_a + R_g)] =$$
$$\frac{1}{2} [0.60(1 + 0.15)1.40 + 0.97(0.50 + 0.70)] = 1.06 \text{ cal cm}^{-2} \text{ min}^{-1}.$$

From Figures 13 and 14:

$$T_l = 36.5° \text{ C}$$
$$LE = 0.28 \text{ cal cm}^{-2} \text{ min}^{-1}$$
$$E = 4.8 \times 10^{-4} \text{ g cm}^{-2} \text{ min}^{-1}$$

If the leaf resistance to transpiration were 9.0 sec cm^{-1} rather than 3.0 sec cm^{-1}, the leaf temperature and transpiration would be:

$$T_l = 43.5° \text{ C}$$
$$LE = 0.17 \text{ cal cm}^{-2} \text{ min}^{-1}$$
$$E = 2.9 \times 10^{-4} \text{ g cm}^{-2} \text{ min}^{-1}$$

If there were no transpiration, the leaf temperature would be 51° C. It is clear that transpiration is extremely important for reducing leaf temperature. Figure 15 shows the departure of leaf temperature from air temperature as a function of the diffusion resistance for values of the absorbed radiation of 1.2, 1.1, 1.0, 0.9, and 0.8 cal cm^{-2} min^{-1} at an air temperature of 30° C and relative humidity of 60 percent. Because there are so many variables involved in the description of the environment that enter into Equation 4, it is clear that it would take a multidimensional plot to present the detailed behavior of leaf temperature as a function of all the variables. Hence, a diagram such as Figure 15 is simply one section through such a multidimensional space. A change in resistance from 3.0 to 9.0 sec cm^{-1} doubles the difference of leaf temperature from air temperature.

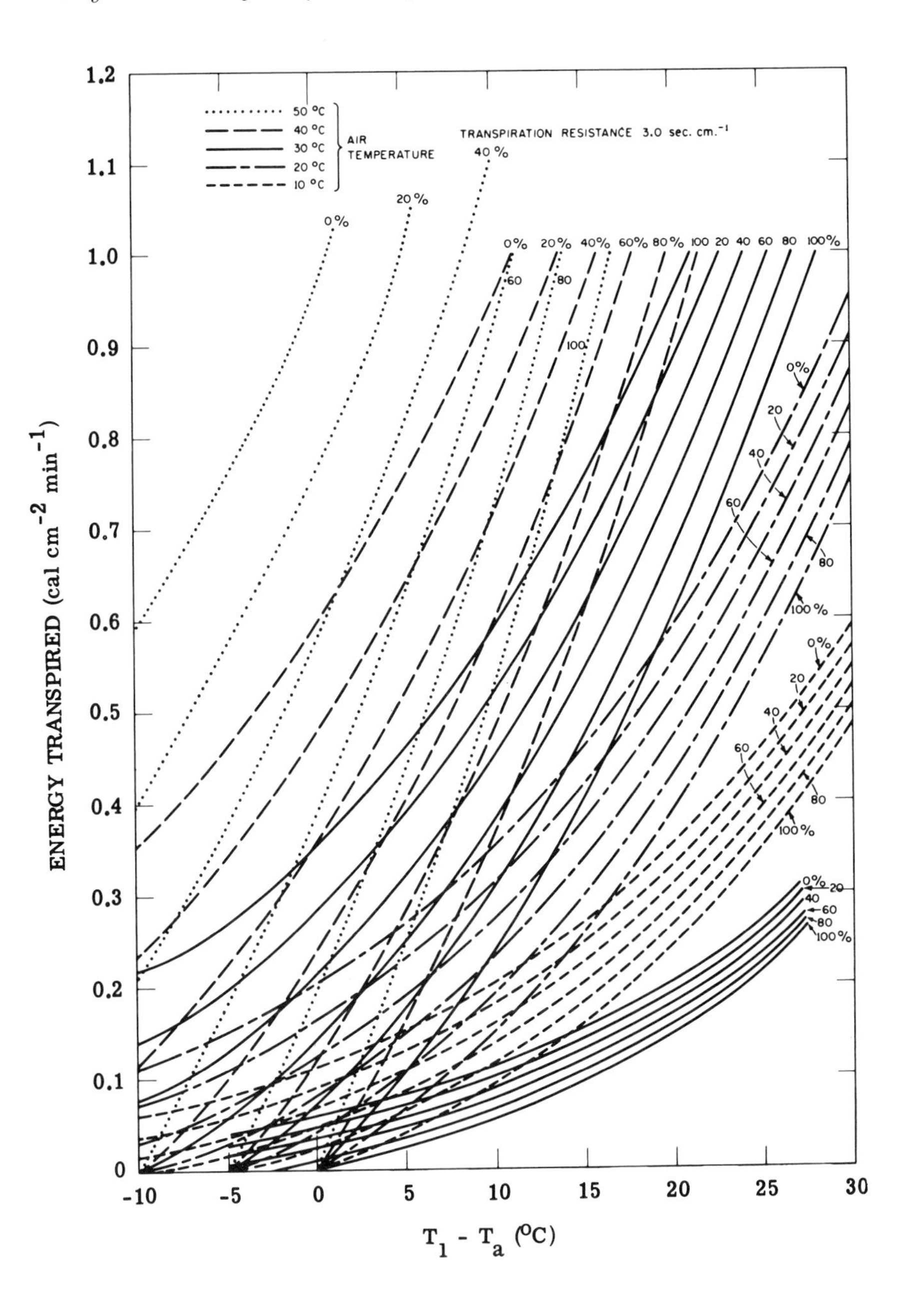

FIGURE 14 Graphical solution of the diffusion equation for transpiration.

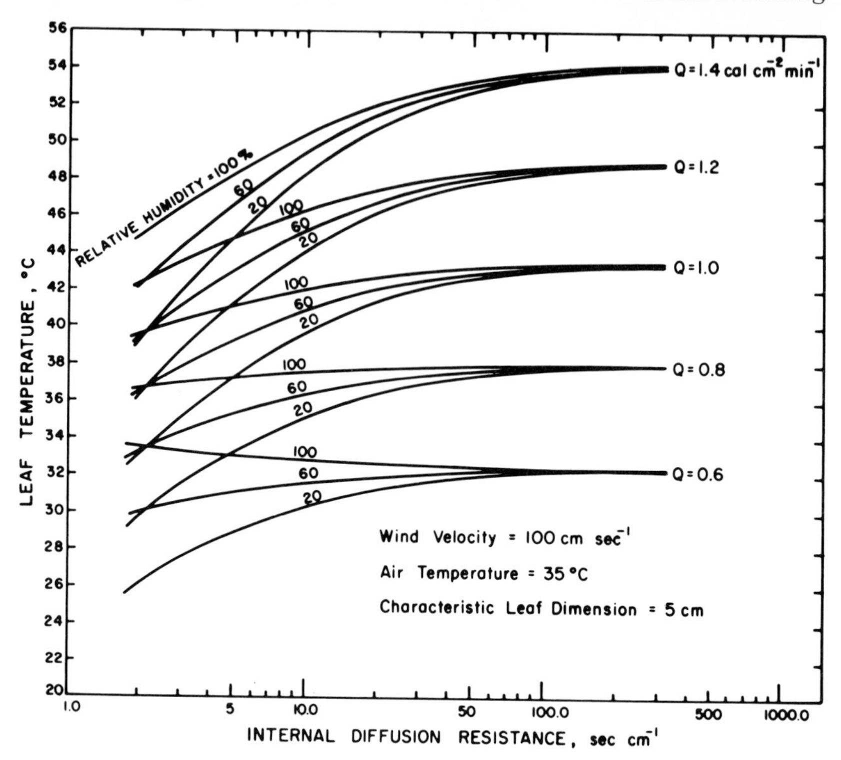

FIGURE 15 Graphical expression of leaf temperature as a function of diffusion resistance and absorbed radiation.

When leaf temperatures become higher than about 40° C, photosynthesis diminishes drastically, respiration increases, carbon dioxide concentration in the leaf guard cells increases, turgor pressure diminishes, stoma close, and transpiration reduces. As the transpiration reduces, the leaf becomes warmer, photosynthesis further diminishes and respiration further increases, and the process repeats. In fact, a type of thermal avalanche or cascade occurs and a thermal catastrophe ensues, resulting in wilting and, possibly, death.

Temperature manifestations of leaves with a crop canopy are as follows. Some leaves will be in full sunlight and will absorb a total amount of radiation between 1.00 and 1.20 cal cm^{-2} min^{-1}. Other leaves will be shaded and will absorb radiation of 0.63 to 0.73 cal cm^{-2} min^{-1}. The stomata of leaves in full sunlight will normally be fully open, except as they become too warm. From Figure 15 it is seen that a leaf with a diffusion resistance of 5 sec cm^{-1}, if absorbing 1.06 cal cm^{-2} min^{-1}, will have a temperature of 43° C when the air temperature is 30° C. However, this leaf temperature is sufficiently high for the thermal avalanche to occur, and the stomata begin to close. If the diffusion resistance of sunlit leaves increases to 20 sec cm^{-1}, the leaf temperature will rise to 47.5° C and with completely closed stomata to 51° C. Leaves in the shade may have stomata partially open and sometimes fully open. Generally the diffusion resistance for shade leaves will be greater than 15 or 20 sec cm^{-1}. From Figure 15 it is seen that in this instance the temperature of shade leaves will be within ±2.0°C of air temperature. At night the radiation absorbed by a leaf is usually less than 0.6 cal cm^{-2} min^{-1}, the stomata are closed, diffusion resistance is very high, and leaf temperature is at or below air temperature. The exposed leaves, which in the daytime are sunlit, at night are the coldest and are as much as 5°C below air temperature.

A radiometer scanning bare soil and vegetation-covered soil will detect a strong difference between the two at noon on a clear day. A dark, dry loam in full sunlight may have surface temperatures as high as 50° C when the air temperature at a height of 2 m is 30° C. A stand of vegetation nearby will have leaf temperatures near or below air temperature. The densely vegetated surface may radiate 0.67 cal cm^{-2} min^{-1} or less, whereas the radiation from the soil may be 0.86 cal cm^{-2} min^{-1}. At midnight, the exposed bare soil will be substantially cooler than the vegetation and the air at the vegetation height. During the early evening, a crop canopy may be cooler than the bare exposed soil.

For a given stand of vegetation, if there is a dense opaque canopy, one may ask whether differences in the amount of water available to various parts of the stand would be detectable as variations in long-wave radiation. Changes in the available water are equivalent to changes in the diffusion resistance. If a dense canopy is absorbing 1.00 cal cm^{-2} min^{-1} and, with ample water available the diffusion resistance is 10 sec cm^{-1}, then from Figure 15 it is seen that, in the absence of wind, leaf temperature will be 41.5° C when the air temperature is 30° C. If the vegetation is not transpiring at all, then the leaf temperature would be 47.5° C. At the lower temperature the canopy would radiate 0.77 cal cm^{-2} min^{-1} and at the higher temperature 0.83 cal cm^{-2} min^{-1}; hence, 0.06 cal cm^{-2} min^{-1} is the maximum difference. Normally the differences will be much less, more on the order of 0.03 cal cm^{-2} min^{-1}, and such small differences are somewhat difficult to detect by radiometers operating from aircraft. A sensitive radiometer used close to the plant canopy can detect temperature differences less than 0.5° C, which are equivalent to 0.005 cal cm^{-2} min^{-1}. Modern optical-mechanical scanners operating in the band from 8 to 14 μ are capable of detecting changes on the order of 0.5° C. One other factor causes serious disturbance of the plant canopy temperature, and that is wind. A very slight breeze, less than 1 mph, is very effective in reducing leaf temperature to near air temperature. Despite the fact that only small differences in radiation from wet and dry canopies may exist, there is evidence that such differences have been detected by infrared scanners for a grass-covered surface. For bare soil with varying moisture content, differences in surface radiation also could be detected readily. For plant canopies that are partly open and for which there is considerable vertical extent, it is doubtful that moisture differences can be detected except in extreme cases, unless the vegetation is viewed obliquely.

Chlorosis and Chemical Deficiencies

The chemical status of plants determines normality or abnormality of growth. A chemical excess or deficiency for a plant may cause chlorosis, premature yellowing and abscission of leaves, burning of leaf tips, bronzing, wilting, mottling, necrosis, water stress, cupping of leaves, flower-color changes, or other abnormalities. Each of these departures from normal is detected visually by the well-trained person. In principle, if these abnormalities occurred in a stand of plants, most of them could be detected by remote means (photometric, radiometric, spectrophotometric, or photographic). Not only do optical properties of the leaf surface (color, shape, size, pubescence) change, but also the leaf temperature relation to the incident energy absorbed and to the transpiration rate changes. An abnormality affecting the absorptivity to incident sunlight will affect the energy budget and hence the leaf temperature. A chemical excess or deficiency may affect the leaf mesophyll or stomatal development and thereby affect the transpiration rate and therefore the leaf temperature. Changes in temperature are detected radiometrically. The author compared the leaf temperature of alfalfa growing in saline soil with alfalfa grown in normal soil and found the saline-soil plants to be 2° to 3° C warmer than the normal plants. He found the temperatures of potassium-deficient sugarcane leaves to be 0.5° to 1.5° C warmer than normal leaves exposed simultaneously to the sunlight. The size, shape, and orientation of the leaves of a plant canopy will determine the reflectance to radiation of the canopy and also affect the temperature of the canopy. The canopy structure will govern the degree to which the underlying soil is viewed from above and hence the photometric property of the soil-plant surface.

The chemical status of a plant will show up in many forms in its pigmentation, cell structure or turgidity, transpiration rate, leaf temperature, etc. New diagnostic techniques are available for detecting chemical deficiency or excess, the full spectrum available to spectrophotometry and radiometry being used. A thoroughly comprehensive compilation of information concerning diagnostic criteria for the chemical status of plants is given by Chapman.[3] It is interesting to note that the book makes no reference whatsoever to spectrophotometry or radiometry as useful diagnostic techniques. This is largely because the bibliography ended with the year 1960. It may also indicate a certain lack of awareness, until very recently, of the importance of spectrophotometry for the detection of the chemical status of a plant. Some examples of plant abnormalities caused by their chemical status follow.

Excess boron causes necrosis of the leaf, beginning with chlorotic yellowing at the edges and progressing toward the midrib.

Leaves from terminal buds of tobacco are light green, and the tissue at the base of young leaves tends to break down.

Calcium deficiency in clover is seen when the leaf tips die, followed by leaf cupping; petioles of younger leaves suddenly collapse, allowing the leaves to droop. Calcium deficiency causes leaf rolling and discoloration in cabbage and cauliflower, and chlorosis in young grape leaves.

The universal symptom of iron deficiency in green plants is chlorosis. Often a chlorotic change may be difficult to identify with the deficiency of a specific element, and in this respect careful spectrophotometry of the leaf may be very helpful. As chlorosis develops, green color is first absent from the finest veins, then from the larger veins, until the leaf is devoid of chlorophyll. On a time-sequential basis the morphology of chlorosis will permit one to distinguish between an iron deficiency and a zinc or manganese deficiency.

Magnesium deficiency shows up visually as a loss of green color between the veins, first of older leaves and then of younger leaves, followed by intense chlorosis and the development of brilliant leaf color. In some plants, magnesium deficiency causes leaf curl.

Phosphorus deficiency often results in an off-color green with purple venation. Sometimes root development is poor and the plant growth is stunted.

Potassium deficiency for many plants shows up first on mature leaves and only later on juvenile leaves. Usually, chlorosis will proceed from light yellow to tan, becoming brown, and finally a "scorch" develops. Legumes show a white speckling of the leaves.

Seleniferous soils are often detected through chlorosis and usually occur only in arid or semiarid regions.

The symptoms of sulfur deficiency are very similar to those of nitrogen deficiency in the form of various degrees of chlorosis; however, with sulfur the new leaves are affected most, whereas with nitrogen the old leaves tend to turn yellow. (Although nitrogen is a constituent of the chlorophyll molecule and sulfur is not, both are essential to the formation of chlorophyll.)

SOILS

Two properties of a soil surface are of general interest in remote sensing. The first has to do with its reflectance of incident radiation, and the second is concerned with its long-wave emittance and surface temperature. The reflectance of a soil surface depends upon its coloration, texture, roughness, moisture content, mineral and chemical composition, angle of illumination, and degree of shadowing by plants, buildings, etc. Examples of the spectral reflectance properties of soils are given in Chapter 7. The long-wave emittance at wavelengths greater than 4 μ is wavelength-dependent and is equal to the long-wave absorptance, which is 100 percent minus the long-wave reflectance. Examples of long-wave reflectance are given in other chapters in this volume. The surface temperature of the soil is important for the radiation emitted from the surface, which is $\epsilon\sigma T_s^4$, where ϵ is the emittance, σ is the Stefan-Boltzmann constant, and T_s is the surface temperature in °K. The surface temperature of the soil is dependent upon the energy budget of the soil surface. The energy budget depends upon the following factors:

$$Q_{abs} = \epsilon\sigma T_s^4 \pm Q_{conv} \pm Q_{cond} \pm LE,$$

where Q_{conv} is the energy exchanged by convection between the surface and the air. If the air is cooler than the surface, the surface loses energy to the air (positive sign), and if the air is warmer, heat is delivered to the surface (negative sign). The rate of convective transfer depends upon the roughness of the surface (bare soil, short grass, tall grass, crops, trees, etc.) and upon the wind-speed profile. The conduction term represents heat conducted from or to the surface within the soil (positive or negative sign, respectively). The conductivity of the soil depends upon its compactness, particle size, moisture content, rock content, etc. The last term on the right-hand side of the equation represents energy lost by evaporation of moisture (positive sign) or energy given up to the surface by the condensation of dew (negative sign).

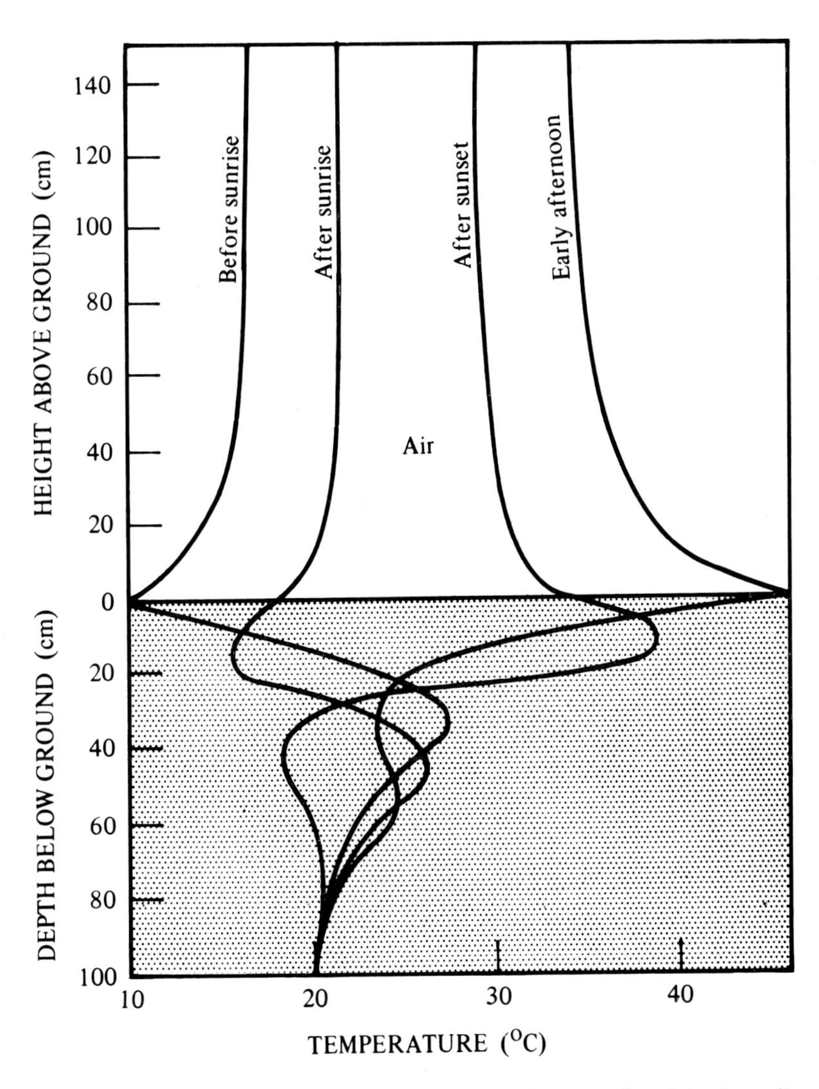

FIGURE 16 Temperature profiles in the air near the ground and in the soil for a typical clear midsummer day and night at temperate latitudes.

The surface soil temperature is influenced by the temperature profile of the air above the surface and of the soil below the surface. The temperature of the soil at any depth varies with date; during the winter months temperatures increase with depth, whereas during the summer months temperatures decrease with depth. The transition period is in March to April and September to October at temperate latitudes. The annual and diurnal amplitude of temperature variation decreases with depth. The annual variation becomes zero at about 12 m, and the diurnal amplitudes become zero at about 75 cm. A general rule of thumb is that the depth of zero annual range is about 19 times the depth of zero diurnal range. Figure 16 shows the temperature profiles in the air near the ground and in the soil for a typical clear midsummer day and night at temperate latitudes. It is evident that the greatest excursion of temperature occurs at the surface and that those organisms living on the surface all the time, such as plants, are subjected to the most harsh environment. A radiometer scanning the surface features will detect these great differences in temperature. A grass or vegetation cover will greatly reduce the daytime rise of temperature, but may, in fact, produce cooler nighttime temperatures. Vegetation cover will considerably reduce the amplitudes propagated into the soil.

References

1. Coulson, L. 1966. Effects of reflection properties of natural surfaces in aerial reconnaissance. Appl. Opt. 5:905–917.
2. Gates, David M., and W. Tantraporn. 1952. The reflectivity of deciduous trees and herbaceous plants in the infrared to 25 microns. Science 115:613–616.
3. Chapman, H. D. [ed.]. 1966. Diagnostic criteria for plants and soils. University of California, Division of Agricultural Sciences. 793 p.

6

Soil, Water, and Plant Relations

VICTOR I. MYERS South Dakota State University, Brookings
Contributing Authors: M. D. HEILMAN, R. J. P. LYON, L. N. NAMKEN, D. SIMONETT, J. R. THOMAS, C. L. WIEGAND, J. T. WOOLLEY

REFLECTANCE STUDIES ON PLANTS AND SOILS AND THEIR APPLICATIONS

Rapid progress in the development and use of new scientific instrumentation for remote sensing has provided new research tools for the agricultural soil and water field, as well as for many other earth resource disciplines. Interpretation of the signals received by these sensors may make it possible to identify many soil and water factors influencing agricultural production. As yet, however, little is known about the specific effects of plant and soil environmental factors on the reflectance characteristics of land cover. Laboratory spectrophotometer and field spectrometer studies (see Chapter 7), along with fundamental plant physiology studies, are the means by which these characteristics will be defined.

Numerous spectral reflectance and transmittance studies of plant leaves have been made[1-7] (see also Chapter 5). Until recently, most spectrophotometer studies have covered the wavelength region 0.4–0.9 μ. This was due largely to the availability of instrumentation with sensitivity in this particular range. Figure 1 shows spectrophotometer reflectance curves for single leaves from some agricultural crops, extending over the wavelength region 0.5–2.5 μ. These curves illustrate that several portions of the electromagnetic spectrum can be exploited in detecting reflected energy from plants. It should be pointed out

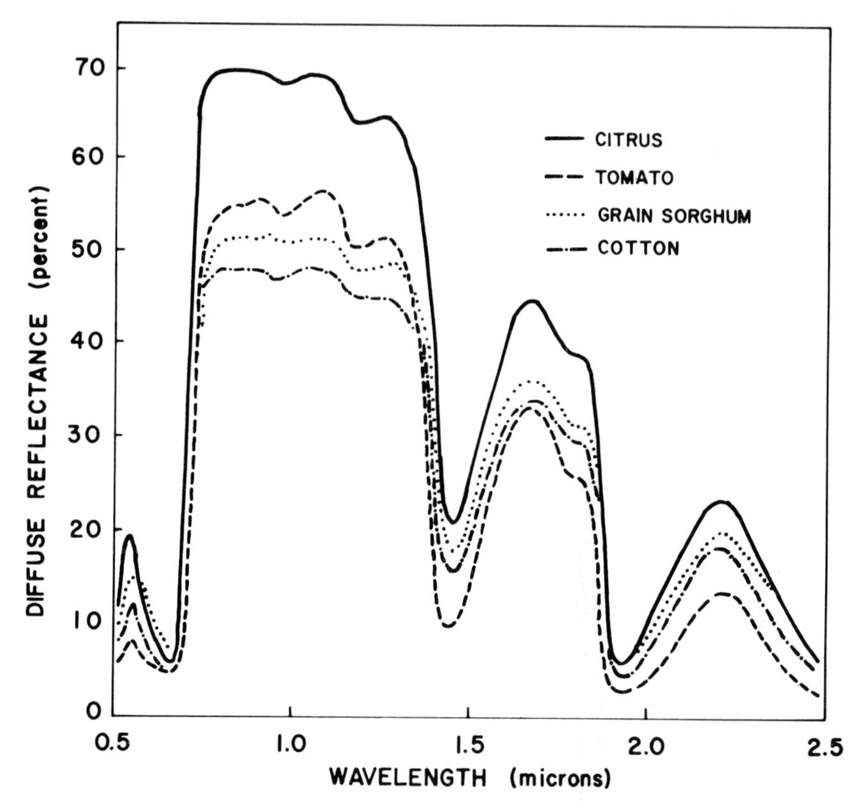

FIGURE 1 Light reflectance from leaves representing four agricultural crops. Instrument used was a DK-2A spectrophotometer.

that not all spectra from leaves of different crop species show the fortuitous anomalies that appear in Figure 1.

One of the known causes of variation in reflectance from a given area of cropland is leaf density or leaf area index (LAI) of the plants. Leaf area index is the ratio of leaf area to soil area. A leafy, densely spaced crop may have an LAI of 6. Prior to crop emergence, when soil reflectance predominates, LAI is zero. The plant reflectance increases with plant development to a maximum when the full plant canopy has developed.

The importance of the transmittance of infrared light through upper leaves and multiple reflection from lower leaves has been demonstrated. Figure 2 shows the results of diffuse reflectance measurements from combinations of cotton leaves stacked on one another, up to six deep, corresponding to an LAI of 6. The reflectance measurements were made on a DK-2A Beckman spectrophotometer.

There were no differences in reflectance in the visible wavelengths from any combination of stacked leaves, indicating that reflectance of visible light from leaf surfaces is from the topmost exposed leaves (or conversely, that the top two layers absorb all the remaining inci-

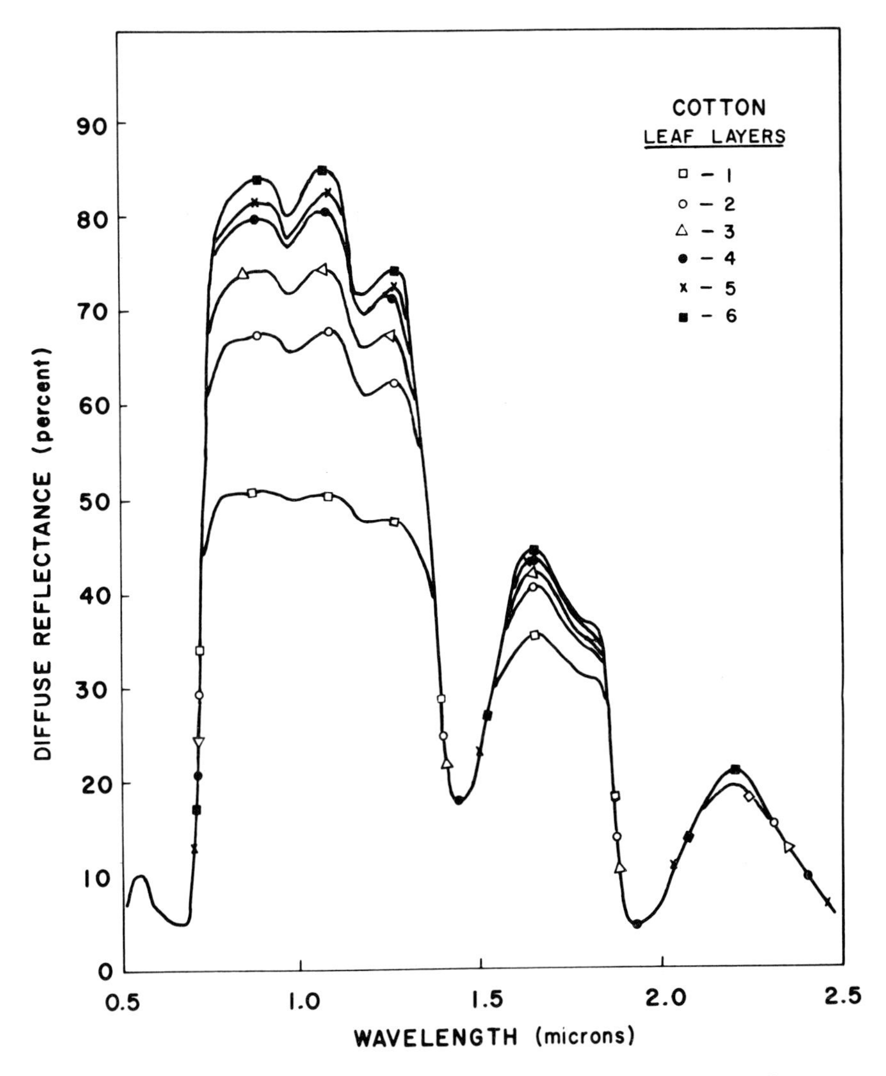

FIGURE 2 Diffuse reflectance from combinations of cotton leaves stacked on one another, up to six deep.

dent visible light.) In the near infrared, however, reflectance of two leaf layers increased about 17 percent over that for one layer. Each succeeding additional layer added diminishingly smaller increments of reflectance. At a wavelength of 0.9 μ, the reflectance for one layer was 50 percent, and the reflectance for six layers was 84 percent. Leaf stacking did not affect reflectance in the 1.45-μ and 1.93-μ water-absorption bands. However, plateau reflectance at about 1.60 μ was increased nearly 10 percent by stacking six layers. Kodak Ektachrome Infrared Aero film (Plate 17) is sensitive in the wavelength interval 0.4–0.9 μ and, therefore, capable of recording variations in reflectance associated with plant vigor and LAI.

In further verification of the phenomenon described, Figure 3 shows reflectance and transmittance curves for a cotton leaf. Transmittance of light through the leaf was substantial at the infrared wavelengths, up to the 2.5-μ limit of measurement (except in the 1.93-μ absorption

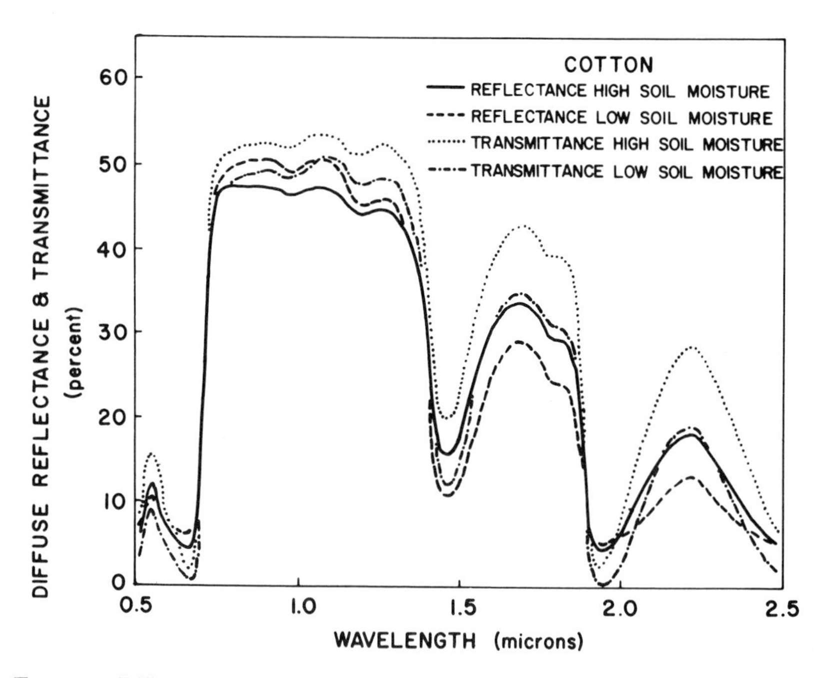

FIGURE 3 Diffuse reflectance and transmittance for leaves from cotton plants at two moisture levels. Differences at each wavelength between 100 percent and the sum of percentages of reflectance and transmittance equals absorption by the leaf.

band) but amounted to only about 2 percent in the visible band. From 0.75 to 1.3 μ, the sum of reflectance and transmittance accounts for 98 to 100 percent of the incident energy; hence, plant leaves are virtually transparent in this wavelength interval.

The data of Figure 3 show that in the 0.75–1.3-μ interval the transmittance is greater for the leaf grown under high soil moisture conditions. The opposite is true for reflectance. There are some crossovers of both transmittance and reflectance at other wavelengths, particularly at water-absorption bands. The ages of leaves sampled may have some bearing on the differences described.

A model study was conducted to determine if the leaf-stacking experiment, the transmittance study, and photographs of fields involved related phenomena. A framework was constructed with eight removable trays. Each tray was made of wire netting so that a layer of leaves could be spread on it. The top tray of leaves was uniformly lighted with tungsten floodlamps, simulating sunlight on a plant canopy. Pictures were then taken on panchromatic film, a black-and-white infrared film with a dark red (Wratten No. 89A) cutoff filter (see Chapter 2 for filter characteristics), of various combinations of numbers of trays stacked in the framework. Figure 4 shows reflectance from eight combinations of stacked trays of Turk's cap (*Malvaviscus grandiflorus*) leaves. Reflectance measurements were made from above the canopy also with an ISCO spectroradiometer sensitive in the wavelength region 0.4–1.5 μ.

Reflectance in the visible wavelengths, as recorded on the panchromatic film and measured by the spectroradiometer, did not change when more trays were added. Figure 2 shows that visible-wavelength reflectance measured on the spectrophotometer did not change with multiple stacking of leaves. However, infrared reflectance increased considerably with an increase in numbers of stacked trays. This result would be predictable from Figure 2.

The leaf-stacking experiments, with the spectrophotometer and model studies, show that near-infrared light transmitted through the top leaves of plants is partly reflected from and partly transmitted through lower leaves. The lower leaves of the plant canopy then act as the trays of leaves in the model in reflecting and transmitting light, with the upward component of the reflected light from the lower leaves reinforcing the reflected light from the surface leaves. The net result, then, is a much brighter infrared reflectance from a field than would be predicted from reflectance curves from individual leaves.

Reflectance from a crop canopy, also discussed in Chapter 5, depends not only on stacking and related geometric phenomena but also on

FIGURE 4 Near-infrared reflectance from eight combinations of stacked trays of leaves. Black-and-white infrared film was used with a dark red (No. 89A) filter.

various aspects of plant physiology. Successful use of interpreted spectra, aerial photography, and scanned imagery from particular crops and/or problems in areas where little or no ground truth exists depends on an understanding of the plant characteristics influencing spectral reflectance from the crop canopy. Leaf structure, pigments, and water status are illustrated here by Figures 5 through 11 to show the influence of these characteristics on reflectance and transmittance of light with respect to cotton leaves.

Most leaves reflect more visible light from the lower (dorsal) surface than from the upper (ventral) surface. This is confirmed by Figure 5 and can be explained by the structure of the leaf. The palisade tissue presents a high concentration of chlorophyll, which absorbs energy in the visible part of the spectrum. The spongy mesophyll tissue usually contains much less chlorophyll and presents more air–water or air–cellulose interfaces because these cells are in less intimate contact with each other than are the palisade cells. The interfaces scatter radiant energy.

Between 0.75 and 1.35 μ, the lower surface of the leaves reflects less than the upper. The leaf is very nearly transparent to radiation in this band (see Figure 3, which shows that 98 to 100 percent of the

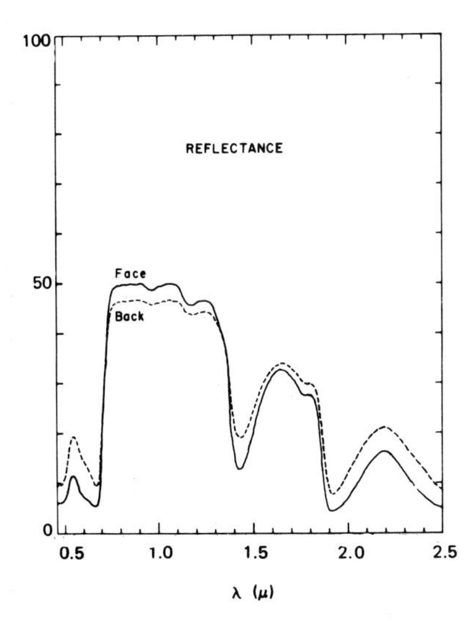

FIGURE 5 Total reflectance of face (ventral side) and back (dorsal side) of a single cotton leaf.

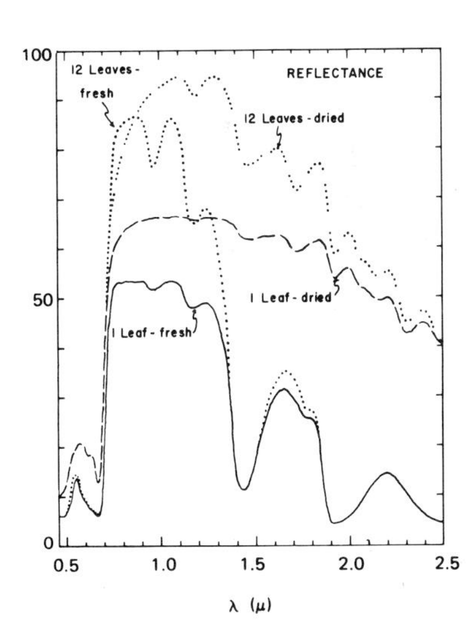

FIGURE 6 Total face (ventral side) reflectance of fresh cotton leaves compared with that of dried cotton leaves.

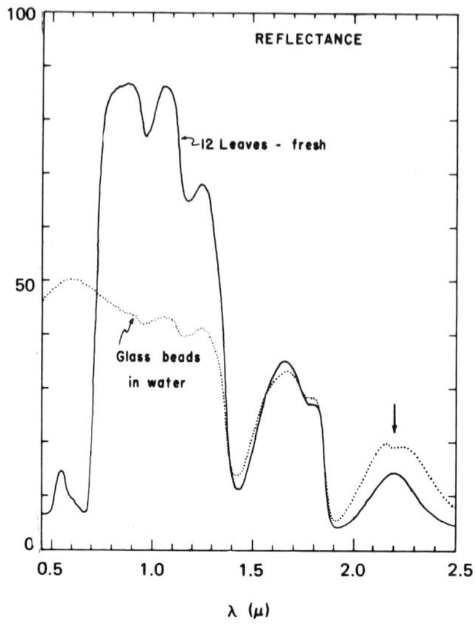

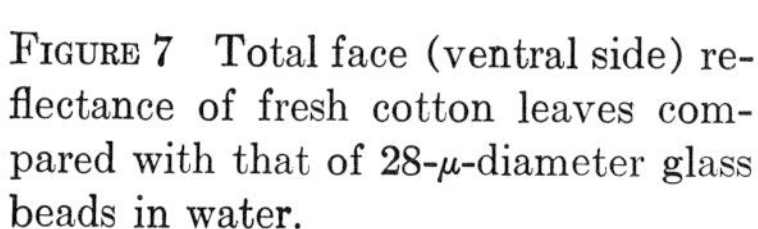
FIGURE 7 Total face (ventral side) reflectance of fresh cotton leaves compared with that of 28-μ-diameter glass beads in water.

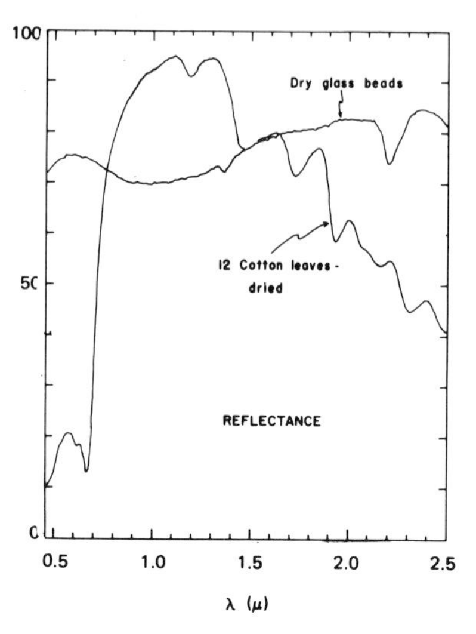

FIGURE 8 Total face (ventral side) reflectance of dried cotton leaves compared with that of 28-μ-diameter glass beads in air.

radiation here is either reflected or transmitted). Absorption throughout the reflected spectrum differs, however, between the upper and lower surfaces of most leaves because reflectance and transmittance also vary. Figure 2 confirms the indication that there is little absorption in the very-near-infrared band, since even stacked leaves having essentially infinite optical thickness have about 86 percent reflectance.

Figures 6–8 suggest that the shape of the reflectance curve at wavelengths greater than 1.35 μ is governed largely by the water in the leaf. The spectrum for very small glass beads in water is shown to be about the same as the fresh leaf spectrum (Figure 7), but the curves for dried leaves and for glass beads in air (Figure 8) are completely dissimilar.

During the drying of leaves, reflectance increases at all wavelengths, probably because of two factors: (a) the shrinking and consequent increased density of the leaf, and (b) the changing of the refractive index discontinuities within the leaf as the water is lost. A further consequence of drying is the appearance of some material that absorbs energy at about 0.81 μ. This can be seen in Figure 6 as a rounding of the reflectance curve for a single leaf and as a decrease in reflectance for a stack of leaves.

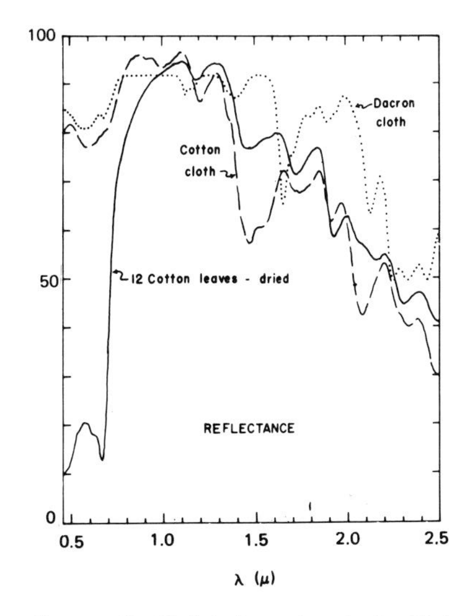

FIGURE 9 Total face (ventral side) reflectance of dried cotton leaves compared with that of white cotton cloth and white Dacron (polyester) cloth.

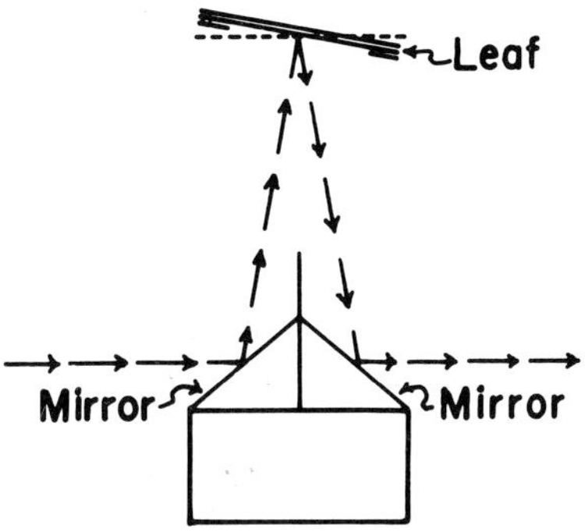

FIGURE 10 Diagram of the apparatus used to measure the specular reflectance of a leaf.

The spectral characteristics of dried leaves at wavelengths between 1.0 and 2.5 μ are similar to those of cotton cloth, but are different from those of Dacron cloth (Figure 9). This suggests that this portion of the dried leaf spectrum is probably due to cellulose.

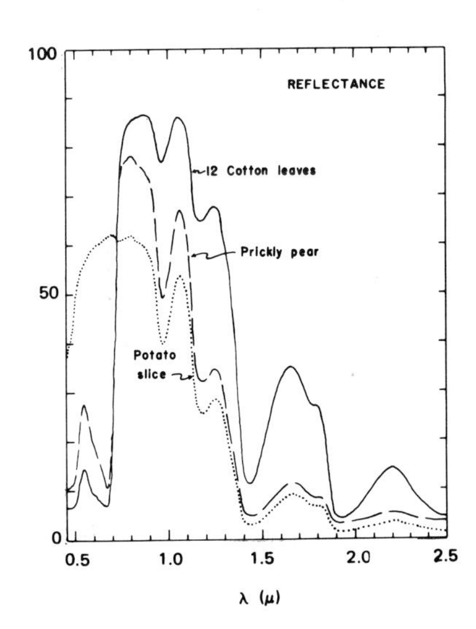

FIGURE 11 Reflectance of a stack of fresh cotton leaves compared with that of two examples of thick plant tissues—an intact prickly pear (*Opuntia* sp.) "leaf" and a 5-mm-thick slice of potato tuber tissue.

Plants needing water seem often to have a "dull" appearance. To determine whether this dull appearance came from a decrease in the specular portion of the reflectance, a piece of leaf was mounted in the device shown schematically in Figure 10. The angle of the leaf can be adjusted in this device so that the specular component can be either included or excluded when reflectance is measured, but the device cannot be used to obtain a valid measure of total reflectance because it lacks an integrating sphere. Young cotton leaves do exhibit a decrease in specular reflectance as they lose water, and this decrease is associated with wrinkling of the epidermal cells as they lose turgidity. This phenomenon does not seem to be evident with other crop leaves, however.

Figure 11 illustrates an interesting aspect of spectral interpretation at wavelengths greater than 0.9 μ. The absorption bands in this region are those of water, but the reflectance of light is caused by scattering at places where the refractive index changes. The prickly pear and the potato tuber have much larger cells than the cotton leaf and therefore have fewer changes of refractive index for a given quantity of water. This causes the shapes of the reflectance curves to be different. Such phenomena may be very useful in explaining some spectral curves or in determining structural characteristics of some types of tissues.

Reflectance and transmittance of light from plants in the wavelength region 0.5–2.5 μ is influenced by at least three phenomena. At visible wavelengths to about 0.7 μ, chlorophyll and carotene absorption most strongly influence the magnitude of reflectance. At near-infrared wavelengths to 1.3 μ, in the absence of chlorophyll and carotene absorption, physiological structure of plants results in high values of reflectance and transmittance. Beyond 1.3 μ, absorption of electromagnetic radiation by water contained in plant tissue influences the magnitude of reflectance and transmittance. Also, the presence of cellulose probably influences reflectance in these wavelengths beyond 1.3 μ.

Radiance spectra acquired in the field with a spectrometer have a general similarity to reflectance spectra measured from individual leaves on a spectrophotometer in the laboratory. There are important differences, however, that result from the following factors: (a) absorption and scattering by gas molecules and atmospheric aerosols, such as dust, reduce incoming solar radiation in certain wavelength bands[8]; (b) illumination from the sun varies in intensity with numerous conditions; (c) radiance from crops in the field is affected by crop geometry, background soil reflectance, and other factors; and (d) radiance from soils in the field is difficult to duplicate in the laboratory with disturbed soils.

A comparison between the spectra from a cotton field measured with an ISCO spectroradiometer, and those from a cotton leaf measured in the laboratory with a DK-2A spectrophotometer are shown in Figure 12. The field reflectance was obtained from a height of 300 cm over the crop canopy.

Identification of crop varieties from aircraft or from spacecraft may have important applications for economic planning and crop census-taking. Plate 17 shows that Kodak Ektachrome Infrared Aero (color-infrared) film produces a wide variety of diagnostic color signatures that may be useful for crop identification. Much research is still needed to establish the characteristic color signatures that will permit identification of crop varieties and the color tone of each, which might be affected by various soil and crop conditions.

In summary, reflectance and transmittance measurements from a spectrophotometer are valuable for controlled laboratory studies of plants and soils and indicate the portion of the spectrum where reflectance anomalies can be expected to occur in field studies. The absolute

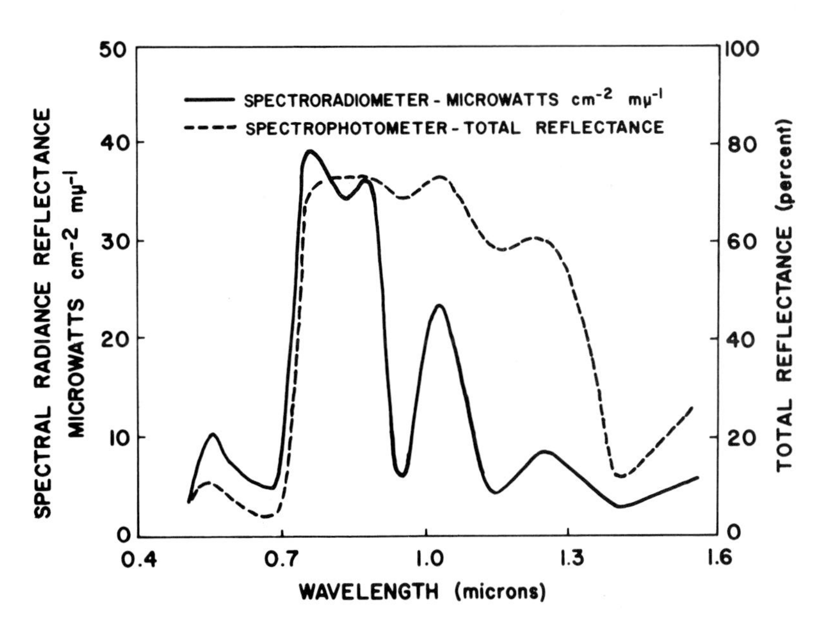

FIGURE 12 Comparison of reflected energy from cotton, measured in the field with a spectroradiometer, and total reflectance from a cotton leaf, measured with a Beckman DK-2A laboratory spectrophotometer.

energy value of the spectrophotometer curves is difficult or impossible to determine, however, and the spectral distribution of the sources differs. Field spectra, which have a solar energy source, involve such factors as reflectance from multiple leaf layers, soil background radiance, and atmospheric absorption and scattering of certain wavelengths. The field and laboratory spectra for cotton shown in Figure 12 can be compared, wavelength by wavelength, to illustrate their similarities and differences.

Reflectance Measurements of Nutrient-Deficient Leaves

Measurement of the diffuse or specular reflectance of developing plant foliage promises to be a reliable, nondestructive technique by which changes in the structure, pigmentation, and moisture content that are induced by changes in the plants' environment can be followed. Nutrient deficiencies in plants, for example, may affect the color, moisture content, and internal structure of leaves, and as a result their reflecting power will also change.

Sweet pepper leaves exhibiting different degrees of nitrogen deficiency are shown in Plate 18. Leaves with a nitrogen deficiency contain relatively little chlorophyll and tend to be pale green; they are generally smaller and contain more water per unit of dry matter than normal leaves.

The reflectance spectra of all nitrogen-deficient leaves are similar to those of normal leaves (Figure 13). However, the severity of the nitrogen deficiency affects the percentage reflecting power of the leaf. Reflectance in the visible region of the spectrum (0.38–0.7 μ) increases as the nitrogen-deficiency symptoms become more pronounced. This is associated with the lower chlorophyll content of the nitrogen-deficient leaves. As has been shown by Thomas,* concentration of chlorophyll in the leaf may be estimated from a measurement of leaf reflectance at a wavelength of 0.54 μ. Calibration curves are required for each plant species.

Leaf morphology is related to the reflectance phenomena of nutrient-deficient leaves. The reflecting power of leaves in the infrared region of the spectrum (0.7–1.0 μ) appears to be associated with the leaf age and size. Reflectance increases exponentially as leaf thickness increases. The increase in reflectance as the leaves become more nitrogen deficient

*Personal communication, 1967.

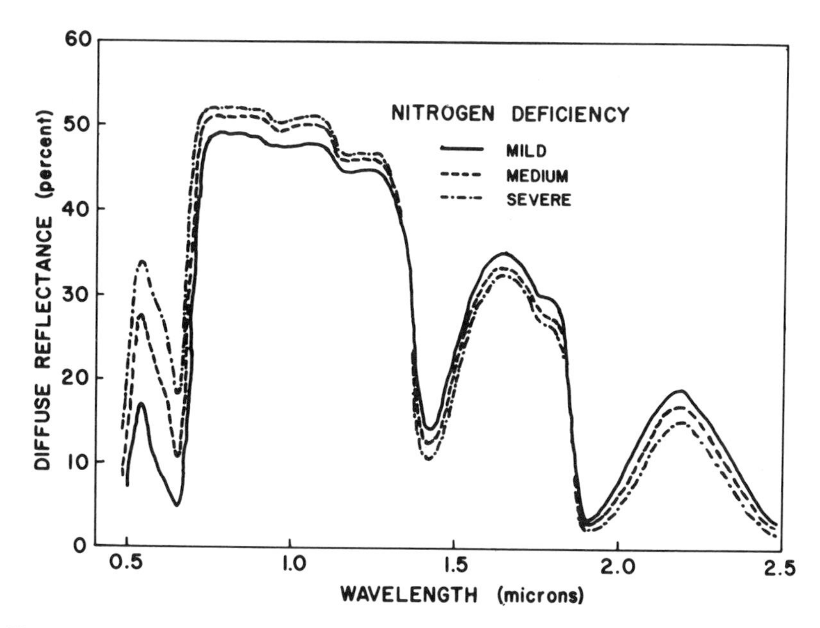

FIGURE 13 Effect of nitrogen deficiency on the diffuse reflectance of sweet pepper leaves.

suggests that the deficient leaves are thicker. This may be due to smaller and fewer cells within the leaf.

The influence of moisture status of leaves on reflectance must be considered in interpreting any reflectance phenomena. The shape of the reflectance curves in the wavelength interval 1.3–2.5 μ is due mainly to water. A fundamental water-absorption region occurs from 2.6 to 2.8 μ. Overtones from this absorption region occur in plants at 1.93 μ and 1.45 μ. Reflectance measurements at these two wavelengths and at 1.65 μ and 2.20 μ are significantly related to the total amount of water in the leaf and to the leaf-moisture deficit. The reflectance curves of Figure 13 suggest that nitrogen-deficient leaves contain more water than normal leaves. The moisture content of the mildly deficient leaf was approximately 3 percent lower than that of the severely deficient leaf.

Small absolute differences in reflectance can be accentuated by use of the differential reflectance technique in which a normal leaf is used instead of magnesium oxide as the reference standard. Differential reflectance can be measured with a Beckman Model DK-2A spectrophotometer over a wavelength interval of 0.38–2.35 μ, although the interval 0.75–2.35 μ is of principal interest here.

Differential reflectance curves are shown in Figure 14 for sweet pepper leaves deficient in nitrogen. If the reference and the test leaf were identical in all respects, the resulting reflectance curve would be a straight line passing through the zero reference point. It is evident from a comparison of these curves that nitrogen deficiency strongly affected both the amount of water within the leaf and the morphology of the leaf.

Photographic Detection of Salinity

Many arid areas of the world are affected by high water tables and resultant soil salinity. Detection of the saline areas and of the degree of salinity in the rooting profile is of considerable interest to agricultural workers involved in reclamation of these soils. Early detection of saline areas may permit preventive measures before significant crop damage is apparent. Furthermore, rapid detection of saline areas by advanced remote-sensing methods and procedures can greatly accelerate initiation of reclamation processes.

Plants are frequently good indicators of conditions that occur below the soil surface. The conditions are manifested in plant appearance and spectral reflectance from leaf surfaces. The root systems of plants explore a rather large soil volume. A plant sample, therefore, is more representative of the site conditions than a single soil sample.

Spectrophotometer studies have shown that plant leaves affected by salinity exhibit increased reflectance, which is particularly evident in the near-infrared wavelengths. Individually, salinity-affected cotton leaves have decreased transmittance; however, the net effect is that the plant canopy in a saline area shows decreased reflectance.

Cotton was used as an indicator plant in a study relating the salinity in the 0- to 1.5-m (0 to 5-ft) soil profile at some reference sites where the salinity is known to that of a number of prediction sites where the salinity is unknown.[9]

In an early study, black-and-white infrared film was used with an 89A dark red filter to absorb the visible light. This film–filter combination emphasizes the high infrared reflectance from healthy vegetation compared with the lower reflectance from plants affected by salinity. It was possible to delineate, by appearance, six different con-

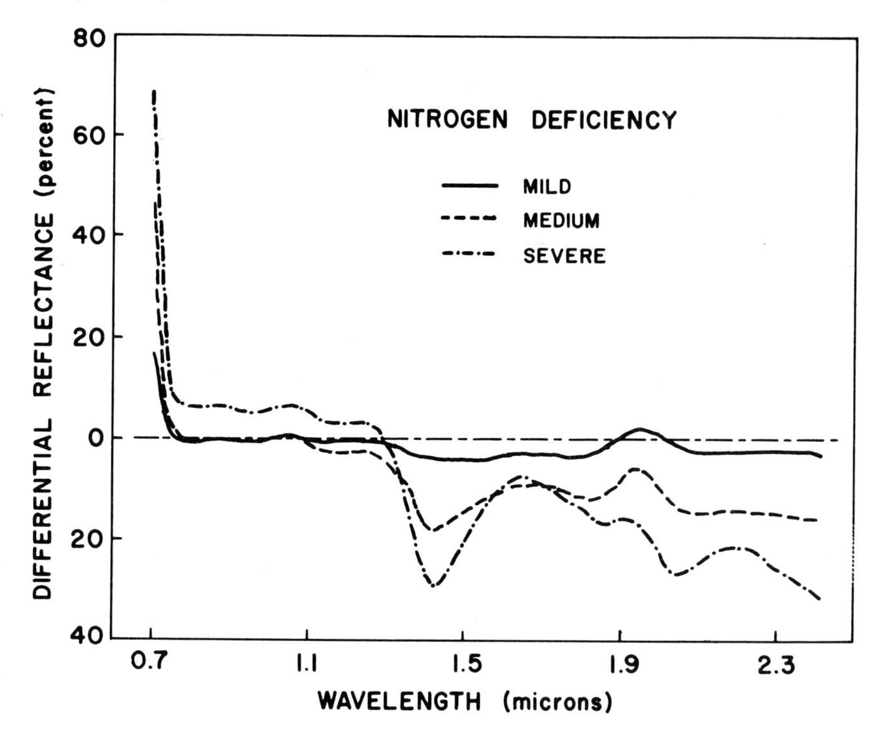

FIGURE 14 Differential reflectance spectra of nitrogen-deficient sweet pepper leaves with respect to a healthy leaf.

ditions of the crop on the film. That these corresponded to different soil salinity ranges was verified by analysis of soil samples collected within the areas delineated on the photographs.

Multiband photography has been used in detecting salinity. Figure 15 is a panel showing pictures taken with a nine-lens camera (see also discussion in Chapters 2 and 4). Insofar as salinity is concerned, and also many other phenomena that influence the high infrared reflectance of plants, the near-infrared wavelengths used for panels with wavelengths 0.67 μ-*F*,* 0.75 μ-*F*, and 0.775 μ-*F* exhibit the best tonal contrasts for detecting the extent of salinity as registered by plants. Panels with wavelengths 0.53–0.63 μ, 0.49–0.58 μ, 0.58–0.66 μ, and 0.63–0.70 μ, on the other hand, are superior for detecting contrasts in soils. Operationally,

*F = full film sensitivity.

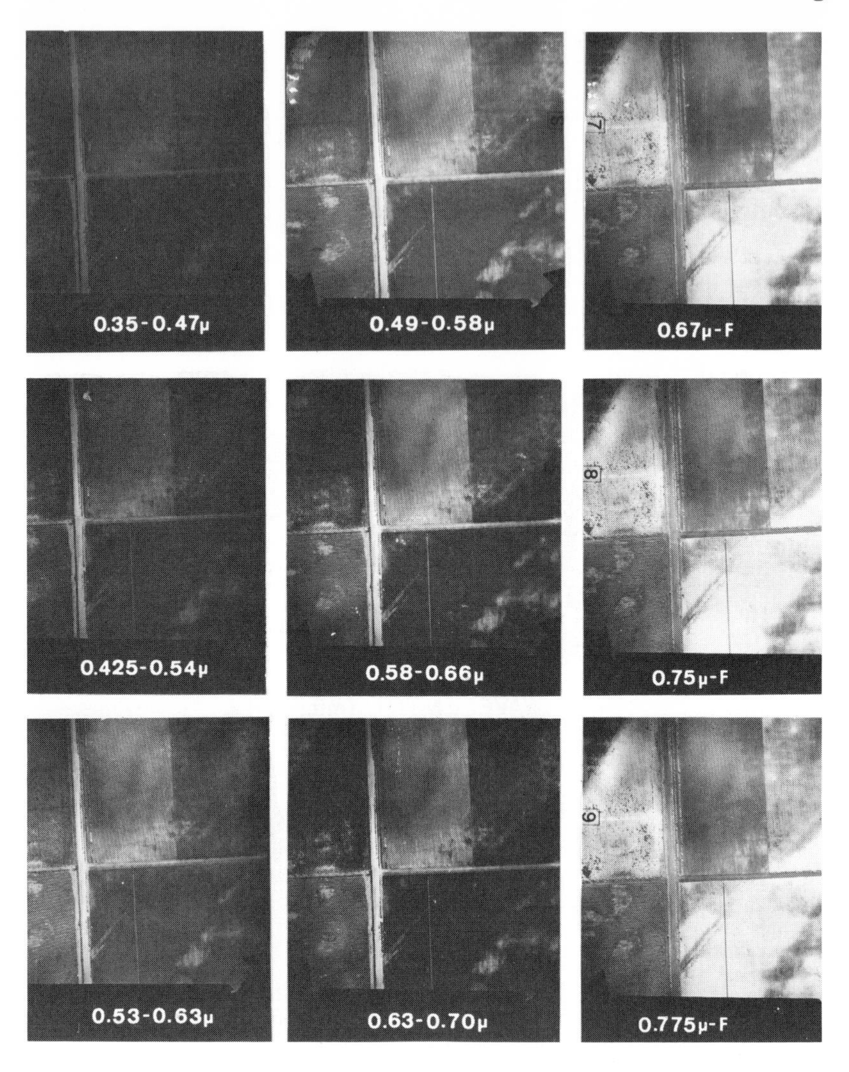

FIGURE 15 A composite of nine pictures, taken with a nine-lens camera, showing an area affected by salinity and cotton root rot. (Wavelength intervals are given for each picture. *F* means full film sensitivity.)

the specific usefulness of the nine-lens camera is yet to be determined. As a research tool, however, it may prove useful in isolating the specific wavelengths where reflectance contrasts can be detected.

Kodak Ektachrome Infrared Aero film has been used in an application of photo interpretation in estimating the severity and extent of known salt-affected areas in cotton fields on nonirrigated farms in the lower Rio Grande valley of Texas (Plate 19). This film, one layer of which is sensitive in the near-infrared wavelengths, emphasizes the infrared reflectance of healthy green vegetation, which appears bright red or pink on the photographs. Cotton plants affected by salinity appear as darker shades of red and, when seriously affected, very dark. White areas on the photo are accumulations of salt on bare soil surface areas.

Ektachrome Infrared photos permit making a distinction between vegetation and soil, which is not always possible from black-and-white photos. It is especially useful where vegetation is sparse, since the bright color of vegetation as it appears on the photos simplifies the identification of living plants in contrast to the background. On black-and-white photos, the background can be of approximately the same tone as the plants, making photo interpretation difficult.

It is significant that spectrophotometer reflectance curves from cotton plants affected by various degrees of moisture stress are different at wavelengths less than 1.35 μ but are nearly identical at longer wavelengths. It has been shown[7] that leaves from different cotton plants affected by various amounts of salinity have reflectance curves with substantial differences, between 0.5 and 2.5 μ, throughout the entire spectrum. Such contrasting reflectance characteristics may provide the means for remote detection and identification of these phenomena.

Reflectance changes in infrared wavelengths associated with moisture deficit may be the result of changes in the size and shape of cells and intercellular spaces as moisture stress progresses. Salinity, which undoubtedly causes similar plant structural changes, also results in an increase in solute concentration within the cell cytoplasm, which may be associated with reflectance contrasts at wavelengths longer than 1.35 μ.

The relation between film optical density, which is an indicator of reflectance from both plant and soil surfaces, and several factors that affect the magnitude of this reflectance was investigated by Thomas *et al.*,[7] using simple and multiple regression analyses. Plant height, percentage of ground cover, relative leaf turgidity, soil salinity, and total soil moisture stress were significantly correlated with the density of black-and-white infrared film. Myers *et al.*[9] reported similar correlations between leaf temperature and these variables. The degree of effectiveness in expressing film density as a function of any one of these factors depended upon the stage of plant development. Cotton plant maturity appeared to be critical for black-and-white infrared film. No significant relations between the variables listed and image density on black-and-white infrared film were found within 3 weeks of harvest, and only relative leaf turgidity was significantly related to film density at a period within one week of the first harvest.

Plant height, percentage of ground cover, soil salinity, and total soil moisture stress contributed significantly to the color density of Kodak Ektachrome Infrared Aero film transparencies, as measured on a color densitometer. In order to minimize the background effects of soil on reflectance, the color density was measured using various filters. The relation between color density and the various ground truths was improved by the use of filters. However, their effectiveness changed as the crop matured.

In establishing a correlation between photographic color contrast and the average salinity in the soil profile, using color-infrared film, a depth increment that is fairly representative of the rooting depth must be selected. Also, timing of the measurement must be correlated with the maximum contrasts of plant color and height. In the case of cotton in south Texas, the ideal timing is usually just before the cotton bolls begin to open. In a reasonably uniform deep soil, cotton plants at the boll stage are utilizing moisture and nutrients to a depth of about 1.5 m. Photographs of an area affected by salinity can be studied without special equipment, and a great deal can be deduced from the observations. It is desirable, however, to take advantage of instrumentation and techniques that have been developed to automate the procedures and of the precision offered by instrumentation. Desirable equipment for such studies includes a densitometer for measuring film density, a stereoscope, and a film-viewing table.

Background reflectance from exposed soil that shows through crops or in furrows between rows can greatly influence the apparent reflectance from crops as observed on photographic film. If color transparencies are available for densitometer studies, color filters inserted in the optical train permit selective color scanning of a cropped area. Since the high infrared reflectance causes increased red dye development in infrared color film, a red narrowband interference filter passes only red light and accentuates high infrared reflectance.

It is readily apparent that a number of factors that influence growing cotton plants, such as available soil moisture and soil salinity, can change from year to year. The significance of these changes is minimized, however, by bench-mark reference sites where the salinity is known, as a basis for predicting the salinity in other areas in the same season. Three years of experience have verified the efficacy of this procedure.

The procedures described for photographic detection of salinity have possible application throughout the world on millions of acres on which cotton is grown. Studies of cotton grown in saline soils at Weslaco, Texas, indicate that yields are depressed by moderate soil salinity, even though cotton is considered to be very salt-tolerant. Rapid detection of such conditions could greatly facilitate implementation of corrective measures and could be expected to result in increased yields.

Thermal Detection of Salinity

The measurement of plant leaf temperature offers another possible technique for detecting the occurrence and extent of soil salinity.[9] If dissolved salts are present in the soil solution the thermodynamic activity of the water is reduced.[10] Consequently, the soil water is rendered less available to plants. A reduction in the rate of water uptake by roots and a decrease in the water content of the stalks are among the first detectable plant responses to salinity. Thus the transpiration rate is reduced.[11]

In addition to symptoms of moisture stress, plants growing in saline soil are usually retarded. Studies at the U.S. Salinity Laboratory, Riverside, California, show that salinity affects cell division. Nieman[12] found that the area of bean leaves increases in direct proportion to the increase in cell numbers and that the difference in size between normal and salt-stunted plant leaves is due to a difference in number of cells. Salinity also slows down cell enlargement so that a longer time is required to reach full size. Chloride salinity tends to prolong cell enlargement, especially of palisade cells. This effect may account for the greater thickness of leaves often encountered under saline conditions.

The many effects of salinity on plants (pigmentation; leaf thickness, size, and structure; hydration; and transpiration) can affect the conversion of electromagnetic energy into thermal energy and the dissipation of this energy. Thus, plant leaf temperature will correlate with many factors—the salinity of the soil, soil moisture, relative turgidity of the plant, plant height, crop ground cover, etc.[13]

The relationship of cotton leaf temperature to the salinity of the soil is presented in Figure 16. The data are expressed as electrical conductivity of saturated soil extracts (EC_e) and are adjusted[14] to remove variations from the mean value of solar radiation on the two sampling days. The data indicate that the range in leaf temperature associated with variations in soil salinity from 0.5 to 15 millimhos cm^{-1} was 2.7° C on June 2 and 5.4° C on June 16. The displacement of the curves from

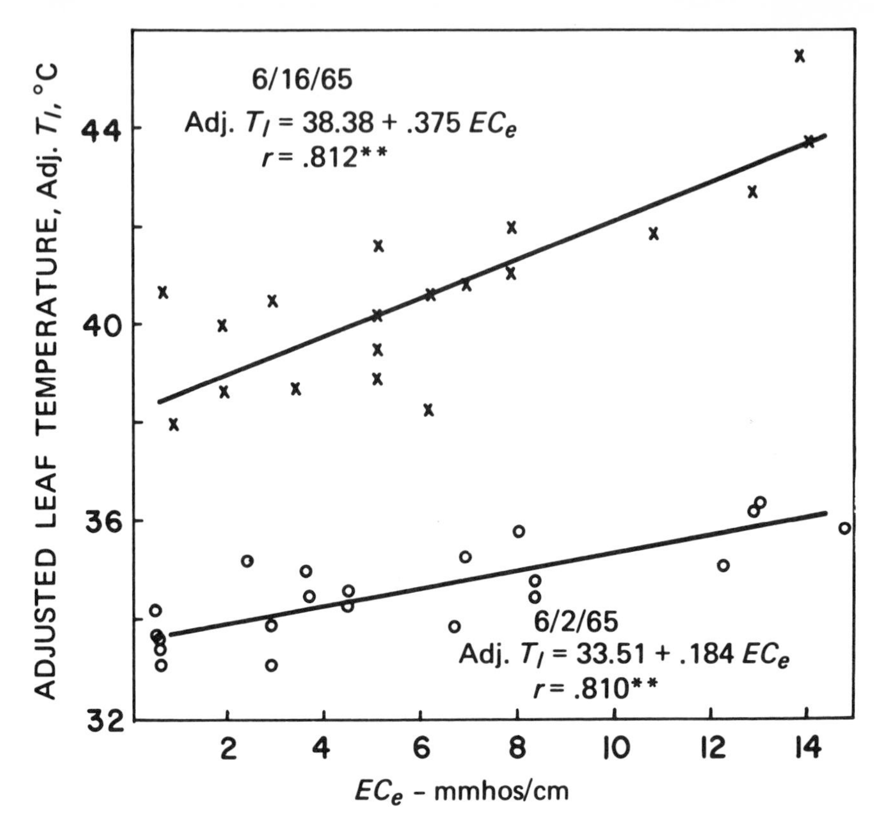

FIGURE 16 Effect of salinity of the soil (expressed as electrical conductivity of saturated soil extracts, EC_e) on cotton leaf temperature adjusted to mean solar radiation (Adj. T_l) during the measurement period of two sampling days. Simple correlation coefficient (r) and regression equations as given on the figure. 0–4-ft profile. ** indicates significance at the 1 percent level.

each other is explained by a higher ambient temperature (34° C versus 32° C), greater direct solar plus diffuse sky radiation load on the plants (1.04 cal cm^{-2} min^{-1} versus 0.86 cal cm^{-2} min^{-1}), and greater soil moisture stress (1.1 atm versus 2.6 atm) on the June 16 date compared with the June 2 date.[15] Equilibrium leaf temperature would differ by the ambient temperature difference.[16] Since leaf temperature increases about 1° C per each 0.1 cal cm^{-2} min^{-1} of solar radiation,* the radiation

*C. L. Wiegand and L. N. Namken, "Cotton Leaf Temperature." Presented at the National Meeting of the American Society of Agronomy, Columbus, Ohio, 1965.

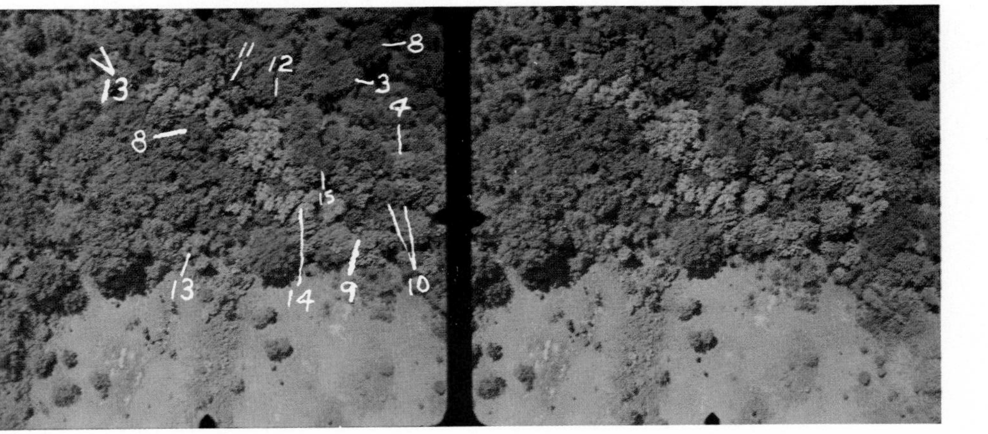

PLATE 14 Two Anscochrome D/200 stereograms of forested areas in the Hudson River valley near Poughkeepsie, New York, taken June 18, 1965, at a scale of approximately 1 : 1,580. Top: A well-drained upland area. Bottom: A poorly drained bottomland area. (*Courtesy Ralph Croxton and U.S. Forest Service.*) (For major discussion, see Chapter 4, specifically, pages 198–199 for key.)

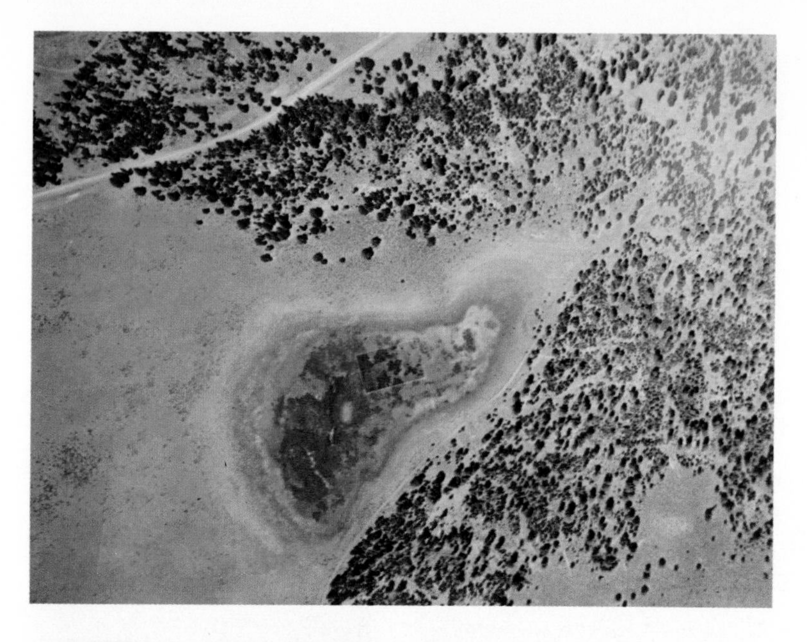

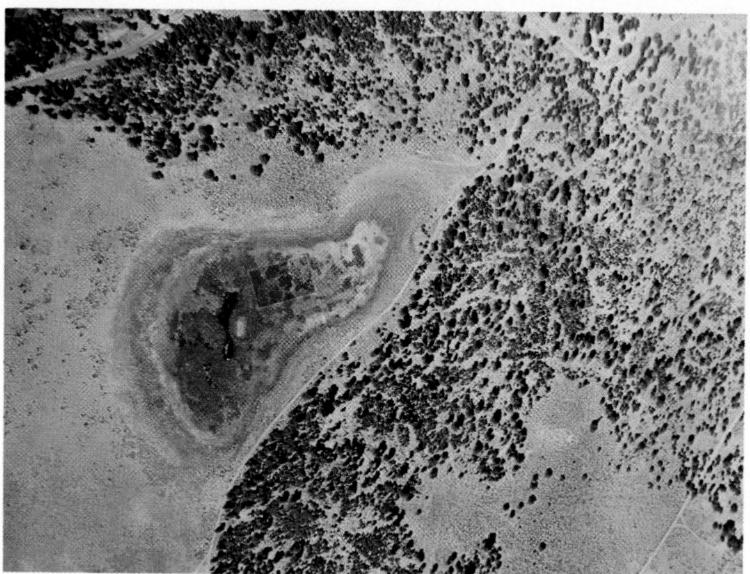

PLATE 15 Portion of NASA Range Resources Test Site, Harvey Valley, California. *A* is Ektachrome and *B* is Infrared Ektachrome, vertical aerial photos. (For major discussion, see Chapter 4, Figure 34.)

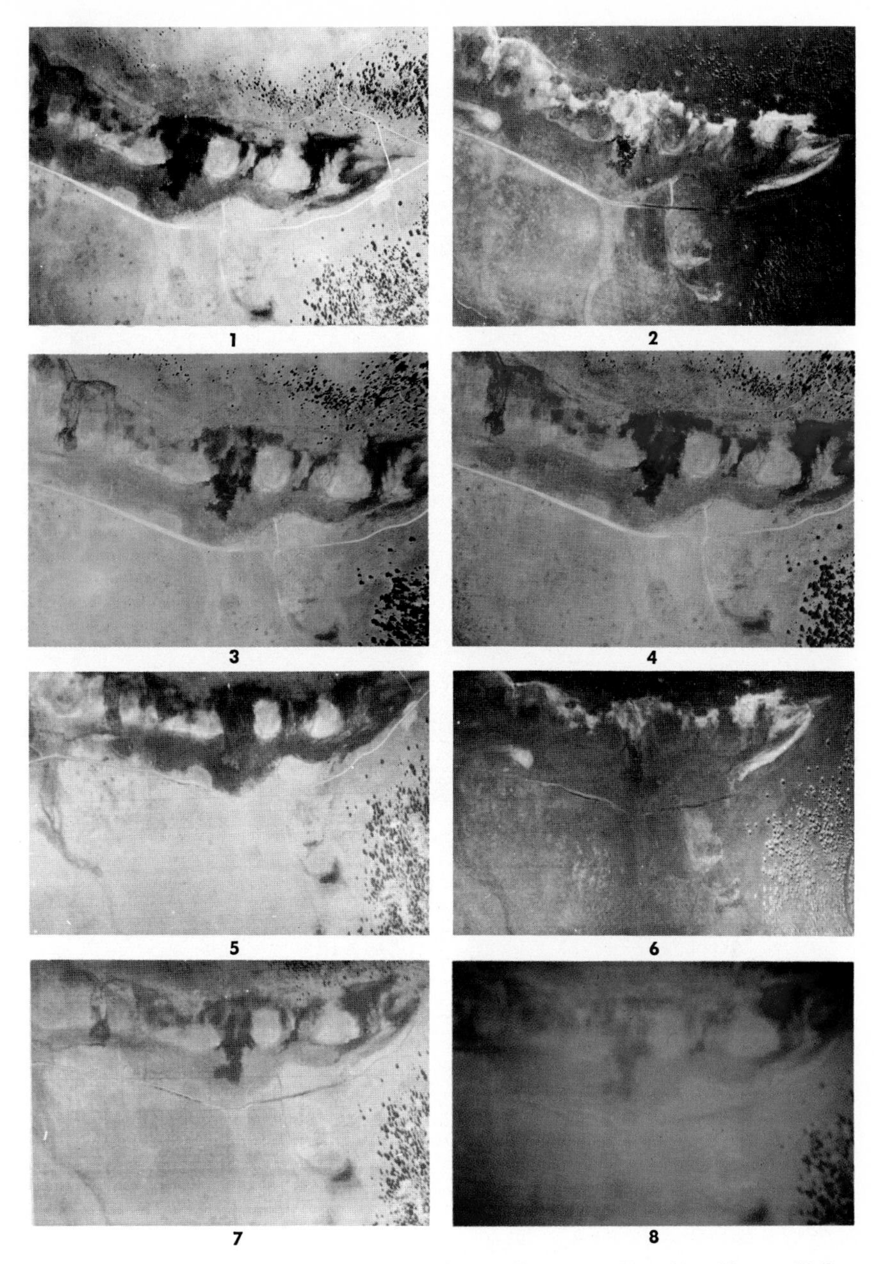

PLATE 16 Multiband photography of NASA Range Resources Test Site, Harvey Valley, California. (*Courtesy David Carneggie, University of California; imagery courtesy The University of Michigan.*) (For major discussion, see Chapter 4, p. 221.)

1: 0.5–0.7 μ
2: 0.7–0.9 μ
3: 0.4–0.5; 0.5–0.6; 0.6–0.7 μ
4: 0.5–0.6; 0.6–0.7; 0.7–0.9 μ
5: 0.32–0.38 μ
6: 0.8–1.0 μ
7: 1.5–1.8 μ
8: Color composite made from black-and-white images of 5, 6, and 7 by means of an "optical combiner"

PLATE 17 Kodak Ektachrome Infrared film produces a wide variety of diagnostic color signatures. Picture taken at 10,000 ft. 1: mature lettuce; 1a: young lettuce; 2: cabbage; 3: oats; 4: peppers; 5: onions; 6: carrots; 7: parsley; 8: mature broccoli; 8a: young broccoli; 9: bare soil. (For major discussion, see Chapter 6, p. 256.)

PLATE 18 Sweet pepper leaves with different degrees of nitrogen deficiency. (For major discussion, see Chapter 6, p. 264.)

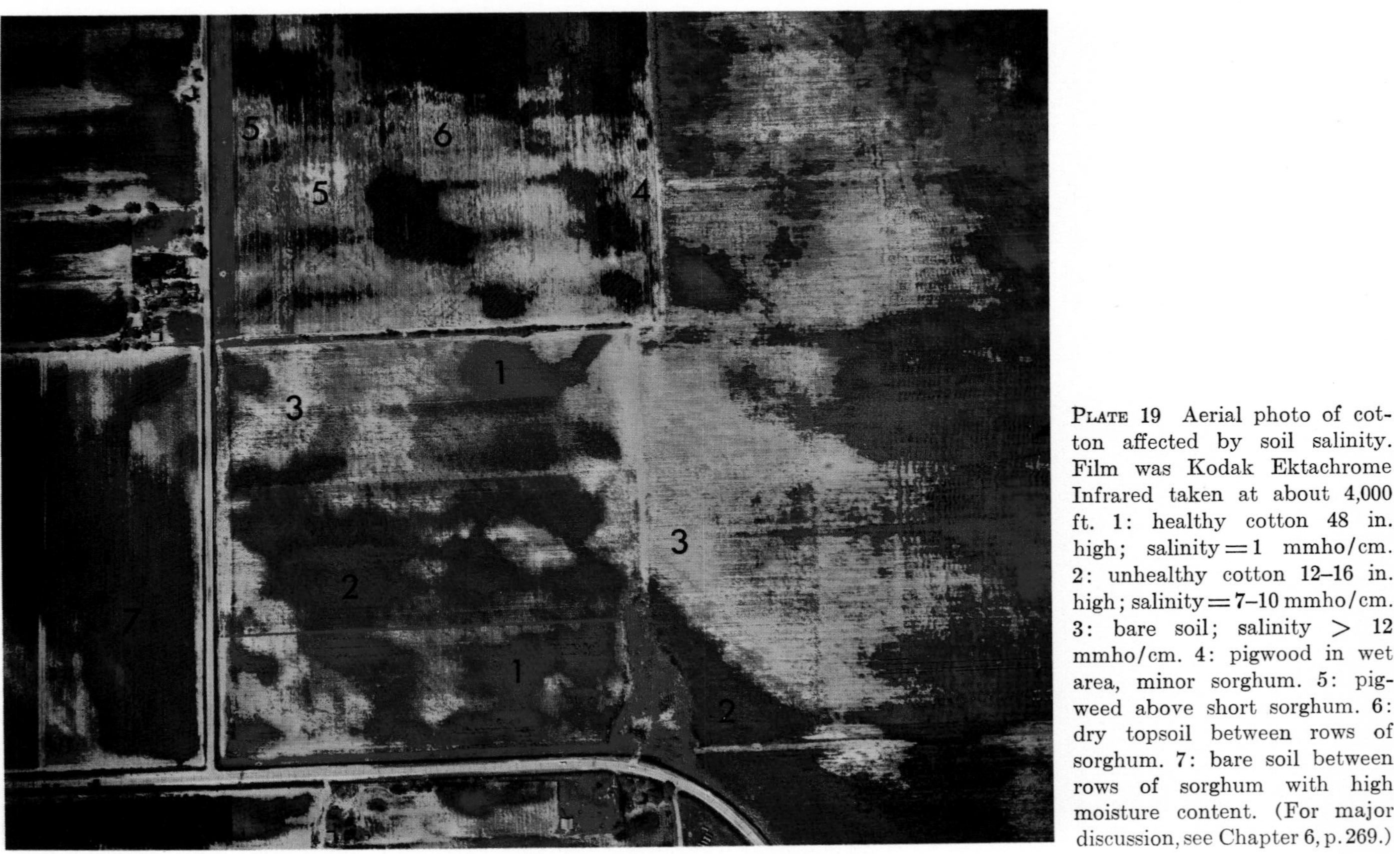

PLATE 19 Aerial photo of cotton affected by soil salinity. Film was Kodak Ektachrome Infrared taken at about 4,000 ft. 1: healthy cotton 48 in. high; salinity = 1 mmho/cm. 2: unhealthy cotton 12–16 in. high; salinity = 7–10 mmho/cm. 3: bare soil; salinity > 12 mmho/cm. 4: pigwood in wet area, minor sorghum. 5: pigweed above short sorghum. 6: dry topsoil between rows of sorghum. 7: bare soil between rows of sorghum with high moisture content. (For major discussion, see Chapter 6, p. 269.)

PLATE 20 Ektachrome Infrared photograph showing soil moisture conditions. Unstressed cotton is bright red; cotton stressed for moisture is darker. Field *A* has distinctive soil contrasts. Dark areas in fields *B* and *C* and the plots to the left of *D* are stressed for moisture. (For major discussion, see Chapter 6, p. 275.)

PLATE 21 Ektachrome Infrared photograph of Moon Lake agricultural area, Lower Rio Grande valley, Texas. (For major discussion, see Chapter 6, p. 289.)

PLATE 22 Agricultural area near Davis, California. See Figure 8, Chapter 9, for description of conditions. *A:* Simulated panchromatic photograph (with K-2 yellow filter), 0.52–0.55 μ, 0.55–0.58 μ, 0.58–0.62 μ. *B:* Simulated Infrared Aerographic photograph, 0.8–1.0 μ, positive transparency. *C:* Simulated color photograph; blue = 0.44–0.46 μ, green = 0.52–0.55 μ, red = 0.62–0.66 μ. *D:* Simulated Infrared Ektachrome photograph; blue = 0.52–0.55 μ, green = 0.62–0.66 μ, red = 0.8–1.0 μ. *E:* Color-wavelength translation in the ultraviolet and visible; blue = 0.32–0.38 μ, green = 0.40–0.44 μ, red = 0.52–0.55 μ. *F:* Color-wavelength translation in the infrared; blue = 0.72–0.80 μ green = 2.0–2.6 μ, red = 8.0–14.0 μ. *G:* Color-wavelength translation in the ultraviolet, visible, and infrared; blue = 0.32–0.38 μ, green = 0.66–0.72 μ, red = 4.5–5.5 μ. (*University of Michigan.*) (For major discussion, see Chapter 9, p. 380.)

PLATE 23 Agricultural area near Davis, California. See Figure 8, Chapter 9, for description of conditions. *A:* Thermal contours, 8.0–14.0 μ (band). *B:* Quantized view, 8.0–14.0 μ (video). *C:* Color-coded apparent temperature. Temperature decreases through the sequence violet (highest), blue, green, yellow, orange, red (wine), brown, black (lowest). *D, E, F, G:* Recognition pictures of four crops. *H:* Colored recognition picture; red = relatively mature green rice, blue = immature rice, green = safflower, black = bare earth. (*University of Michigan.*) (For major discussion, see Chapter 9, p. 381.)

PLATE 24 Aerial color photographs of portion of Purdue University agronomy farm near Lafayette, Indiana (cf. Figure 14, Chapter 9), taken at 2,000 ft. *A:* Conventional color film, 9:38 A.M., June 30, 1966. *B:* Infrared Ektachrome (camouflage-detection) film, 9:38 A.M., June 30, 1966. (For major discussion, see Chapter 9, p. 387.)

PLATE 25 Time-of-year differences in appearance of agriculture materials on Infrared Ektachrome (camouflage-detection) photographs, taken at 2,000 ft, of the area shown in Plate 24. *A:* May 6, 1966, 2:00 P.M. *B:* June 30, 1966, 9:38 A.M. (For major discussion, see Chapter 9, p. 391.)

difference would account for approximately an additional 2° C separation of the best-fit regression lines. Soil matrix suction would reduce the availability of water to the plants during the rainless period between sampling dates and contribute another degree or so separation of the regression equations.[17]

As stated earlier, leaf temperature should integrate all the variables affecting transpiration rate and energy transfer between the leaf and its environment. In the example given, ambient air temperature, intensity of solar radiation, and soil water availability in the root zone—as well as soil salinity—contributed to equilibrium leaf temperature. The variables ambient temperature and solar radiation will vary only gradually over a cloud-free crop-producing area; however, the soil water will be either rather uniform (general rains) or different from the salinity pattern (scattered showers or irrigation). Thus, identification of the salinity pattern does not require that all these variables be measured in detail.

Irrigation water always contains salts that are crystallized out and deposited in the soil when the water is evaporated. Unless excess irrigation water is periodically applied to leach these salts from the soil profile, and unless soil drainage is adequate, salts accumulate in the root zone. These requirements are seldom met, so the worldwide soil salinity problem can be expected to worsen.

Remote-sensing techniques should improve the diagnosis and delineation of saline areas. In nature, the pattern of salinity is generally very erratic, the saline and the nonsaline soil being interspersed. A further complication is that an abrupt change from unaffected to barren soil may occur over a lateral distance of a few feet or a few meters, whereas in other instances, such as encroachment of salts from a gradually rising water table, the lateral gradations in salinity are very subtle. Simultaneous detection of both aspects requires both high resolution and sensitivity in the techniques used.

Another factor to consider in interpretation is the wide range in tolerance of salinity among crops. The pattern of detectable salinity and the indicated severity may be influenced by the crops grown. Barley, cotton, and sugar beets are very salt-tolerant. Most of the other cereal and forage crops are moderately salt-tolerant. Fruit crops are salt-sensitive.[18]

Plant-temperature measurement by remote-sensing techniques is facilitated if there is a fully developed plant canopy. Under these conditions if the ground is nonsaline it will likely be shaded, whereas in the saline areas, shorter plants and fewer and smaller leaves may permit the ground to show through the canopy. These conditions enhance the signal difference between salt-affected and salt-unaffected areas, unless one views the scene obliquely.

Experience has also indicated that remote sensing of plant temperatures is most successful when some physiological moisture stress has developed, especially under high evaporative demand conditions. Hence, differences between salt-affected and salt-unaffected crops will be more evident if one delays measurements for a week or more following a soaking rain or irrigation and if the measurements are made around noon or in the early afternoon, when incident solar radiation and air temperatures are near their maximal daily values.

In making such measurements, one should always take enough on the ground to verify that the observed thermal or image pattern is due to salinity and not to topographical undulations (which would cause high areas to suffer moisture stress first and might result in crops drowning out in low areas), to soil texture or soil series patterns, to variations in management of different fields, or to other causes not apparent from the spectral signatures alone.

The spatial resolution of thermal imagery is poorer than that of photographic imagery. Thus the fine detail in photographic coverage can be used to verify stand conditions resulting from uneven plant emergence; to verify disease, insect damage, or nonuniformity of fertility that would affect stand and plant size; to verify percent ground cover differences due to differences in planting date and species; and to verify other features that could complicate the interpretation of the thermal imagery. For these reasons, simultaneous infrared-film photographic coverage of the area to be thermally scanned is recommended.

Photographic Detection of Moisture Stress

Plant moisture stress changes the reflectance characteristics of a crop canopy, making it possible to relate photograph density to available soil moisture.[7] In the case of irrigated crops, this is a potentially useful tool for determining when to irrigate. The detection of moisture-stressed nonirrigated crops would indicate drought conditions. However, as previously mentioned, other factors can affect the infrared reflectance of crops. Moisture stress apparently causes physiological changes in leaves that influence the reflection of solar radiation from the canopy. This is evidenced from the plant canopy by a decrease in infrared reflectance with an increase in moisture stress. Reflectance differences between moisture-stressed and moisture-nonstressed plant canopies in the visible wavelengths are not as great as in the infrared.

One study of spectral reflectance from cotton subjected to moisture treatments illustrates problems encountered in remote sensing of moisture stress in plants. The study involved three moisture treatments in which 40.4 cm, 28.2 cm, and 18.1 cm of irrigation water were received. All plots received 7.1 cm of rain. Leaf samples were taken on June 16, when the average soil moisture per foot of depth in the 1.5 m profile was 8.9, 6.3, and 5.1 cm for the three treatments, and average plant height was 81.4, 69.9, and 64.9 cm, respectively.

Spectrophotometer reflectance curves from leaf samples are shown in Figure 3 and discussed briefly in the first section of this chapter. Near-infrared reflectance from individual leaves of plants subjected to various degrees of moisture stress show about the same or slightly higher reflectance than those not subjected to moisture stress, although near-infrared reflectance from a crop canopy as registered by black-and-white or color-infrared film is less for the stressed cotton. This phenomenon is seen in Plate 20. Site *A* has distinctive soil contrasts: the dark areas need irrigation; however, the bright red area in the middle of the field is not stressed because of soil with a higher moisture-holding capacity. The dark areas in fields *B* and *C* and the research plots to the left of site *D* are stressed for moisture.

Observations and measurements indicate that the leaves of cotton plants affected by moisture stress undergo physiological changes that substantially affect reflectance and transmittance characteristics. Kramer[19] states that loss of turgor and wilting of leaves is accompanied by measurable reduction of leaf area and thickness, and that the dwarfing effect of drought on plants is due largely to decreased cell enlargement and earlier cell maturation. The many changes that take place in leaves subjected to prolonged or intermittent drought undoubtedly all influence spectral characteristics to some degree. Some of these changes described by Kramer are thicker cuticle, more compact structure, thicker cell walls, better developed vascular system, and increased concentration of cell sap.

The physiological factors that influence near-infrared reflectance result in leaf reflectance characteristics that are not immediately changed by the addition of water to the soil. This is because water in the leaves is not the phenomenon influencing this reflectance. Color, apparent in the visible spectrum, does change with addition of water to stressed cotton plants. Cotton plants under moisture stress continue to exhibit low near-infrared reflectance until a new flush of top growth appears, usually within several days after water is added.

In summary, a number of factors, some of which remain unidentified, influence reflectance from crop canopies. However, three principal wavelength regions are associated with distinct phenomena. In visible wavelengths, from about 0.3–0.7 μ, chlorophyll and carotene absorption predominates. This absorption is strong at 0.38 μ and 0.65 μ but is weaker at 0.55 μ, the typical green vegetation peak. Moisture and salinity stresses, and addition of nitrogen, decrease reflectance in visible wavelengths. These decreases are associated mainly with changes in chlorophyll and carotene.

Reflectance in the shortest of the near-infrared (0.7–1.35-μ) wavelengths is caused by the lack of pigment absorption and by the lack of absorption by liquid water. The negligible absorption in these wavelengths is a distinctive feature of vegetation. Reflectance changes are associated primarily with changes in the size and shape of cells and intercellular spaces, and, perhaps, with other physiological changes in leaf structure.

Beyond 1.35 μ, water absorption predominates. Particularly strong liquid-water-absorption bands appear at 1.45 μ and 1.95 μ.

Other factors, such as crop geometry, background soil radiance, and number of reflective surfaces, influence the net reflectance from crop canopies.

The relationships described have been verified for cotton only and must be established for other crops.

Thermal Detection of Moisture Stress

Both the absolute temperature of leaves and plant canopies and the temperature difference between leaves and the ambient air are of interest. Plant leaf temperature is of primary interest in considering biochemical reaction rates and the water relations of plants, the latter due to the influence of temperature on water vapor pressure. Leaf temperature minus air temperature is frequently used for comparing treatment effects at varying ambient temperature and is required for convective energy-transfer calculations.[20,21]

Curtis[22] pointed out that it is possible to control experimental conditions so that any one of the more important factors such as convection, reradiation, transpiration, or angle of incident radiation may control plant leaf temperature. Leaves normal to the sun in midday sunlight have temperatures up to 15° to 20° C above the ambient air temperatures.[22-24]

Curtis,[22] Clum,[25] and Smith[26] found that reduction in transpiration caused by withholding water resulted in leaf temperature increases of 2° to 5° C. Pallas and Harris[27] reported leaf temperature of cotton to be highly correlated with transpiration under most conditions, and that leaf temperature falls below ambient air temperatures when light intensity and relative humidity are low. Tanner[28] reported a 0° to 3° C temperature difference between irrigated and nonirrigated potatoes. He estimated that a 10 percent decrease in transpiration of a full cover of alfalfa evapotranspiring at 0.76 mm hr^{-1} would cause a temperature increase of approximately 1° C when the associated windspeed at the 1-m height was 3.1 m sec^{-1} and the net radiation was 0.95 ly min^{-1}. Gates[20] calculated the energy balance for single leaves and concluded that, for each 0.10 mm hr^{-1} of transpiration, the effective radiation load on a leaf is reduced 0.10 ly min^{-1}. Compared with no transpiration, transpiration at these levels would lower the leaf temperature 5°, 2.5° and 1° C below the nontranspiring leaf at windspeeds of 1, 5, and 15 mph, respectively. Mellor[29] concluded that leaf temperature and transpiration rate increase linearly with the energy absorbed over a range of energies from 0.14 to 2.79 ly min^{-1}.

Curtis[22] and Shaw[30] present data showing that citrus and tomato leaves are 1° to 2° C colder when radiating to a clear sky than when the sky is cloudy or the leaves are shielded from the sky.

Figures 17 and 18, taken from Wiegand and Namken* depict the influences of solar radiation and plant relative turgidity†[31,32] on leaf temperature of cotton plants. The two sets of data in Figure 17 were obtained on different days when air temperature (T_A) at plant height and relative turgidity (RT) were uniform among all experimental treatments. Air temperature at the time of measurement (2:30–3:00 P.M., CST) differed by 3.5° C, and relative turgidity of the cotton leaves differed by 10 percent on the two dates. The variable radiation was created by intermittent clouds. The radiation indicated was the direct solar and diffuse sky radiation as measured with an Eppley pyranometer. In the study cited, it was found that individual cotton leaves equilibrated with a change in radiation intensity in about 45 sec. Thus the leaf-temperature measurements were deferred until radiation remained steady for a minimum of 45 sec.

The data show that the thermal response of the leaves to changing radiation is linear, and they imply that if radiation conditions are variable it will be necessary to measure the incident radiation simultaneously with leaf temperature. For example, an increase in radiation from 0.6 to 1.6 ly min^{-1} resulted in a 9° to 10° C increase in leaf tem-

*Wiegand and Namken, *op. cit.*

†Relative turgidity (RT) $= 100\ (FW - DW)/(TW - DW)$, wherein FW is the fresh or field condition weight of leaf samples, TW is the turgid weight achieved by floating the leaf samples on distilled water overnight under illumination, and DW is weight of the samples after drying at 60° C.[32]

perature as indicated by the regression coefficient, b. The standard errors of estimate, $S_{y \cdot x}$, indicate that leaf temperatures could be estimated within 0.9° C two thirds of the time.

The variable cotton plant moisture conditions shown in Figure 18 were achieved by timing of irrigation during a rainless period. At the midafternoon measuring time, cotton plant leaves exhibited symptoms of wilt at 70 to 72 percent relative turgidity. Leaves at 60 percent relative turgidity were extremely flaccid. The standard deviations of air temperature and solar radiation associated with these observations were 0.7° C and 0.04 ly min^{-1}, respectively. The regression coefficient indicates that leaf temperature increased 0.15° C per percent decrease in relative turgidity over the relative turgidity range 83 to 59 percent. These and additional data indicate that cotton leaf temperature can vary about 3.5° C with plant moisture stress.

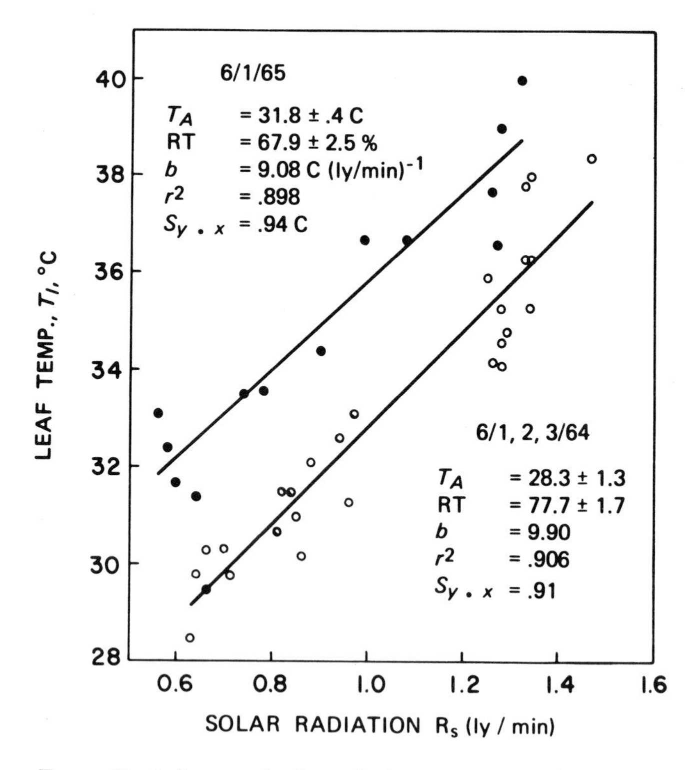

FIGURE 17 Influence of solar radiation on cotton leaf temperature when air temperature at plant height and relative turgidity were approximately constant. *(Courtesy Wiegand and Namken.)*

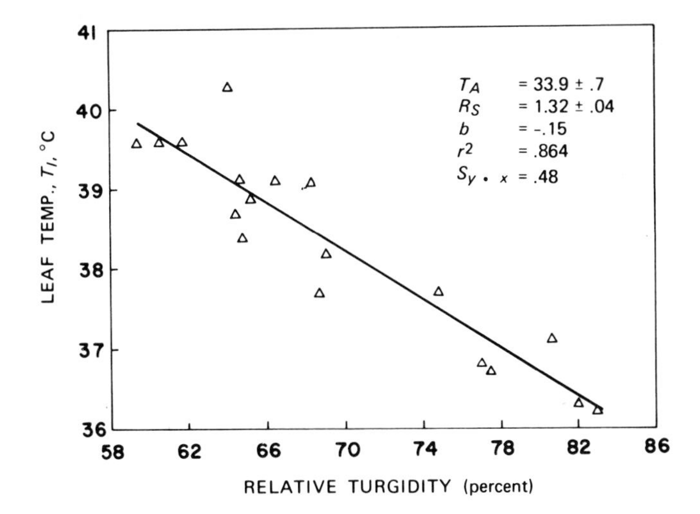

FIGURE 18 The effect of relative turgidity on leaf temperature when air temperature and solar radiation were approximately constant. (*Courtesy Wiegand and Namken.*)

From the data for 37 individual days covering two crop seasons, it was determined that leaf temperature could be estimated with a standard error of 1° C from measurements of relative turgidity, solar radiation, and air temperature at plant height. For these measurements, solar radiation averaged 1.15 ly min^{-1}, and leaf temperature minus air temperature averaged 4° C.*

Most plant-temperature measurements reported in the literature have been made on individual, well-exposed leaves. Figure 19 is a composite of four thermograms obtained with an infrared camera during a study of diurnal plant canopy temperature changes in small, differentially irrigated cotton plots.

The time of day (CST) at which the thermograms were taken is given in the legend. The first was obtained at 5:40 A.M., well in advance of daybreak. The light areas from bottom to top—ignoring the one at the very bottom—are a man kneeling between the plot in the foreground and the center plot, an incandescent lamp in the far plot, and three side-by-side instrument shelters just outside the plots. The other three thermograms depict the same target at later times during the day.

In all these thermograms, the lighter tones indicate warmer targets. Targets within the field of view are matched in tone with one of the

*Wiegand and Namken, *op. cit.*

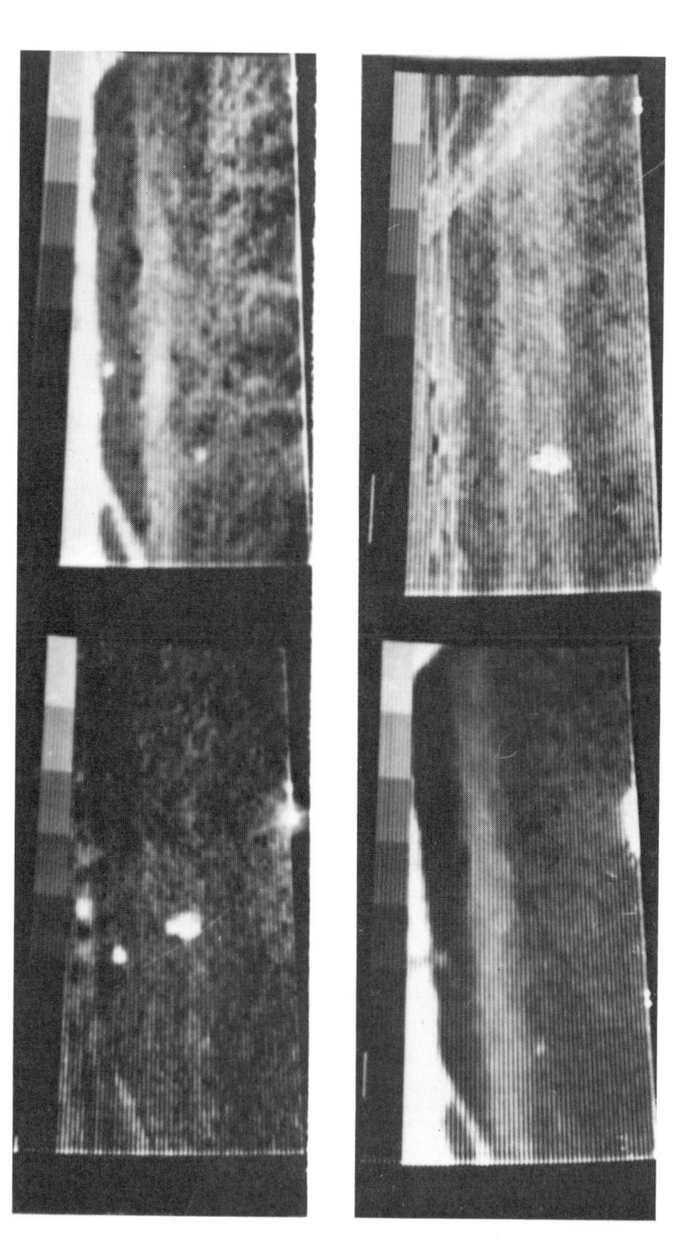

FIGURE 19 Thermograms of cotton plots varying in soil moisture condition taken at four different times during the day of June 10, 1965. Top left, 5:40 A.M.; top right, 9:35 A.M.; bottom left, 3:20 P.M.; bottom right, 10:10 P.M.

eight gray-scale steps printed automatically as the last several scan lines of each thermogram. From the electronic settings used to obtain the thermograms and a parameter corresponding to the gray-scale step, the target radiance may be calculated and then converted to target temperature. It was necessary to vary the electronic settings as the crop surfaces warmed so that temperature differences cannot be compared by visual inspection except within thermograms.

The cotton plot in the foreground of each thermogram and the one in the background were at about the same moisture condition and have the same tone. The calculated temperature difference between dry and wet plots was 0.1°, 0.3°, 2.0°, and 0.2° C at 5:40 A.M., 9:35 A.M., 3:20 P.M., and 10:10 P.M., respectively.

The remote sensing of plant canopy temperatures appears feasible in studying a broad array of situations: assessing the need for irrigation, or conversely, the extent and severity of drought; distribution of rainfall as revealed by vegetative thermal patterns; effect of slope on soil-moisture distribution; site exposure and aspect effects;[33] proximity of water tables to the soil surface; effects of altitude and longitude on plant temperature; plant-temperature variation with distance from coastal or inland waterways; diurnal patterns; ecological studies of plant communities; and crop inventories related to gross thermal patterns.

To be meaningful in a number of the studies suggested, the indicated temperature must be measured with a resolution of at least 1° C. If plants completely cover the ground surface and solar radiation is constant, the temperature range among crops may be only a few degrees centigrade. Where tall and short plants are interspersed and it is desired to observe temperature differences between sunlit and shaded plants, a measurable temperature range of 10° to 12° C is recommended.

If solar radiation is variable, as between cloudy and fully sunlit conditions, a temperature range similar to that allowed for shaded versus sunlit plants is required. Variable radiation, however, requires that the solar radiation be measured simultaneously with the plant temperatures since it has a stronger influence on leaf temperature than any other variable.

Transpiring leaves will be cooler than the ambient air under low insolation intensities, and plant leaves will be 1° to 2° C cooler than the air at night. Thus the lower limit of the temperature range desired will be 2° or 3° C lower than the ambient air temperature at the height of the plants. The usual Weather Bureau local temperature reports suffice for estimating the lower temperature limit required.

A number of factors must be considered in interpreting thermal radiation measurements. Generally, the difference in radiation between a blackbody reference and the target is measured, and temperature is inferred, from the Stefan-Boltzmann equation. Since each measurement yields an equivalent blackbody temperature, the emissivity of the target (and reference blackbody) must be known. The emissivity of plant leaves is indicated to be in the range 0.95–0.98.[21,34] In general, emissivity may vary with both temperature and wavelength. A further complication is that the exponent on temperature in the Stefan-Boltzmann equation is 4 only for the full-spectrum integration from zero to infinity of the Planckian radiation function,[33] whereas measurements are made in short-wavelength segments of the whole range, frequently in the 8–14-μ atmospheric window. Atmospheric attenuation must be considered. Sensor response itself may be nonuniform over the wavelengths of interest.

If plants do not completely cover the soil surface, both plant and soil surfaces contribute to the radiance sensed. Exploratory measurements indicate that under the incident radiation range, approximately 0.8 to 1.5 ly min^{-1}, dry soil will have equivalent blackbody temperature about 20° C warmer than plant leaves.

The limitations and precautions characteristic of other remotely sensed targets are described throughout this book and apply also to plant temperature measurements.

Radar Sensors in Soil Mapping

There is no literature on the utility of radar images in soil mapping other than that of Beatty.[35] Consequently, studies were begun in 1965 by D. Simonett and Dr. James Thorp at the University of Kansas to test the value of different radar wavelengths and polarizations in soil reconnaissance. One study, with a *K*-band system, is reported here.

Figures 20–22 summarize some observations on the identification of soil associations on this imagery in a portion of Woods County, Oklahoma. Figure 20 shows three classes of vegetated and bare sand dunes in the county. Experienced radar interpreters with a good soils-geomorphology background usually can distinguish these three types on the radar film transparencies. Bare dry sand is a good absorber of energy in the radar wavelengths. Hence, it yields very little backscattered return and appears black on a film positive (Figure 23). Vegetated dunes and sand sheets return more energy and have a lighter

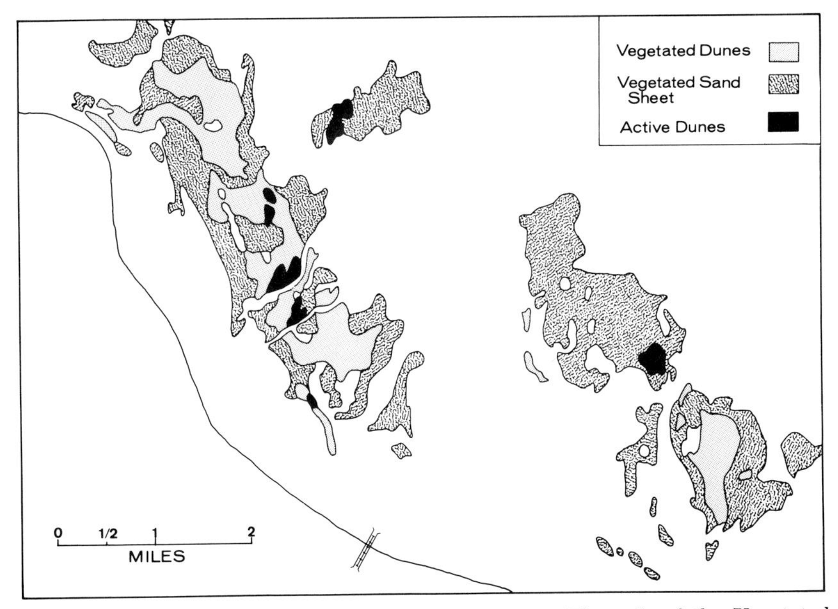

FIGURE 20 Aeolian sands, Woods County, Oklahoma. The soils of the Vegetated Dunes are the Tivoli Fine Sand, dun phase (15–25 percent slopes) and Dune Sand (stabilized). The Vegetated Sand Sheet occurs on the Tivoli Fine Sand, and the Active Dunes are bare sand with 15–25 percent slopes. These three soil groups are usually, but not always, distinguishable on *K*-band radar imagery.

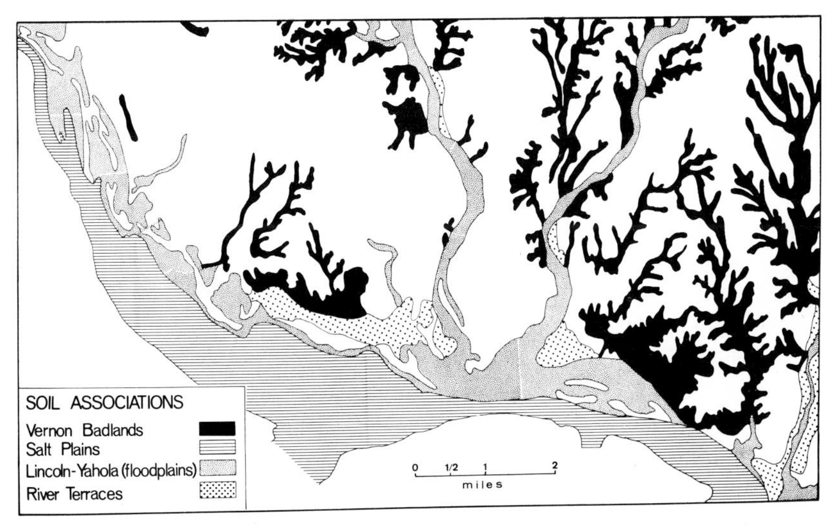

FIGURE 21 Soil associations normally distinguishable on *K*-band radar imagery, Woods County, Oklahoma.

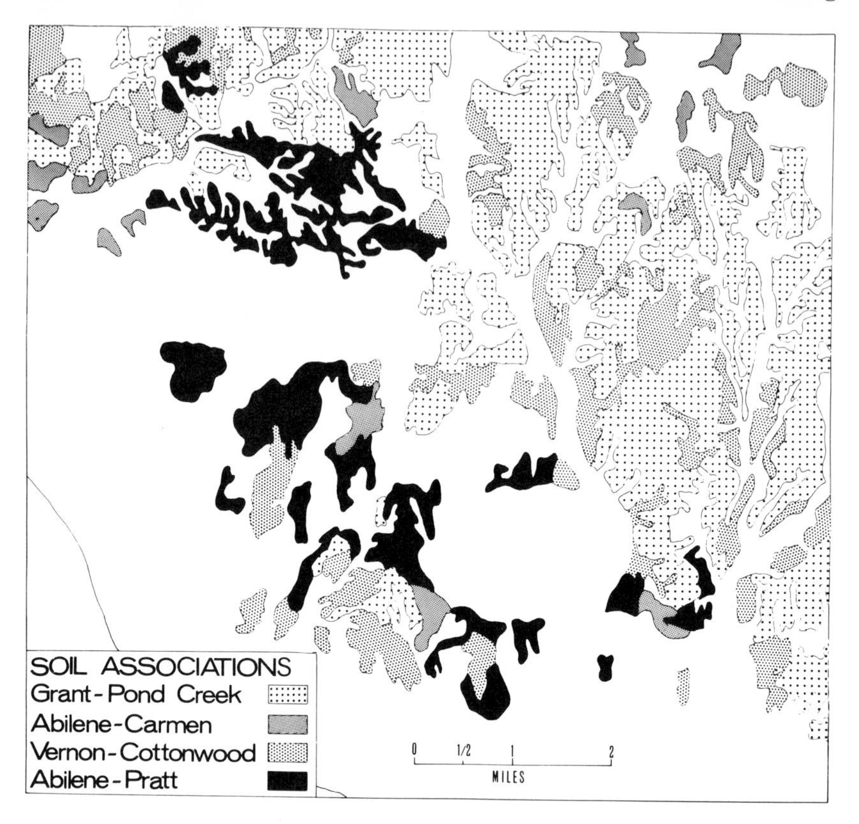

FIGURE 22 Soil associations that cannot be separated on K-band radar imagery, Woods County, Oklahoma.

image tone. They also have a distinctive texture that, along with the topographic undulations of the dunes, aids in their detection.

Figure 21 shows four soil associations in Woods County, Oklahoma: Vernon badlands, salt plains, Lincoln-Yahola (floodplains), and river terraces. The first three are distinguishable with ease on the radar imagery, the fourth with some difficulty. The Vernon badland association typically has a steeply convex valley side plunging into a narrow, almost flat-flooded valley. The rough valley walls, clad variously in small trees, grass, sagebrush, or exposed gypsum, are easily delineated on the radar imagery. The salt-plain association is a bare saline river wash with distinctive light tone on the radar image. The flat Lincoln-Yahola floodplains association is covered with mixed grass, sagebrush, and occasional trees. The association as a whole is easily mapped, but its component soil series cannot be detected. The river terraces association usually can be separated from the floodplains, but again its several soil series cannot be noted. In other regions we have found it more difficult to distinguish terraces from floodplains, for example, in the Kansas and Waukarusa Valleys, near Lawrence, Kansas. The ability to distinguish river terraces on radar imagery varies in relation to the environment and the radar look angle, being easiest at low angles of illumination and where the vegetation differs on the terraces and floodplains.

Figure 22 shows four additional soil associations in Woods County, Oklahoma, that could not be distinguished from one another, although collectively they could be separated from the associations mapped in Figures 20 and 21. The associations shown in Figure 22 include soils that would be classified as Udic Argustolls, Typic Ustochrepts, and Haplic Calcuistolls.

The inability to discriminate between these soil associations on the radar image, let alone the series of which they are composed, is in many ways to be expected in an area so completely given over to cultivation. Lundien[36] and Simonett *et al.*[37] indicate that radar is much less sensitive to soil-texture variations within a field than it is to crop and soil moisture and other differences between adjacent crops. Field patterns are dominant on this radar imagery, and they mask any weakly expressed soil information. In reaching this conclusion, however, it is not proper to deduce further that radar imagery itself is of little utility in distinguishing between any but extreme soil differences of the type indicated in Figures 20 and 21. Radar imagery of grazing land in the Flint Hills of Kansas shows some sensitivity to differences in height and types of grass and weeds. Discrimination might thus be achieved where the native vegetation varies in its seasonal response on different soils. This remains to be studied.

A *K*-band radar image showing a vegetated coastal sand strip backed by sand dunes along part of the Oregon coast (Figure 23) is included to show the close relation between vegetation cover on dune sands and the radar return. Field studies of the coastal sand strip show that those areas that are dark are relatively free of Marram grass and other vegetation, whereas the areas that are lighter have a denser cover. Dry bare sand, as noted above, absorbs almost all the incident radar energy. Within the dune field, bright returns come from clumps of Sitka spruce, Lodgepole pine, and other conifers, and wispy gray tones coincide with pioneer regrowth in flat areas where the water table is near the surface.

To summarize these observations, it is clear that the information

FIGURE 23 K-band radar image on sand dune area south of the Umpqua River, Oregon. 1: Foredune vegetated by European and American dune grass. Brighter areas are more densely covered. 2: Open sand (unstabilized). 3: Stabilized dune surface predominantly vegetated by Sitka spruce and other conifers. 4: Trough between dune ridges vegetated by shrubs and grasses.

obtainable from this radar imagery for soil mapping is distinctly uneven in both distribution and quality, for though it is sometimes possible to make clear distinctions between adjacent soils even at the series level, and more usually at the association level, there are many instances when neither is feasible. Separation of soil groups at the association level is more likely in untilled and subhumic to arid regions than in cultivated or densely forested humid lands. To express these conclusions in another way: Where extreme differences occur in adjoining plant structures, in soil or plant moisture content, in soil texture, in topography, and (in areas of scanty vegetation) in small-scale surface roughness, then discrimination of the radar image of soil units closely tied to these differences will usually be possible. Lesser differences will not be so easily detected, especially in cultivated areas, and careful timing of aircraft flights to coincide with the greatest seasonal vegetation contrasts will be necessary.

Additional studies employing multiple wavelengths, polarizations, and radar bandwidths may show that it is possible to better this level of discrimination of soil units. Even if this is not the case, the over-all perspective provided by radar coverage of large landscape units will be a valuable addition in reconnaissance surveys.

Thermal-Infrared Sensing of Soils

Soil, so important in the production of crops, is extremely complex—in formation, stratification, organic and mineral content, fertility, and many other ways. Remote sensing measures surface phenomena primarily and would at first seem to have limited application in measuring soil properties and transient soil conditions (see Chapter 5, pp. 236, 250). Surface soil conditions, many of which can be detected by remote sensors, frequently are indicative of subsurface conditions. Also, natural vegetation of uncultivated areas and agricultural crops in cultivated areas can frequently be used as indicators of soil properties. Thermal-infrared sensing, however, holds considerable promise for identifying some subsurface soil conditions.

Thermal Properties of Soils

The intake and expenditure of radiant energy at the earth's surface are expressed in the following equation of radiation balance:

$$R = Q - S - U,$$

where R is net radiation retained by the earth, Q is total solar radiation, S is reflected solar radiation, and U is effective radiation by the earth.[38]

The amount of radiation reaching the earth from the sun depends upon absorption and scattering by the atmosphere. In addition, certain factors such as slope, exposure of the land, nature of the soil and vegetation, and elevation and latitude affect the amount of radiation retained.[39]

The effect of latitude on the angle at which the sun's rays strike the earth is well illustrated by differences in temperature between the arctic and tropical regions. Even though days and nights may be of equal lengths in both regions at certain periods of the year, latitude has an enormous effect upon the soil temperature.

The flow of heat through soil involves the simultaneous operation of several transport mechanisms.[40] Conduction is responsible for the flow of heat through the solid materials. Across the pores, conduction, convection, and radiation act together. When water is present, latent heat of vaporization is an additional factor involved in heat transfer. Carson[41] identifies two independent parameters, thermal conductivity and volumetric heat capacity, as the principal factors involved in the thermal behavior of soils. Association of surface temperatures with the principal soil characteristics influencing heat transfer enhances the opportunity for using remote sensing to help identify certain soil factors. Dynamic thermal changes within soil profiles—diurnal, seasonal, and annual—greatly strengthen the opportunity for thermal-infrared sensing, providing the soil factors causing the changes can be identified.

Transfer of Heat

1. Heat may flow toward the soil surface or away from it. The sign of this flux alternates seasonally and with the time of day. The transmission of heat in the soil occurs chiefly through molecular thermal conductivity and by convection and radiation.[42] The greater the temperature difference between two points in a soil profile, the greater the flow of heat between these points.

2. The heating and cooling characteristics of a soil depend on its thermal conductivity (the number of calories flowing in 1 sec through a 1-cm thickness of homogeneous material with an area of 1 cm^2 when the temperature difference is 1° per cm). The thermal conductivity of dry soil is very low, the reason being the limited areas of contact be-

tween soil particles. With increasing moisture content, the conductivity increases rapidly at first and then more slowly.

3. The heating and cooling of a soil also depends on its heat capacity. A distinction must be made between gravimetric heat capacity (specific heat) and volumetric heat capacity. The former is the number of calories necessary to warm 1 g of soil by 1° C; the latter is the number of calories required to warm 1 cm^3 of soil by 1° C. Heat capacity of a soil depends on its water content, porosity (air content), and mineralogical composition. The volumetric heat capacity of sand relative to other soils is highest in dry conditions and lowest in moist conditions.

4. Temperature conductivity, expressed in degrees rather than calories, determines changes of soil temperature in time and depth. The coefficient of temperature conductivity, K, equals the coefficient of thermal conductivity, λ, divided by the volumetric heat capacity of the soil, c_p:

$$K = \lambda / c_p.$$

Temperature sensing of surface soils with an airborne thermal line scanner has been used for detection of certain soil characteristics. An area on a typical alluvial floodplain is shown in Plate 21, which is an Ektachrome Infrared photograph. The soil surface and subsurface vary in texture from loam to sandy loam. The light-colored soils of Plate 21 are sandy loams, and the darker soils are loams.

Diurnally flown thermal imagery of the area in Plate 21 appears in Figure 24. Each of the transparencies was scanned with an isodensitracer (IDT) to determine film densities. Figure 25 is an isodensitracing of line-scan thermal imagery for the mission flown at 7:00 P.M. on June 1, 1966. Film densities are directly related to equivalent blackbody temperatures (see the discussion of thermal detection of moisture stress, page 276).

The IDT automatically scans the film transparency and plots the measured density values in a quantitative two-dimensional equal density tracing of the scanned area. The optical system of the IDT automatically probes the entire specimen area by making a series of closely spaced parallel scans. For each scan of the specimen, a corresponding coded parallel line is recorded, forming a contour map of the scanned area. The code in the recorded lines indicates precisely the amount of density change and shows whether the density is increasing or decreasing.

Microdensitometer traces can be made along selected isodensitracing

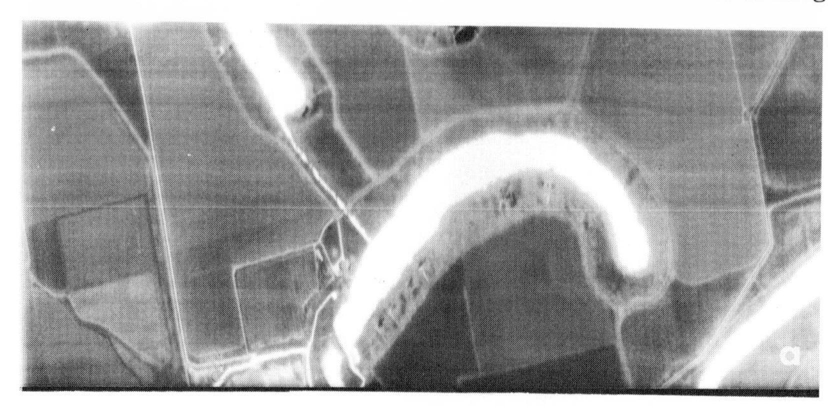

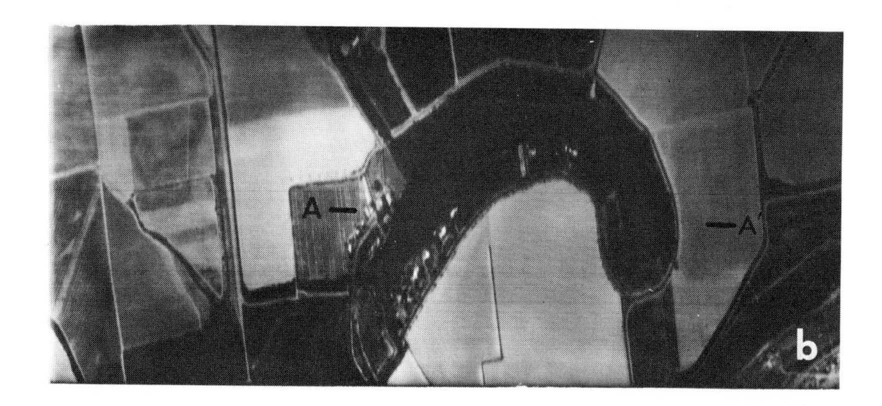

FIGURE 24 Diurnally flown thermal imagery (8–14 μ) of the area shown in Plate 21. June 1, 1966, Weslaco, Texas. *a,* 6:00 A.M.; *b,* 2:00 P.M.; *c,* 7:00 P.M. For microdensitometer traces of *A–A′* and *B–B′*, see Figure 26.

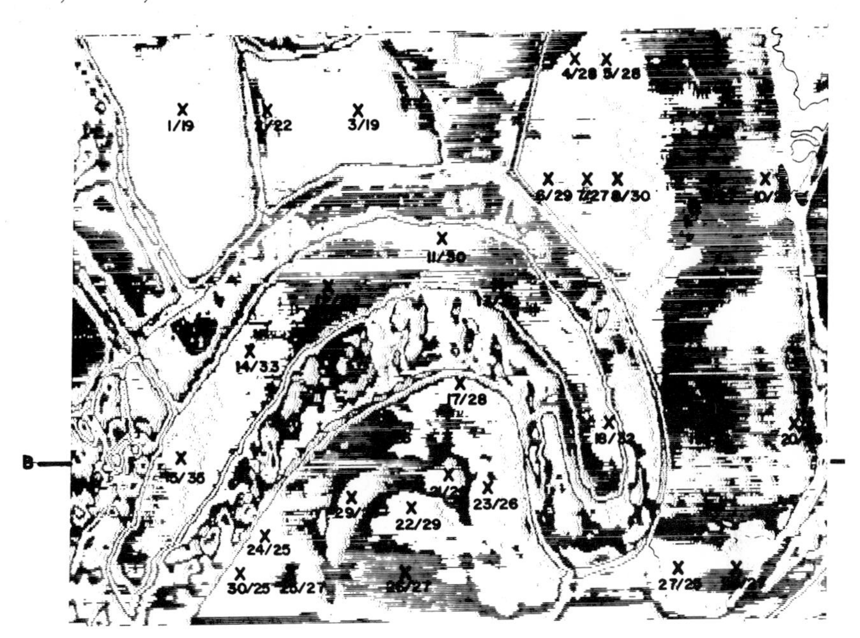

FIGURE 25 Isodensitracing, Moon Lake area, in the lower Rio Grande Valley, Texas, June 1, 1966, 8–14 μ, 7:00 P.M., 2,000 ft. Numbers indicate site number/degree centigrade.

scan lines, and the results can then be presented as *X-Y* coordinate traces of film density along those lines. Figure 26 shows two microdensitometer traces along transects shown on Figure 24.

Temperatures at 30 selected sites were determined and are shown on the isodensitracings of Figure 25. The significance of these temperatures in relation to soils is briefly summarized as follows:

1. Thermal imagery of soils has the unique characteristic of being influenced by subsurface characteristics. Equivalent blackbody temperatures of the surface soil in one of the fields in Figure 24 were shown to be related to profile characteristics at depths of about 0–50 cm.
2. Soil-moisture contrasts in bare soil (Figure 24) produced a 3° C difference between surface soil temperatures as observed on imagery flown at 6:00 A.M.
3. A soil-moisture deficit in part of a cropped area had a temperature higher by 7° C than the portion of the field with adequate moisture.
4. Diurnal surface soil temperatures in a dry field varied by an average of 48° C on June 1, 1966. At the same time, surface tempera-

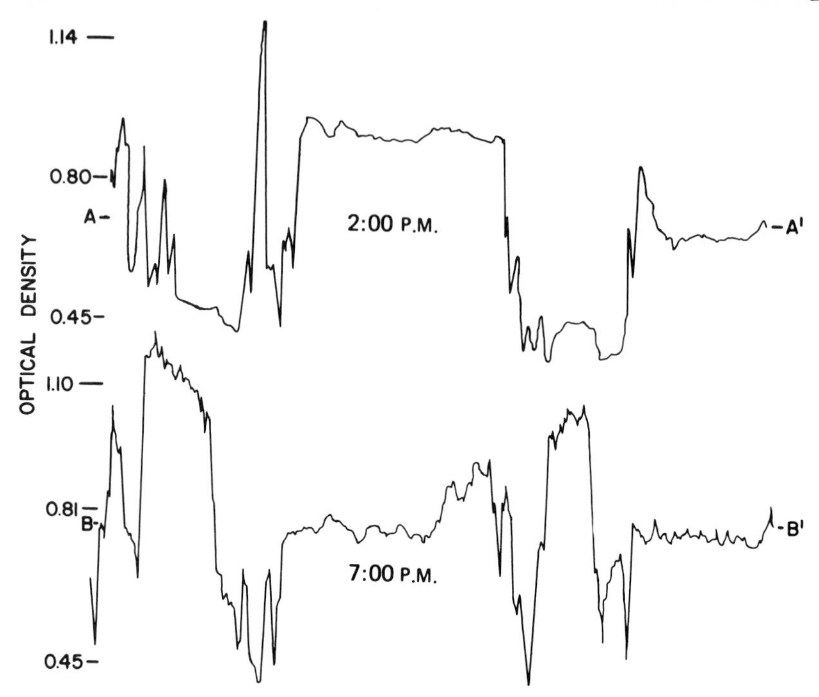

FIGURE 26 Microdensitometer traces for transects *A–A′* and *B–B′* of Figure 24.

tures varied an average of 41° C in a field with a higher soil moisture content.

Factors Affecting Diurnal, Seasonal, and Annual Temperature Changes

1. The size of soil particles has a considerable influence on heat transfer. Experiments conducted by Chudnovskii[42] show a variation of 400 percent in thermal conductivity of three kinds of soil: sand, chernozem loam, and podzolic loam. The more the soil is dispersed, the lower its calorimetric conductivity. (Chudnovskii uses the terms "dispersion" and "particle size" synonymously.) The experiments show that the greater the content of clay particles relative to sand, the smaller the thermal conductivity for the solid phase.

2. Soil moisture and soil air humidity are the most important factors influencing thermal characteristics of the soil. As a soil becomes more moist, its conductivity increases, but the increase becomes gradually less marked as moisture content continues to rise. This is due to the circumstance that changes of temperature conductivity K depend on

simultaneous changes in thermometric conductivity and in volumetric heat capacity c_p. ($K = \lambda/c_p$, where λ is wavelength.)

At low moisture levels, the main factor is the ratio between the calorimetric conductivity of water and soil, which produces a sharp increase in the thermometric conductivity. At higher moisture levels, a second factor appears, namely, the ratio between the specific heats of water and soil. This ratio has no effect at low moisture content. The overwhelming majority of soils in the dry state have a specific heat about 0.2 that of water. Consequently, the higher the moisture content of a soil, the greater its specific heat and the lower its thermometric conductivity. Furthermore, for every type of soil, the specific volume decreases with increase of moisture.

3. As porosity increases, the air content also increases, and soil conductivity decreases, but the calorimetric conductivity of air is a hundred times lower than that of soil materials. Figure 27 gives the generalized results of a number of experiments conducted by Chudnovskii[42] and others. The plotted points show the dependence of conductivity on porosity and indicate the same hyperbolic picture for sand and loam. The solid line is a theoretical curve obtained by using methods described by Chudnovskii[42] (after Bogomolov).

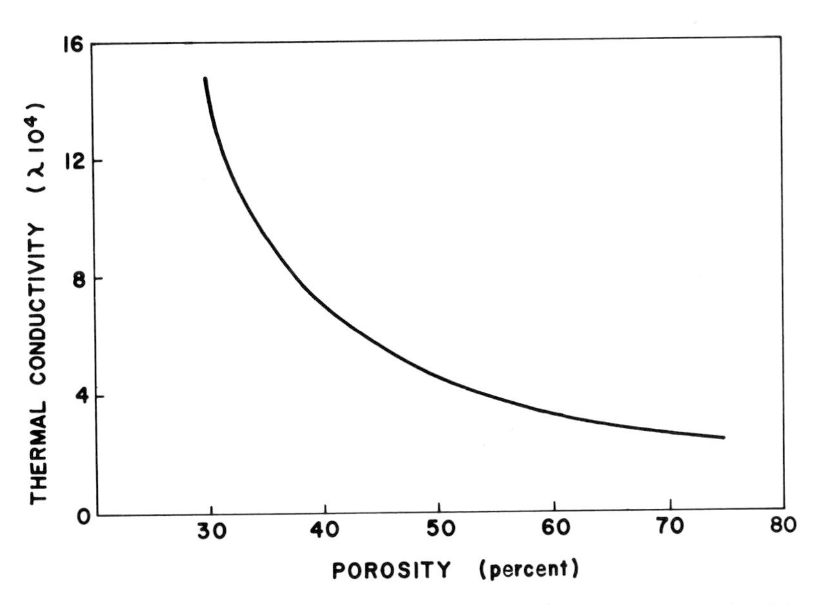

FIGURE 27 Relation between soil porosity and coefficient of thermal conductivity. (*After Chudnovskii.*)

To summarize: porosity of natural soils is high, about 30 to 40 percent in virgin soils and 40 to 60 percent or more in cultivated soils. As can be seen from Figure 27, the variation of porosity in cultivated soils will have an effect on the thermal characteristics, though the influence will not be as great as that of particle size and soil moisture.

Factors Influencing Interpretation

1. The transfer of heat in the soil profile, as well as addition and loss of heat, is complex because a soil's thermal conductivity depends on its moisture content, which changes with depth and time. Furthermore, during the warm season, the distribution of temperatures in the soil is determined by the thermal properties of the upper soil layer, which may be a poor conductor of heat, depending on cultivation practices.

2. In dry weather a high temperature gradient within a superficial surface layer may be due to decreased temperature conductivity of this layer, caused by lesser compactness and lower moisture content.

3. The presence of a crop affects the soil temperature. A grass cover does not greatly alter the mean daily and annual soil temperature, but it does reduce the daily and annual range.

4. In spring, clayey soils are cooler than sandy ones; in autumn, the reverse is true.

Factors Influencing Time of Sensing

1. The soil profile depth of damping of constant daily temperature and that of constant annual temperature are in the ratio of the square roots of oscillation periods. Since these are 24 hours and 1 year, the depth of damping of annual oscillation exceeds the depth of damping of diurnal oscillation by a factor of 19.

2. During daylight the variations of heat exchange follow those of solar radiation. The time at which heat exchange becomes negligible coincides more or less with sunrise and sunset. Thermal sensing of soils at night, in contrast to daytime sensing, shows greatest surface contrasts indicative of profile conditions. After sundown, heat flow is upward and varies with soils, depending on their respective heat-storage capacities and thermal conductivities.

3. The times of maximum and minimum daily temperature are retarded with depth. The retardation is proportional to depth. The amplitudes decrease with depth. As depth increases in arithmetic progression, the amplitude decreases in geometric progression and is damped at a certain depth.

The amplitude of soil-temperature oscillations according to depth depends on the composition of soils. In granite, which has the highest temperature conductivity, the amplitude is least at the surface and greatest at depth 60 cm. In sand, which has a much lower temperature conductivity, the amplitude is relatively greater at the soil's surface.

4. Diurnal oscillations of temperature in a moist soil are less than those in a dry soil. Thermal imagery produced over an area of relatively uniform soil may show temperature differentials at a time after rainfall when there is a moisture differential between well-drained and poorly drained topsoils.

5. The annual course of heat exchange in soil shows certain peculiarities. The greatest positive heat exchange occurs in spring and in early summer; the greatest negative heat exchange occurs in early winter. Imagery produced by thermal sensing in the late fall or early winter, depending on latitude and likely weather conditions, can be most productive for detecting soil conditions by taking advantage of uniform crop and soil-moisture conditions that exist at that time of year over wide areas.

References

1. Colwell, R. N. 1956. Determining the prevalence of certain cereal crop diseases by means of aerial photography. Hilgardia 26(5):223–286.
2. Gates, D. M., H. J. Keegan, J. C. Schleter, and V. R. Weidner. 1965. Spectral properties of plants. Appl. Opt. 4(1):11.
3. Keegan, H. J., J. C. Schleter, W. A. Hall, Jr., and G. M. Haas. 1955. Spectrophotometric and colorimetric study of foliage stored in covered metal containers. U.S. Dep. Com., Nat. Bur. Standards Rep. 4370. Noc. (AD 84 923).
4. Myers, V. I., L. R. Ussery, and W. J. Rippert. 1963. Photogrammetry for detailed detection of drainage and salinity problems. Trans. Amer. Soc. Agr. Eng. 6(4):332–334.
5. Olson, C. E., Jr., R. E. Good, C. A. Budelsky, R. L. Liston, and D. D. Munter. 1964. An analysis of measurements of light reflectance from tree foliage made during 1960 and 1961. Office of Naval Res. NR-387-025. Dep. For., Agr. Expt. Sta., Univ. of Illinois, Urbana.
6. Olson, C. E., Jr. 1964. Spectral reflectance measurements compared with panchromatic and infrared aerial photographs. Univ. of Michigan, Ann Arbor. (AD 603 499).
7. Thomas, J. R., V. I. Myers, M. D. Heilmand, and C. L. Wiegand. 1966. Factors affecting light reflectance of cotton. *In* Proc. 4th Symp. Remote Sensing of Environment, Inst. Sci. Tech., Univ. of Michigan, Ann Arbor. p. 305–312.
8. Monteith, J. L. 1965. Radiation and crops. Exp. Agr. Rev. I. p. 241–451.
9. Myers, V. I., D. L. Carter, and W. J. Rippert. 1966. Remote sensing for estimating soil salinity. J. Irrig. Drain. Div., Amer. Soc. Civ. Eng. 92(IR4):59–68.

10. Taylor, S. A. 1958. The activity of water in soils. Soil Sci. 86:83–90.
11. Ehlig, C. F., and W. R. Gardner. 1963. Relationship between transpiration and the internal water relations of plants. Agron. J. 56:127–130.
12. Nieman, R. H. 1965. Expansion of bean leaves and its suppression by salinity. Plant Physiol. 40:156–161.
13. Myers, V. I., C. L. Wiegand, M. D. Heilman, and J. R. Thomas. 1966. Remote sensing in soil and water conservation research. *In* Proc. 4th Symp. Remote Sensing of Environment. Inst. Sci. Tech., Univ. of Michigan, Ann Arbor. p. 801–813.
14. Snedecor, G. W. 1956. Statistical methods. 5th Ed. Iowa State Univ. Press. p. 129–131.
15. Thomas, J. R., and C. L. Wiegand. In press. Osmotic (salinity) and matric suction effects on relative turgidity and temperature of cotton leaves. Soil Sci.
16. Baker, D. N. 1966. The microclimate in the field. Trans. Amer. Soc. Agr. Eng. 9:77–81, 84.
17. Cox, L. M., and L. Boersma. 1967. Transpiration as a function of soil temperature and soil water stress. Plant Physiol. 42:550–556.
18. U.S. Salinity Laboratory Staff. 1954. Diagnosis and improvement of saline and alkali soils. U.S. Dep. Agr. Handbook No. 60.
19. Kramer, P. J. 1959. Transpiration and the water economy of plants. Plant Physiol. Academic Press. 701 p.
20. Gates, D. M. 1964. Leaf temperature and transpiration. Agron. J. 56:273–277.
21. Raschke, K. 1960. Heat transfer between the plant and the environment. Ann. Rev. Plant Physiol. 11:111–120.
22. Curtis, O. F. 1936. Leaf temperatures and the cooling of leaves by radiation. Plant Physiol. 11:343–364.
23. Gates, D. M. 1963. Leaf temperature and energy exchange. Arch. Meterol. Geophys. Bioklimatol. 12:321–336.
24. Gates, D. M. 1963. The energy environment in which we live. Amer. Sci. 51:327–348.
25. Clum, H. H. 1926. The effect of transpiration and environmental factors on leaf temperature. I. Transpiration. Amer. J. Bot. 13:194–216.
26. Smith, A. M. 1909. On the internal temperature of leaves in tropical insolation. Ann. Roy. Bot. Gard. Peradeniya 4:229–298.
27. Pallas, J. R., Jr., and D. G. Harris. 1964. Transpiration, stomatal activity and leaf temperature of cotton plants as influenced by radiant energy, relative humidity, and soil moisture tension. Plant Physiol. 39:xliii (Abstr.)
28. Tanner, C. B. 1963. Plant temperatures. Agron. J. 55:210–211.
29. Mellor, R. S. 1962. Influence of environmental factors on the temperature and energy transfer mechanisms of plant leaves. Ph.D. Thesis, Colorado State Univ., Ft. Collins (Univ. Microfilms, Inc., Cat. No. 63-1403, Ann Arbor, Mich.).
30. Shaw, R. H. 1954. Leaf and air temperatures under freezing conditions. Plant Physiol. 29:102–104.
31. Namken, L. N. 1964. The influence of crop environment on the internal water balance of cotton. Soil Sci. Soc. Amer. Proc. 28:12–15.
32. Namken, L. N. 1965. Relative turgidity technique for scheduling cotton (*Gossypium hirsutum*) irrigation. Agron. J. 47:38–41.
33. Brooks, F. A. 1964. Agricultural needs for special and extensive observations of solar radiation. Botan. Rev. 30:263–291.

34. Gates, D. M., and W. Tantraporn. 1952. The reflectivity of deciduous trees and herbaceous plants in the infrared to 25 microns. Science 115:613–616.
35. Beatty, F. D. 1965. Geoscience potentials of side-looking radar. Raytheon/Autometric Corp., Alexandria, Va., 90 p.
36. Lundien, J. R. 1966. Terrain analysis by electromagnetic means: radar responses to laboratory prepared soil samples. U.S. Army WES Tech. Report 3-639, Report 2. 55 p.
37. Simonett, D. S., J. E. Eagleman, A. B. Erhart, D. C. Rhodes, and D. E. Schwarz. 1967. The potential of radar as a remote sensor in agriculture: 1. A study with *K*-band imagery in western Kansas. CRES Report 61-21. 13 p.
38. Shul'gin, A. M. 1965. The temperature regime of soils. English translation from Russian. Published for U.S. Dep. Agr. by the Nat. Sci. Found., Washington, D. C., by the Israel Program for Scientific Translation, Jerusalem.
39. Baver, L. C. 1956. Soil Physics. 3rd Ed. John Wiley, New York. p. 363.
40. Jackson, R. D., and S. A. Taylor. 1965. Heat transfer. Methods of soil analysis. Agron. Monograph No. 9, Academic Press, New York. p. 349–360.
41. Carson, J. E. 1961. Soil temperature and weather conditions. Argonne National Laboratory, Argonne, Illinois.
42. Chudnovskii, A. F. 1962. Heat transfer in soil. English translation from Russian. Published for the Nat. Sci. Found. and U.S. Dep. Agr. by the Israel Program for Scientific Translation, Jerusalem.

7

Field Spectroscopy

Roger A. Holmes School of Electrical Engineering, Purdue University, West Lafayette, Indiana

Introduction

Field spectroscopy is the general term describing the techniques, instruments, and considerations necessary to obtain reflectance, emittance, and radiance spectra of vegetation and soils in their natural environment. The last is the salient point: crops are viewed on a large or a small scale as they grow in the field under ambient micrometeorological conditions; soils are viewed as they are, plowed, harrowed, disked, and so on. The measurements are made under natural irradiance with all the variability associated with cloud cover, a point that must be considered in calibration efforts. The advantage of field spectroscopy to the agriculturalist who desires *in situ* measurements is obvious: he cannot transport samples to the laboratory spectrophotometer, mount them there for a specific measurement, and hope to reproduce the many variables of the natural environment. On the other hand, many of the niceties of optical design built into the laboratory instrument and taken for granted are not present in the field—the solar irradiance at the scene of view replaces the lamp as a source, for example. This means that the optics of the data-gathering process must be carefully considered for the data to be meaningful.

The discussion that follows is concerned with the wavelength range from 0.35 to 16 μ. Atmospheric absorption rules out other portions of the optical region electromagnetic spectrum in a practical sense. The instrumentation of radar or microwave spectroscopy is sufficiently different to warrant separate treatment and is discussed in Chapter 3. X-ray and gamma-ray spectroscopy are also special cases.

There are natural divisions within the 0.35–16-μ-wavelength range from a field spectroscopy point of view. From 0.35 to about 4 μ, the scene viewed reflects direct and scattered solar irradiance. From about 4 μ to the end of the 8–14-μ atmospheric window, the scene emits radiation as a graybody, with some emittance variation as a function of wavelength. There is some overlap of reflective and emissive behavior in the vicinity of 4 μ. These aspects pose different calibration and view-angle problems in each spectral region and have a strong influence on instrument design choices.

The discussion of field spectroscopy opens with what may be called exterior parameters, those points to be taken into account assuming an adequate instrument is at hand. This leads naturally to the problems of instrument calibration in the face of the variability of field experiments. This, in turn, requires consideration of the scan time, spectral resolution, and sensitivity of the field spectrometer. These features, together with signal-to-noise ratio properties, are discussed in conjunction with specific wavelength ranges that are of interest in agriculture. Finally, there is a brief review of instrument types available and under design. It should be emphasized at this point that most instruments that have been used in the field to date are not stock catalog instruments—they have been designed specifically for field spectroscopy, often as prototypes. The existence of a well-accepted, time-tested line of instruments, such as in laboratory spectrometers, spectrographs, and spectrophotometers, is still some years and a considerable amount of user demand in the future. In this vein, it is mandatory that an investigator making field spectral measurements have more than a casual acquaintance with instrumentation. As a closing point in the section on instrument types, various forms of data readout are discussed with a view to what the experimenter wishes to accomplish; a balance is sought between the time required to gather data and the time necessary to analyze these data.

Geometrical Parameters of Field Measurements

Several vital features of accurate field spectroscopy can be discussed with only the more gross specifications of instrument performance kept in mind. These specifications are field of view, effective aperture, focusing capabilities, and polarization sensitivity.

Field of view is best defined in terms of the procedure used to measure this quantity. In order to map out the field of view of a particular instrument, a small radiation source is located on the nominal optical axis of the instrument. The source may be a tungsten lamp radiating in the visible and near-infrared portions of the spectrum. From about 3 to 16 μ, a hot object such as a soldering iron behind an aperture plate will suffice, or direct viewing downward onto a surface of known emissivity cooled to 77° K in a small Dewar half-filled with liquid nitrogen will serve the purpose. While the instrument is held fixed and focused on the source, the source is moved in measured amounts in the plane perpendicular to the optical axis, and the relative instrument signal output is monitored. In this way loci of constant instrument response can be found. Simple trigonometry converts loci positions into degrees off the instrument axis. The resulting intensity profiles can be changed into power contours through appropriate calculations. Some typical field-of-view calibrations for instruments used in the field are shown in Figure 1.

"Effective aperture area" is the cross-sectional area, at the instrument's optical entrance, of the bundle of rays emanating from a small source in the field of view. The word "effective" is used to indicate that only those rays that proceed to a detector in the instrument are included. The larger the effective aperture, the more radiation is collected from the sources in the field of view.

The focusing capabilities of an instrument depend on the specific task at hand. General-purpose instruments should have a focusing capability from about 1 m to infinity. As a general rule, however, if the scene is much closer than 1 m (and the field of view is small), it may be more reasonable to transport a standard spectrophotometer to the field location or to design a special-purpose arrangement of detectors and optical components than to insist on stringent short-focus requirements for a general-purpose instrument.

Reflex optics enabling the experimenter to determine both the scene of view and instrument focus just prior to measurement are most desirable.

In most spectrometers there will be some sensitivity to the polarization of the incident radiation (*cf.* discussion in Chapters 3 and 9). Whenever radiation reflects from or refracts through a medium such as a mirror or grating surface or prism, the intensity after this event will generally be a function of the polarization of the incident electromagnetic waves. Since experiments have been reported[1] indicating reasonably strong polarization effects in radiation reflected from vegetation and soils, polarization sensitivity of an instrument is an im-

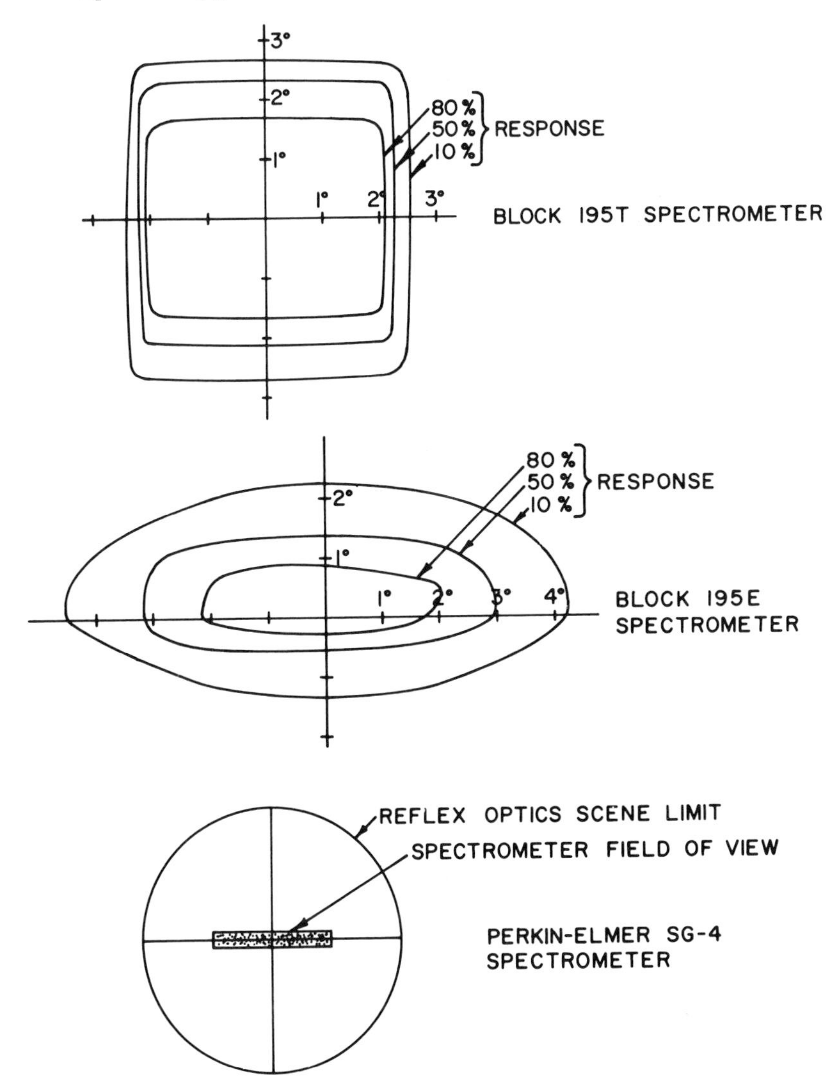

FIGURE 1 Fields of view for three spectrometers currently used in field spectroscopy. For the Perkin-Elmer SG–4, the spectrometer field of view at 10 m viewing distance is a 21 × 3-cm rectangle.

portant parameter in fieldwork. At least one instrument currently in use in field spectroscopy is extremely sensitive to polarization degree and direction. This is not a drawback, provided the instrument's polarization axis is known; in fact, this feature allows measurement of

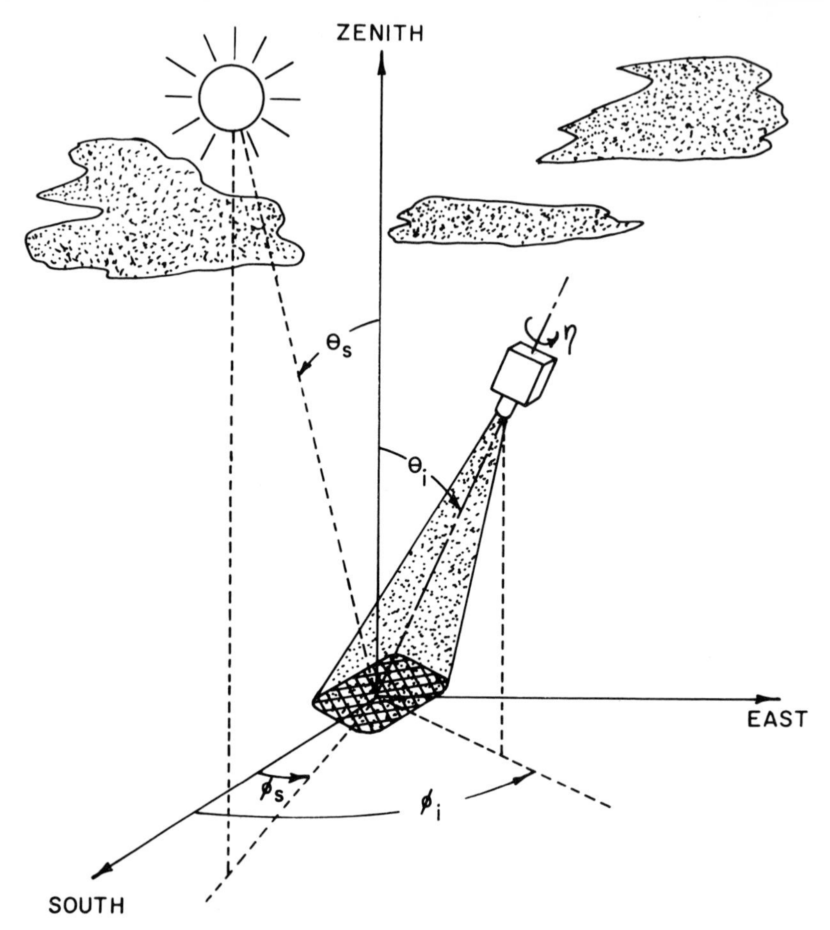

FIGURE 2 Geometrical parameters at the field spectroscopy site. θ_s = angle of sun's altitude; ϕ_s = angle of sun's azimuth; θ_i = angle of instrument's altitude; ϕ_i = angle of instrument's azimuth; η = optical axis.

polarization properties of natural objects, although the experiment is more involved than a simple intensity measurement at a given wavelength. In the visible and near-infrared regions it is not difficult to calibrate an instrument in this respect with a bright source and a commercial polarizer. Polarizers are also available for the emissive range.

The general experimental conditions are drawn in Figure 2. Clouds are included to emphasize the effect they may have on the experiment. The field of view of the instrument at the ground plane is shown by cross-hatching. Important angles are those concerned with the altitude and azimuth of both the sun and the instrument (θ_s, ϕ_s and θ_i, ϕ_i), the orientation angle of the instrument with respect to its optical axis, η, and the lay of any row structure in the crop or soil being observed. Sun angles are measured directly or are computed from the time of observation and latitude–longitude of the test site, using navigational tables.

To emphasize the geometric effects one may observe, and illustrate the need for knowledge of the geometric parameters, curves drawn from data taken in three experiments are shown in Figures 3–5. These data were taken in the summer of 1966 at the Purdue University Agronomy Farm, with a Perkin-Elmer SG-4 rapid-scan grating spectrometer. The field of view of this instrument is extremely small even at large viewing distance, as can be seen by reference to Figure 1. Figure 3 shows the effect of a small field of view on data gathered on a windy day. The spectrometer was set for a fixed wavelength and was held on a fixed scene for many seconds. The "noisy" variations in the response are the result of leaf motion in the field of view about 11.5 m

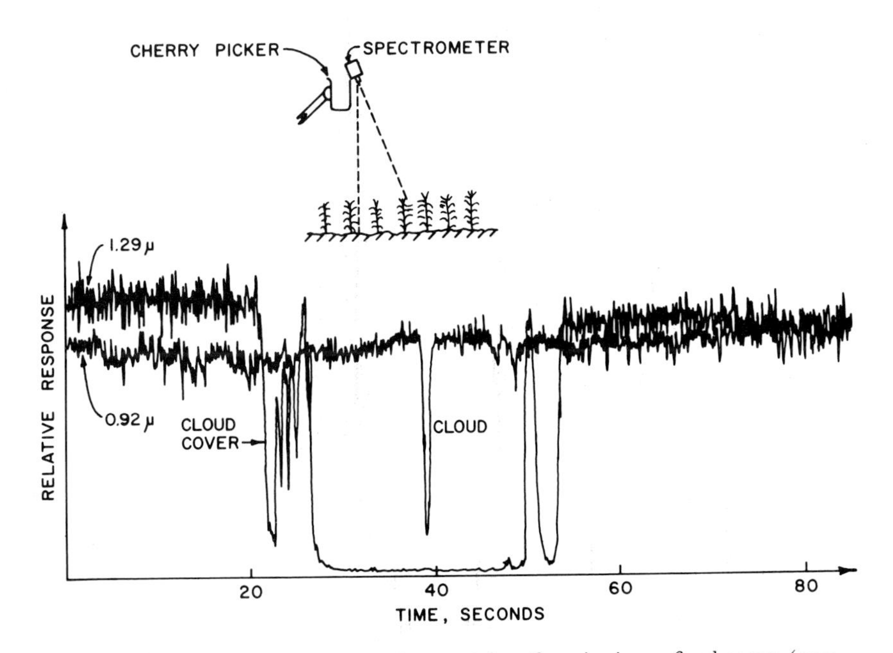

FIGURE 3 Spectrometer response at two wavelengths, viewing a fixed scene (cornfield) for a period of time. $\theta_i = 20°$. Day gusty. Rapid leaf motion in field of view, spectrometer mounted in cherry picker at field's edge.

from the instrument aperture. Further, the gross effects of scattered cumulus cloud cover are manifest. Note that significant change due to cloud cover takes place in a few seconds or less.

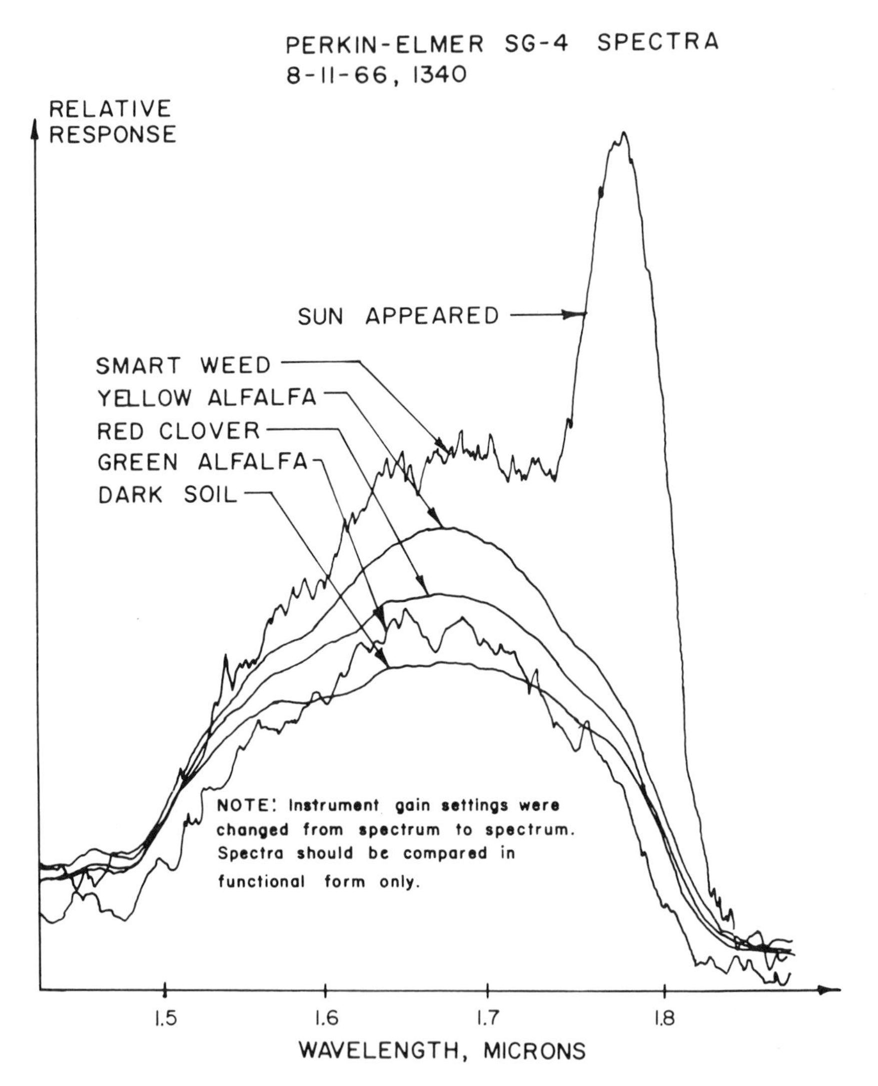

FIGURE 4 Relative spectral radiance of five subjects, showing the effect of solar irradiance variation in the reflective region, August 11, 1966, 1:40 P.M. Instrument gain settings were changed from spectrum to spectrum. Curves should be compared in functional form only.

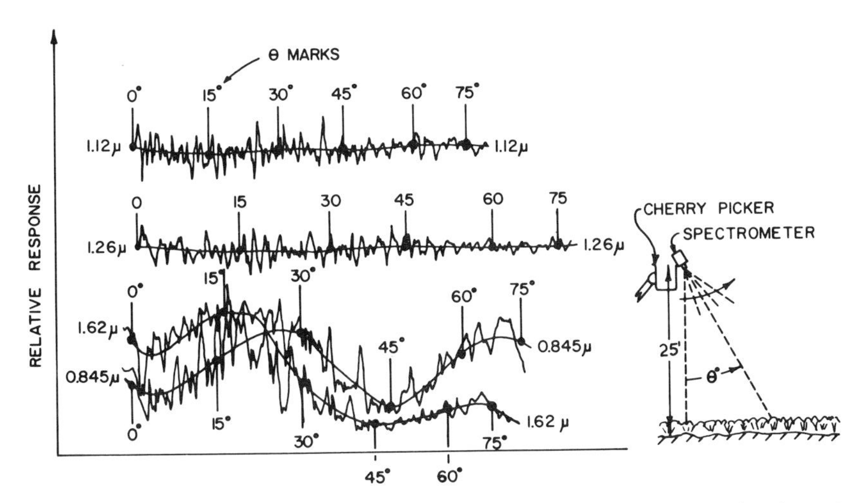

FIGURE 5 Angular sweeps along a soybean row at four fixed wavelengths, August 3, 1966, 1:40 P.M. The spectrometer was manually panned from $\theta_i = 0°$ to $\theta_i = 75°$ at each wavelength setting. Smooth lines through the response curves are drawn in to show average trends.

Figure 4 shows several spectra taken in the 1.5–1.8-μ region. The day was one of light overcast, with the sun occasionally breaking through for short intervals. The total spectral scan time was 15 sec, and the curve shows that even in this region of the spectrum, cloud-cover variability can affect results in such short times as a second or so, to the extent of at least doubling the instrument response.

Figure 5 shows measurements at several fixed wavelength settings of the Perkin-Elmer spectrometer. In this experiment the instrument was manually panned out along a soybean row four times, once at each wavelength setting. The day was clear, so cloud cover was not a factor. Again, the minor variations of the recording are due to the small field of view with respect to the crop's leaf size. At some wavelengths (1.12 and 1.26 μ) the response appears to be independent of view angle; at others, there is significant variation as θ_i is increased.

Most of the instrument parameters can be collected into a concise notation that clarifies what is being measured and how much power is flowing into the instrument aperture. The amount of power flowing into the instrument aperture in the wavelength range λ to $(\lambda + \Delta\lambda)$ is

$$\Delta P = \tau_a(\lambda) N_\lambda A_a \Omega_f \Delta\lambda \qquad \text{(watts)}, \tag{1}$$

where A_a is the effective instrument aperture, Ω_f is the solid angle of the field of view, and τ_a is the transmission in the atmospheric path from the scene to the instrument aperture. The spectral radiance (N_λ) of an incremental surface area is best defined in an operational manner; refer to Figure 6. The radiant power flow from a source of incremental surface area dS into a solid angle increment $d\Omega$ about a given direction from the source surface normal is measured over the wavelength range λ to $(\lambda + d\lambda)$. This radiant power flow measurement is divided by the increment $d\Omega$, the projection of dS in the direction of the measurement, and the wavelength increment $d\lambda$; the result is the spectral radiance N_λ. (See Kingslake[2] for detailed definitions of radio-

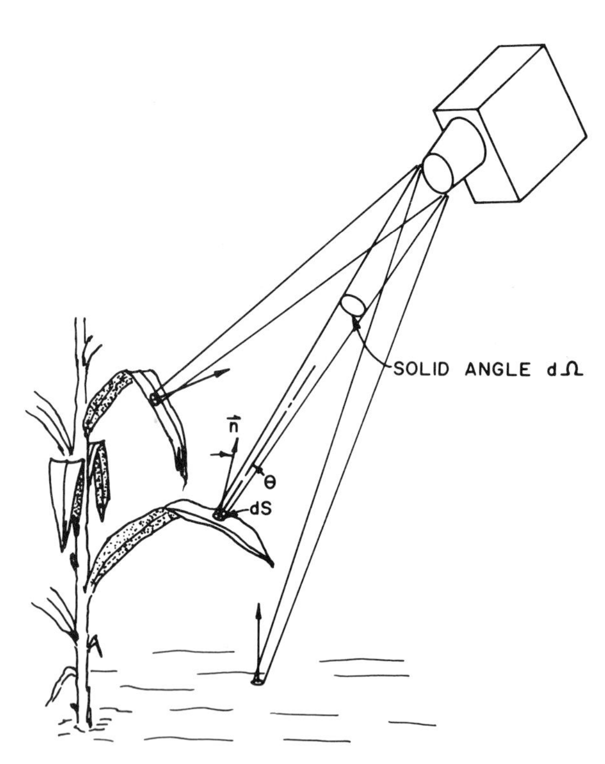

FIGURE 6 Power flow from a surface element dS into a solid-angle increment $d\Omega$ in the wavelength range λ to $(\lambda + d\lambda)$ is $\Delta P = N_\lambda \cos \theta \, dS \, d\Omega \, d\lambda$. The spectral radiance is N_λ, generally a function of θ and λ. The irradiance at the spectrometer aperture comes from a variety of surface types over a broad range of angles with respect to surface element normals, $\vec{n}$.

metric quantities.) The quantity N_λ in Equation 1 is the average spectral radiance over the field of view of the spectrometer. It is averaged in two senses, both over all the sources within the field of view and over the great variety of angles between surface normals and the instrument's optical axis. The relation between $\overline{N}_\lambda$ over a heterogeneous scene in the field of view, such as a one-meter square of cornfield, and normal-incidence reflectance measurements on one corn leaf segment in a laboratory spectrophotometer is not well understood at present. Indeed, the study of the scattering and emission of radiation from rough or heterogeneous surfaces is an active research discipline in its own right.

Still, the experimenter is faced with the fact that the power flow into the field spectrometer is a measure of $\tau_a \overline{N}_\lambda$ simply multiplied by A_a and Ω_f, parameters of the designer's choice. Some general functional aspects of $\overline{N}_\lambda$ can be stated and kept in mind when field spectroscopy data are interpreted. For the reflective region (refer to Figure 2), $\overline{N}_\lambda$ is expected to be a function of θ_s, ϕ_s, θ_i, ϕ_i, atmospheric conditions, and the detailed geometry of the scene viewed. In the emissive region the same variables are important during the day, with the obvious simplification at night. In the reflective region, assuming that direct solar irradiance is dominant, $\overline{N}\lambda$ is the product of the solar power per unit area coming into the scene ($H_{\lambda n} \cos \theta_s$) in the $\Delta\lambda$ range and the bidirectional reflectance for the scene. This bidirectional reflectance is averaged in the same involved sense as $\overline{N}_\lambda$. Then

$$\Delta P = \tau_a H_{\lambda n} \cos \theta_s \, [\rho_\lambda(\theta_s, \phi_s; \theta_i, \phi_i)] A_a \Omega_f \, \Delta\lambda \qquad \text{(watts)}, \qquad (2)$$

where $\tau_a H_{\lambda n} \cos \theta_s$ = incident radiation modified by atmospheric conditions

$\cos \theta_s [\ ]$ = angular effects and detailed scene geometry

$H_{\lambda n}$ = normal direct solar spectral irradiance

ρ_λ = bidirectional reflectance

If effects of sky irradiance are to be included, it is necessary to add an additional term to $\overline{N}_\lambda$ involving a hemispherical integral over the sky radiance distribution.

If the experimenter's major purpose is to correlate spectra with aircraft scanner data, then $\overline{N}_\lambda$ is the term of interest, for this is what an airborne reflective region scanner is sensing. On the other hand, if the experimenter's major purpose is to correlate field spectra with laboratory spectrophotometer data on natural objects, then $\overline{\rho}_\lambda$ is the term of interest, for it is the hemispherical integral of this quantity that is measured in the laboratory. Equation 2 shows clearly that in order to interpret field spectra in terms of reflectance, it is essential to measure concurrently the direct solar irradiance $H_{\lambda n}$ and the sky irradiance if that term is added.

In the emissive region an approximate form for $\overline{N}_\lambda$ can be written, based on radiation theory. Thus, the power flow into the instrument aperture is

$$\Delta P = \tau_a 1.19 \times 10^{-4} \lambda^{-5} \overline{\epsilon(\lambda, T) [\exp(14{,}338/\lambda T) - 1]^{-1}} A_a \Omega_f \Delta\lambda, \tag{3}$$

where λ is in microns and T is absolute temperature in degrees Kelvin. The constants can be derived from the standard Planck equations. If the scene were an isothermal Lambertian radiator of single emittance value, the quantity $\overline{\epsilon(\lambda, T)}$ would be the same as the conventional emittance, ϵ, of the radiator. Even for nearly idealized flat isothermal radiators, there is a variety of emittance quantities such as total, hemispherical, normal, and relative (angular) emittances.[3] In a heterogeneous, nonisothermal scene of a variety of sources at many angles, it is best to report the average radiance $\overline{N}_\lambda$. The writer prefers to report $\overline{N}_\lambda$ in terms of an "equivalent blackbody temperature," T_{BB}, defined by

$$\overline{N}_\lambda \equiv \frac{1.19 \times 10^4}{\lambda^5 [\exp(14{,}338/\lambda T_{BB}) - 1]}.$$

The equations just given suffice to estimate the order of magnitude of power flowing into an instrument, in addition to providing concise statements of the variables.

It was pointed out that the reflectance or emittance of a natural object as measured in the laboratory is difficult to apply in the prediction of radiance spectra from a heterogeneous field scene, and vice versa. This fact places a burden on the experimenter in the field to describe accurately both the field conditions and the instrument parameters in reporting his data. The description should include all the angles of Figure 2 and the obvious instrument variables; in addition, well-selected close-up and instrument-range photographs of the detailed scene geometry are mandatory for adequate spatial description of the experiment. It is only through the correlation of many sets of such carefully defined field spectra that behavioral trends will emerge that will ultimately correlate with laboratory experiments and aid in the interpretation of aircraft scanner flight data.

INSTRUMENT CALIBRATION

Calibration has long been a subject of investigation by workers in spectroscopy and radiometry. The power flowing into the aperture of a field spectrometer generally passes through foreoptics into the wavelength-determining elements of the instrument. Following some form of wavelength discrimination, the power in the spectral range λ to $(\lambda + \Delta\lambda)$ falls on a detector that transduces the optical power to an electrical signal. Electronic amplification of this signal follows and finally results in an instrument response or output, usually a voltage signal. The first step in instrument calibration is to determine how the response is related to power flow into the aperture.

Standard quartz–iodine lamps of known spectral irradiance are commercially available. For such a source, the number of watts per unit area per unit wavelength increment at a certain distance from the lamp is given by the supplier. Thus, the instrument to be calibrated by viewing a standard lamp is set up in the laboratory, and the power flowing into the aperture in the wavelength range λ to $(\lambda + \Delta\lambda)$, call it $\Delta P(\lambda)$, is known. The instrument can vary its wavelength selection, and the response, $V(\lambda)$, typically in volts per watt, is recorded as a function of wavelength. It is now reasonable to define an instrument responsivity $R(\lambda)$, by

$$[V(\lambda)]_{\text{std lamp}} = R(\lambda)\ [\Delta P(\lambda)]_{\text{std lamp}}. \tag{4}$$

Now, having the responsivity, if the response from a scene is recorded in the field, the result is

$$[V(\lambda)]_{\text{scene}} = R(\lambda)\ [\Delta P(\lambda)]_{\text{scene}}; \tag{5}$$

and simple algebra shows that

$$[\Delta P(\lambda)]_{\text{scene}} = \frac{[\Delta P(\lambda)]_{\text{std lamp}}}{[V(\lambda)]_{\text{std lamp}}}\ [V(\lambda)]_{\text{scene}}. \tag{6}$$

Each quantity on the right-hand side of Equation 6 is known; hence $[\Delta P(\lambda)]_{\text{scene}}$ is determined.

This simple process will suffice for calibrating the instrument to determine $\overline{N}_\lambda$ only if one major assumption is made: that the instrument is invariant under the environmental change from the air-conditioned, low-humidity laboratory to the ambient conditions in the field. The full import of this statement has been detailed by Nicodemus.[2] At least three different spectrometers used in the field at present require this assumption to be made since they possess no internal calibration source. Spectra reduced to relative spectral radiance curves by this laboratory-lamp comparison method are shown in Figures 7–9. The relative spectral radiance is found because gain settings in the signal amplifier must be different for the bright lamp source as compared with the lesser field radiance. The responsivity is changed by a constant factor for all wavelengths as gain is changed: the quantity $[\Delta P(\lambda)]_{scene}$ of Equation 6 is multiplied by this constant factor. For absolute measurements of the scene radiance, all instrument settings should be identical for both laboratory and field, or they should be determined precisely for both laboratory and field and taken into account in the reduction arithmetic.

A serious problem prevents the use of this laboratory calibration

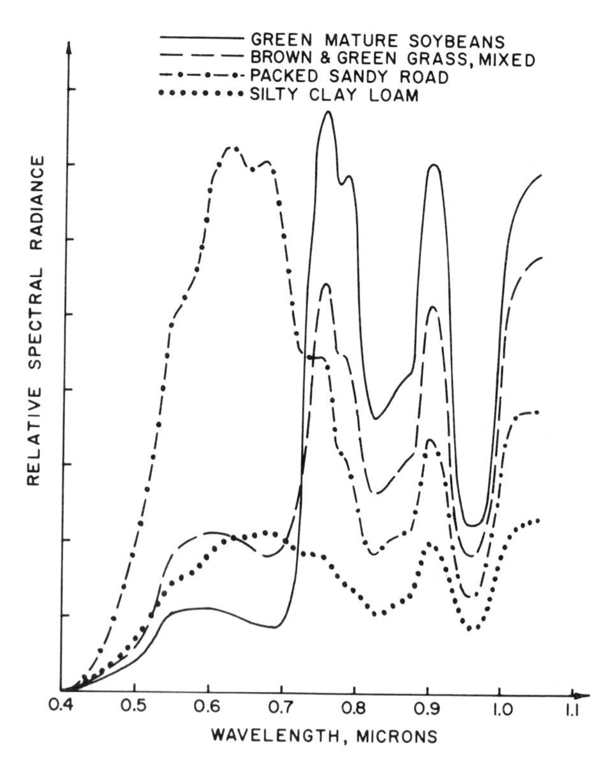

FIGURE 7 Relative spectral radiance spectra for agricultural scenes in the range 0.4–1.05 μ, August 30, 1966, 11:50–11:56 A.M. Curves should be compared in functional form only; gain settings were changed between spectra.

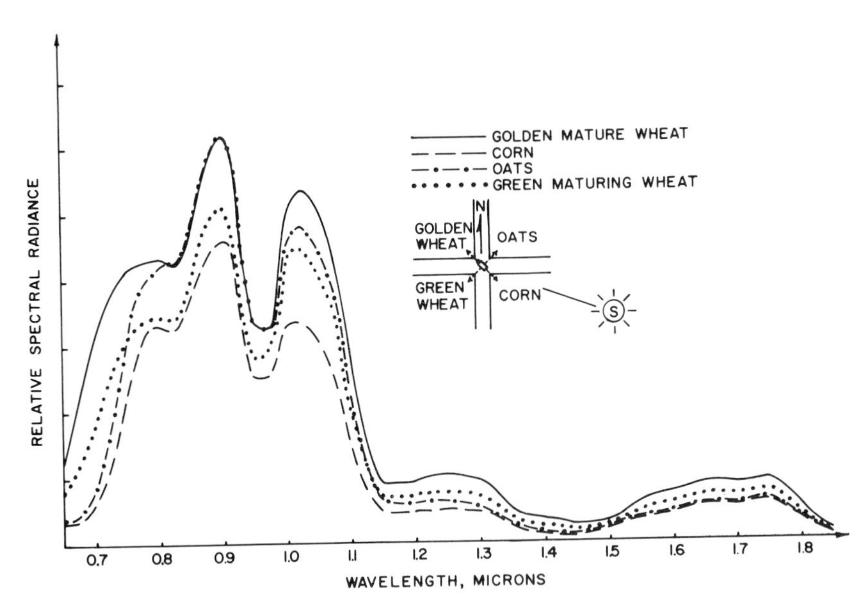

FIGURE 8 Relative spectral radiance spectra for agricultural scenes in the range 0.65–1.85 μ, June 30, 1966, 10:15–10:20 A.M. Curves should be compared in functional form only; gain settings were changed between spectra.

technique in the emissive range unless special precautions are taken. Although standard blackbody sources are commercially available to provide the emissive-range equivalent of the spectral lamp in the laboratory, the environmental changes from laboratory to field and the details of instrument design play significant roles in the spectrometer output. The spectrometer detector will "see" (interchange radiation with) the instrument optics and mechanical parts as graybodies at ambient temperature. Through careful design the detector can be made to view only a limited area of surface of known temperature monitored by thermistor sensors. Without going into instrument details at this point, suffice it to say that an emissive-range spectrometer will always contain a graybody that competes with the scene viewed in exciting the detector. This effect may be exploited by including a self-contained blackbody calibration source within the spectrometer. Finally, it is possible to employ a standard calibration source at the scene location; this was done in presenting the data shown in Figure 10.

The next calibration technique is related to the discussion of the previous paragraphs and is of interest when the calibration source is built into the spectrometer. The detector alternately views the scene and the calibration source at a rapid rate, on the order of hundreds of cycles per second. The resultant detector signal is electronically demodulated to yield a response that measures the difference between scene response and source response:

$$[V(\lambda)]_{\text{scene}} - [V(\lambda)]_{\text{source}} = R(\lambda)\{[\Delta P(\lambda)]_{\text{scene}} - [\Delta P(\lambda)]_{\text{source}}\}. \quad (7)$$

For the reflective range, $R(\lambda)$ can be found by simply covering the aperture of the spectrometer:

$$R(\lambda) = \frac{[V(\lambda)]_{\text{source}}}{[\Delta P(\lambda)]_{\text{source}}}. \quad (8)$$

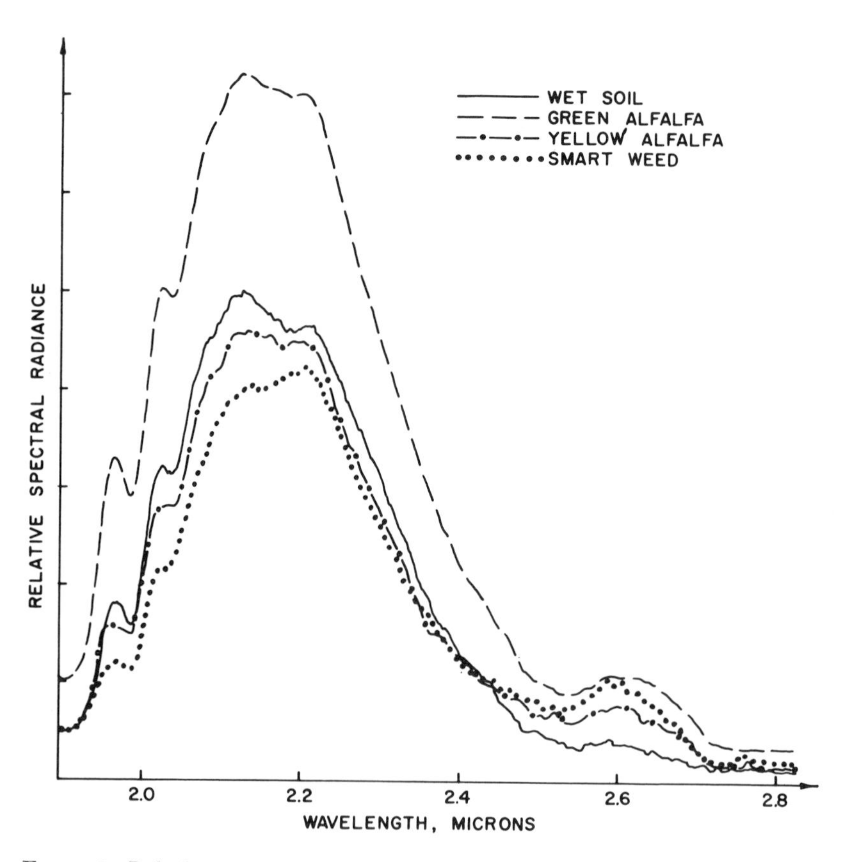

FIGURE 9 Relative spectral radiance spectra for agricultural scenes in the range 1.9–2.8 μ, August 11, 1966, 12:04–12:15 P.M. Curves should be compared in functional form only; gain settings were changed between spectra.

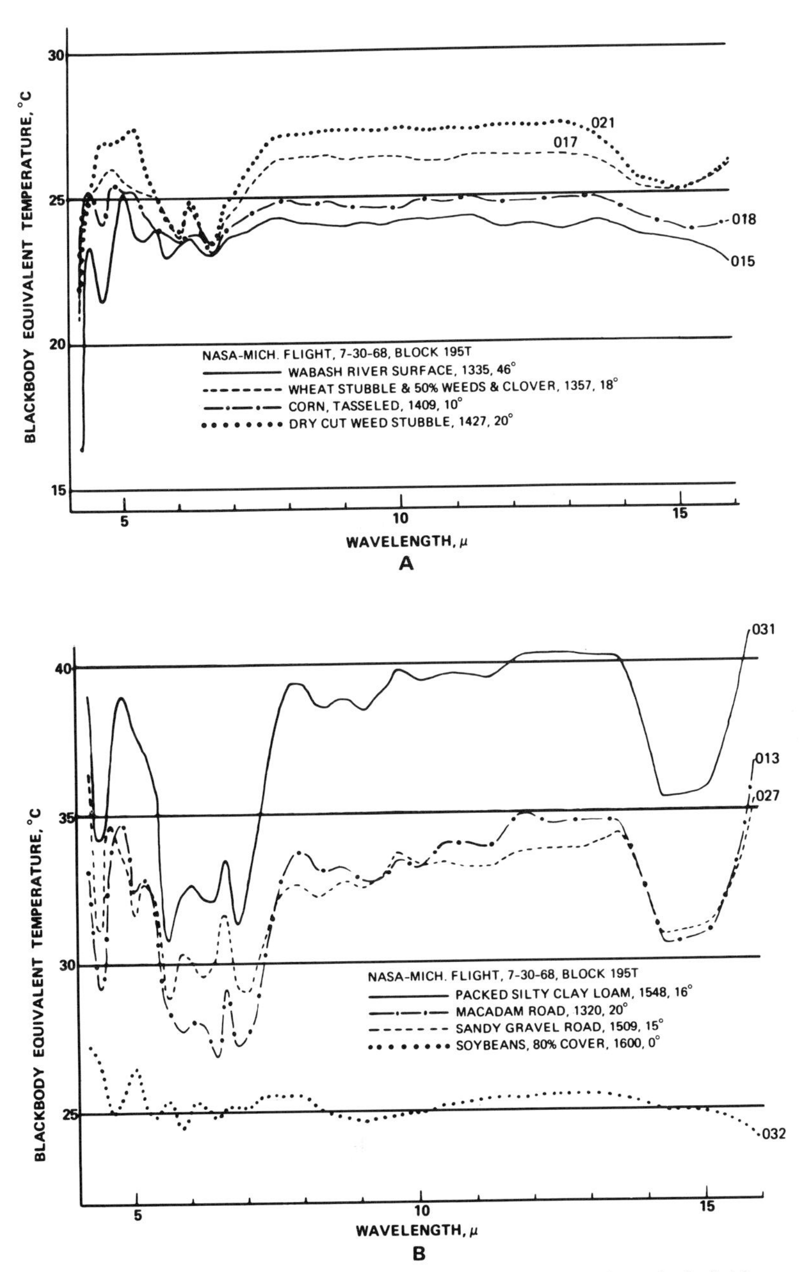

FIGURE 10 Blackbody equivalent temperature spectra of typical field scenes. Angles indicated are θ_i of Figure 2. Note, in *B*, the reststrahlen structure in the vicinity of 8.5–9.3 μ for mineral composite targets.

If Equation 8 is substituted into Equation 7, the result is

$$[\Delta P(\lambda)]_{scene} = \frac{[\Delta P(\lambda)]_{source}}{[V(\lambda)]_{source}} [V(\lambda)]_{scene}, \qquad (9)$$

which shows that this method is formally identical with the laboratory-source calibration. The important practical distinction of this method, however, is that the instrument is viewing the source and the field scene in the same environment. If the spectrometer transfer function is changed, either for source or for scene viewing, the effect is subtracted out by the differential character of the measurement.

The internal source is always present in the emissive-range instrument, whether designed into the spectrometer or not. It is customary design practice to have the detector "see" alternately the scene radiance and the radiance of an internal blackbody of controlled, or at least measured, temperature. Equation 9 applies; thus a differential measurement is again employed, removing instrument component variation effects. The resultant output, when reduced, gives the radiance of the scene and intervening atmosphere times the atmospheric transmittance. Even if the atmospheric transmittance is known, the measurement does not give the emittance-like variable, $\overline{\epsilon(\lambda, T)}$, unless the average surface temperature is known for the objects in the field of view. The average surface temperature is one of the most difficult measurements to make correctly, even for a homogeneous scene of uniform material. The reasonable course at the present time is to report apparent scene radiance if the atmospheric path effects are known; otherwise, aperture irradiance is reported.

A third calibration method consists of placing a material of known reflectance or a blackbody source of known temperature near the scene. If the spectral scanning rate of the instrument is sufficiently fast, a spectral scan of the scene is followed immediately by a scan of the reflectance standard or the blackbody. In the reflective region this amounts to a comparison of scene reflectance with the standard, providing the solar and skylight irradiance does not change during the total time required to take both spectra. In the emissive range this method essentially eliminates the effect of atmospheric transmission between the scene and the spectrometer. For both the reflective and emissive ranges, one of the major difficulties of the technique is the required size of the standards to fill the field of view. The effect of too small a field of view in gathering spectra of field vegetation is shown in Figures 3 and 5. If the field of view is large enough to average fine-grained detail, the standards will have to be of corresponding size.

A second difficulty with the reflectance standard or the blackbody is the difficulty of assuring the quality of the standard in the field environment. A large blackbody source must be maintained at uniform temperature in the face of variable solar irradiance, wind velocity, humidity, and so on. The reflectance standard is subject to subtle surface changes due to interaction of the surface material with ambient atmospheric gases. Further, to avoid time-consuming calibration of the reflectance standard in a spectrophotometer, it is desirable to have a standard that is stable and easily repaired when damaged. Standards tested in the summer of 1966 included white poster board, 3M* white velvet paint on poster board, barium sulfate slurries in a variety of vehicles, and flowers of sulfur. The flowers of sulfur standard proved to be workable. A panel was constructed on a 30 × 30-in. piece of quarter-inch wallboard. Alternate coats of 3M white velvet paint as a binder, and flowers of sulfur were applied to a thickness of 0.25 in., the last sulfur layer being quite thick. The entire surface was flat pressed in a large hydraulic press. A protective layer of wallboard over the sulfur surface and normal care in handling the panel resulted in a standard that held up well in the field over the summer. The panel reflectance was nearly that value reported in a test of sulfur as a standard,[4] and was measured with respect to magnesium oxide on a DK-2 spectrophotometer. A spectrum reduced with respect to this panel is shown in Figure 11.

One modification of the reflectance-panel method is feasible. A small reflectance standard can be placed close to the instrument aperture; its spectrum is then measured immediately following measurement of the scene spectrum. This method has the advantage of small panel size and the disadvantage of not including atmospheric transmission effects over a distance of approximately twice the scene-to-spectrometer length. The disadvantage is relatively minor and the method is the only one practical for an instrument with a large enough field of view to average out vegetation scene details.

The reflection-standard calibration process results in data suitable for comparison with laboratory spectrophotometer data. Aircraft or satellite data give a measure of irradiance at the scanner aperture, and these data include both solar irradiance and reflectance. At the present stage of remote sensing, ideal field spectroscopy should include both scene radiance measurements calibrated against a built-in source and measurements from a reflectance panel or an independent spectrometer measuring solar irradiance separately.

*Minnesota Mining and Manufacturing Co., St. Paul, Minn.

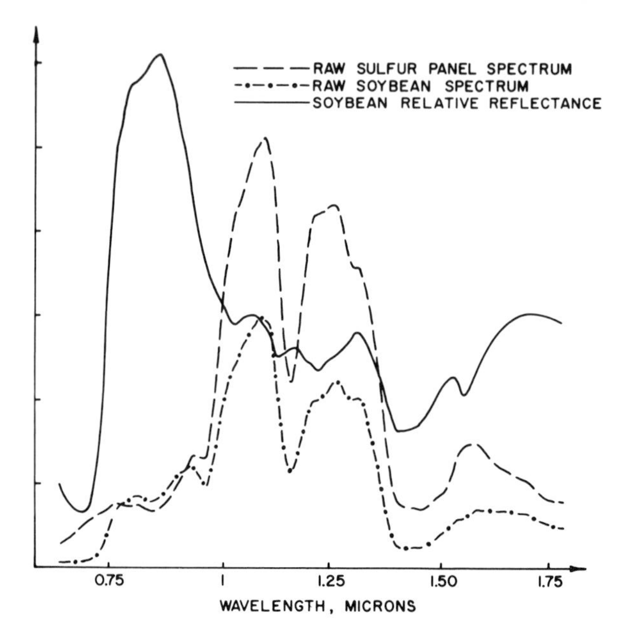

FIGURE 11 Reflectance of soybeans, August 3, 1966, 11:06 A.M. The two raw spectra include the instrument transfer function, which cancels in taking the ratio of crop to standard panel response over the wavelength range. Instrument gain was reduced for the sulfur panel spectrum.

For further information on calibration, see the broad bibliographies in the *Handbook of Military Infrared Technology*.[5]

INTERNAL INSTRUMENT PARAMETERS

In addition to the geometrical parameters already covered, any field spectrometer design will have to include other important characteristics: sensitivity, dynamic range, signal-to-noise ratio, resolution, and spectral scan rate. Although these details concern the engineer, it is important for the field experimenter to have an understanding of how they affect his work and his choice of instrument.

Sensitivity is mainly a function of the choice of detector for a given optical system of a certain aperture and field of view. The voltage, current, or conductance variation from a detector per unit of electromagnetic power incident must be sufficiently large to overwhelm noise signals inherent in any detector. Applicable discussion of specific detector types is given in Chapter 3. In a practical vein, the relation between sensitivity and field spectroscopy comes to a focus on the annoying but essential requirements of detector cooling. As a general rule, rapid-scan spectroscopy of reasonable resolution will involve liquid nitrogen or high-pressure gaseous nitrogen for cooling in the range 3–5.5 μ, and liquid helium cooling in the range 5.5–16 μ, although $Hg_xCd_{1-x}Te$ detectors, which have recently become available, will cover much of the range 5.5–16 μ at liquid nitrogen temperature. For the visible and near-infrared regions, uncooled detectors often suffice. Cooling problems are not insurmountable, but they do impose the requirements of coolant availability and handling techniques.

The *dynamic range* of an instrument is a measure of the ability to take spectra of both bright and dim scenes. One usually wants to make maximum use of the linear ranges of the detector signal amplifier and the recording instrumentation (X-Y plotter, tape recorder, oscilloscope, and so on). For most common field scenes on a clear day, the gain of the spectrometer amplifier is set to a fixed value such that the irradiance from the brightest scene gives a signal just below the saturation (overload) level of the amplifier and recording instruments. If the day is cloudy, this method of manual gain control requires repeated resetting of the gain. Electronic automatic gain control can be built into the spectrometer circuitry, but field experience to date has shown that dynamic range problems are not severe, and manual gain control is sufficient for the normal variety of agricultural scenes.

Signal-to-noise ratio, resolution, and *scan rate* are interrelated parameters. Each parameter represents a desirable property that the experimenter would like to have on as large a scale as possible. A single example is sufficient to show the interrelation and the resultant constraints. Figure 12 shows the structure of the Perkin-Elmer SG-4 spectrometer, and is self-explanatory. The signal-to-noise power ratio, S/N, is proportional to the width of the entrance slit, w, and inversely proportional to the square root of the bandwidth of the detector signal amplifier, Δf. Thus,

$$S/N = \text{Constant} \times w/\sqrt{\Delta f}. \tag{10}$$

The resolution of this instrument is limited by the entrance slit width rather than the grating size, considering the normal slit widths used in the field. The resolution, r, is the ratio of wavelength, λ, to the wavelength difference, $\Delta\lambda$, of two spectral lines at λ that can just be distinguished from one another. In the spectrometer of Figure 12,

$$r = \text{Constant}/w, \tag{11}$$

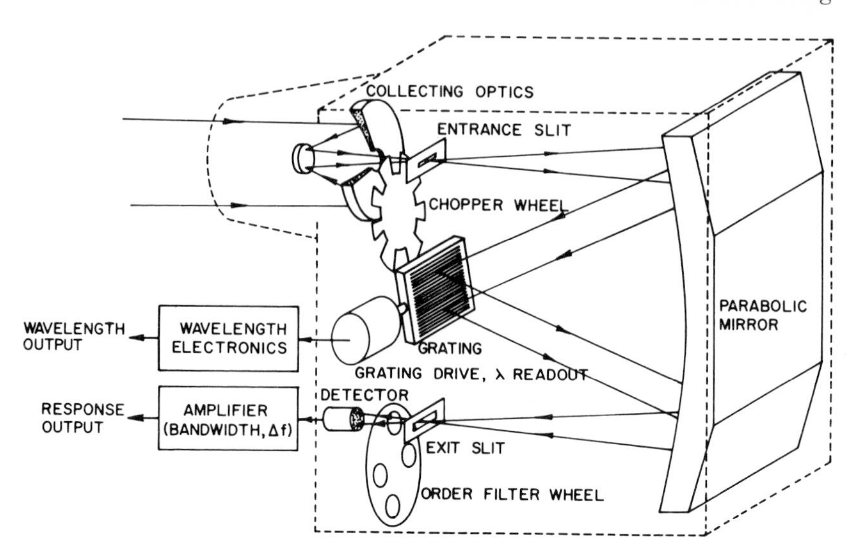

FIGURE 12 A typical field spectrometer system. The diagram is based on the Perkin-Elmer SG-4 spectrometer, with several details of practical importance omitted.

which holds only for slit-limited resolution. The scan rate, F, is limited by the ability of the detector amplifier to respond to sudden changes in detector irradiance, and this ability is determined by the amplifier bandwidth, Δf. Therefore,

$$F = \text{Constant} \times \Delta f. \tag{12}$$

When Equations 10, 11, and 12 are combined, it is clear that

$$(S/N)\,(r)\sqrt{(F)} = \text{Constant}. \tag{13}$$

This relation shows that for a given design one may trade resolution for scan rate, or signal-to-noise ratio for resolution, and so on. Since there is usually a minimum acceptable signal-to-noise ratio, Equation 13 really amounts to a design constraint between scan rate and resolution.

The demonstrated rapid variability of the spectral radiance of an agricultural scene warrants emphasis on high spectral scan rate at the expense of resolution. This is consistent with the irradiance-limited low resolution of aircraft scanners. Spectrometer manufacturers tend to emphasize resolution at the expense of scan rate for the demands of general spectroscopy.

Wavelength Range

Thus far it is apparent that field spectroscopy calls for a rapid-scan spectrometer with a large field of view; two scans per second and a 15° field of view are current requests in instrumentation procurement. In considering the wavelength ranges that field experience has shown to be of interest, the reader should refer to Figures 7–10 for graphic support. Further, a data compilation of both laboratory and field measurements is available.[6]

The range 0.4–1.1 μ has a fascinating variety of spectra from agricultural scenes. In the near infrared, the high reflectance of turgid vegetation compared with soils stands out clearly. The physiological variable, color, is contained in the spectral curves from 0.4 to 0.7 μ. The entire range certainly contains information of value in remote sensing and should be studied with as high a resolution as possible that is consistent with scan-rate requirements.

The range 1.1–2.0 μ may contain information in the amplitudes of the pronounced peaks between the 1.14-, 1.38-, and 1.87-μ water-absorption bands. Spectral curve shape, on the other hand, seems to be the same for a large variety of targets. This is in contrast to the curves for the 0.4–1.1-μ region. It is reasonable to assume that a very gross spectrometer of small spectral resolution capable of measuring only the total irradiance between each absorption band will be necessary here.

Spectral character in the 2.0–2.6-μ window and the 3.0–4.1-μ window is similar to that in the 1.1–2.00-μ region. Details of spectral shape within these windows do little to enhance target identification, though there may be information in the total power in each window.

Spectroscopic capability in the 4.5–5.5-μ window was not available to agricultural researchers until late autumn 1966 for this wavelength range. One reported spectrum on grass indicates typical graybody emitter behavior, as expected.[7] Total power measurements in this window are available on natural objects that confirm this general tendency to appear as graybodies. Even if significant fine spectral structure of the effective scene emissivity $[\overline{\epsilon(\lambda, T)}]$ were known to be present, there is a large amount of atmospheric band structure in this window range that may obscure interpretation. Recent low-spectral-resolution (0.1-μ) data in this range are shown in blackbody equivalent temperature form in Figure 10.

Total power measurements are common in the 7–16-μ wavelength range, to be interpreted as radiation emanating from a graybody. There are recent data[8,9] showing that reststrahlen absorption bands can be detected in minerals; these may be of value in soil type evaluation. Still, the major application for information on radiance in this window is expected to be surface temperature radiometry for both crops and soils. This is borne out by the data of Figure 10. Interpretation of apparent temperatures so measured is discussed in Chapter 5.

Instrument Types and Data Readout

The specific spectrometer system of Figure 12 illustrates the general features of a broad class of field instruments. First, there are collecting optics or foreoptics, which usually determine the field of view (in conjunction with an entrance slit or field stop) and the effective aperture. Details of collecting optics are available in standard optics texts. Next, there is a wavelength-selecting unit, for example, the grating monochromator. The radiation passing through the wavelength-selecting portion goes on to the detector, possibly through a detector-optics system. Detector types are similar for most instruments and are discussed in Chapter 3. Finally, there are the electronic signal-processing units, details of which are outside the scope of this chapter.

The manner of wavelength selection is the distinguishing feature of field spectrometer designs. The most common monochromators employ a grating or prism as the dispersing element. These monochromators are amply discussed in standard spectroscopy texts such as that by Sawyer.[10] In fieldwork over the range 0.35–16 μ, several gratings or prisms are required to cover the spectrum. If a single grating-mount or prism-mount instrument is utilized, it is necessary to change dispersing elements several times in the field environment in order to cover the spectral range on a chosen scene. One expensive way around this difficulty is an instrument with several gratings (or prisms) and dichroic mirrors to steer gross spectral ranges to the appropriate gratings. An alternative solution is the utilization of rapid-change modular monochromators with convenient plug-in gratings. Such monochromator units are commercially available (Bausch & Lomb), but the collecting optics and detector arrangement must be tailored to the unit.

Another common wavelength-selection method is the use of interference filters that pass only a narrow spectral range on to the detector. A large selection of fixed-range filters may be mounted on a wheel and alternately placed in the optical path, essentially sampling the spectrum at the wavelengths of the filters. An interesting variation on this idea is the continuous filter, with wavelength selection determined by the location of the optical path on the filter surface. One instrument, the ISCO Model SR spectroradiometer, utilizes a continuous filter laid down in a line on a glass substrate for the spectral range 0.35–1.05 μ. Continuous filters laid along the circumference of a wheel are available from Optical Coating Laboratories, Inc. These filter wheels cover the spectral range 1.2 μ to well beyond the 16-μ limit discussed in this chapter. Several instrument proposals based on the filter wheel are being studied by remote-sensing researchers, and one filter wheel instrument for aircraft use was designed and built at Goddard Space Flight Center of the National Aeronautics and Space Administration. Filter wheels below 1.2 μ are currently under development.

Still another form of wavelength selection utilizes interferometer techniques, commonly based on the Michelson or Fabry-Perot interferometers, or Soleil compensator–polarizer combinations. In this scheme, different wavelengths of the incoming spectrum are converted to different audio frequencies at the detector, the amplitude of each frequency depending on spectral power inflow at the corresponding wavelength. The output might be called a "spectral symphony"; mathematically it is the Fourier transform of the input spectrum. This places on the experimenter the burden of performing the inverse Fourier transform to regain the original spectrum, either at general computational facilities or in an available special-purpose computer. Three Block Engineering, Inc., interferometer spectrometers are currently in use in agricultural field spectroscopy.

Whatever the particular design features of an instrument, the agricultural experimenter is always faced with deciding how to record his data in the light of subsequent reduction problems. Most instruments have outputs suitable for connection to conventional recorders such as *X-Y* plotters or strip charts or oscilloscope–camera combinations. This recording mode is sufficient when a relatively small number of spectra are to be studied. The labor of manually reducing these data is minor compared with the expense and debugging effort necessary to establish a more sophisticated data-analysis system. On the other hand, if the experimenter intends to record field spectra over a growing season on a wide variety of crops and soils in order to establish statistical trends, the result will be hundreds or thousands of spectra. In this case, the more sophisticated data-handling system becomes a necessity, not a luxury. Analog or digital tape recording in the field is the mode of data gathering, and facilities are required for later playback and analysis of the tapes. An interesting report on data-handling problems met by geophysical research workers in remote sensing is available.[11] It is good policy to gather data on a modest scale when one is first experimenting with a new field spectrometer, carefully checking for data consistency, calibration reliability, and proper instrument functioning with a reasonable number of simply recorded spectra.

Future Field Spectroscopy Plans

The field spectroscopy techniques discussed thus far result in a single value of instrument response (R_λ) for a given scene and given wavelength interval, or equivalently, one spectrum for the entire area covered by the field of view. As pointed out, this response can be reduced to an average scene radiance, a most involved average in terms of the multiplicity in variety and angular variation of the objects in the scene.

In the research work in the field, the prime question is how the overall reflectance or emittance of the scene in, say, a 2×2-m patch can be interpreted in terms of what is there, including geometry, spatial layout, and variety of surfaces. This, in turn, requires a knowledge of the location in the scene of the various levels of radiance, N_λ, for wavelengths of interest. For example, a cornfield scene will include tassels, brightly irradiated top leaves, dimly illuminated lower leaves, stalks, and soil between rows, all of which contribute to the integrated radiance measured by the spectrometer.

One type of spectral instrument that by its very nature will yield a spatial map of the scene is the multilens camera. This instrument takes several simultaneous photographs of the same scene on a single plate, each through a bandpass filter of chosen wavelength limits. The resulting photos give a good qualitative view of the distribution of spectral radiance levels over the scene, but the problems of readily analyzing this distribution are significant, requiring accurate film calibration and a flying-spot scan of the photographic plate. In addition, the experiment is limited to the spectral range covered by the emulsion sensitivity.

A field spectrometer can be made that will have a relatively small field of view of about 5 mrads or less. The area covered by this field of view can be scanned over a scene such as the 2×2-m patch by a rotating mirror system in a rectangular, televisionlike raster. The outputs in various spectral bands would be detected and recorded in parallel, simultaneously, on magnetic tape. This tape could then be played back to create a digital line-printer image in the same manner that digital printouts from airborne scanners are now produced at the Laboratory for Agricultural Remote Sensing at Purdue University. In fact, the field spectrometer necessary to perform this imaging-type measurement could be a reasonable modification of an airborne scanner system. Such a scanning field spectrometer is as technologically feasible today as a simple field spectrometer. What is needed is the impetus, the time, and money necessary for successful development. The research necessary to put remote sensing in agriculture on an operational basis has provided the impetus for field spectrometer development.

References

1. Coulson, K. L. 1966. Effects of reflection properties of natural surfaces in aerial reconnaissance. Appl. Opt. 5:905–917.
2. Kingslake, R. [ed.]. 1967. Applied optics and optical engineering. Academic Press, New York. See especially Vol. 4, Chap. 8, Radiometry, by F. E. Nicodemus.
3. Rutgers, G. A. W. 1958. Temperature radiation of solids. *In* Handbuch der Physik, Springer-Verlag, Berlin. Vol. XXVI. p. 129.
4. Kronstein, M., R. J. Kraushaar, and R. E. Deacle. 1963. Sulfur as a standard of reflectance in the infrared. J. Opt. Soc. Amer. 53:458–465.
5. Wolfe, W. L. [ed.]. 1962. Handbook of military infrared technology. Office of Naval Research, Library of Congress Catalog Card No. 65-62266.
6. Earing, D. Target signature analysis center: data compilation. Willow Run Labs., Inst. Sci. Tech., Univ. of Michigan, Ann Arbor. (A continuing series under Projects 7850 and 8492, beginning July 1966.)
7. Fredrickson, W. R., and N. Ginsberg. No date. Infrared spectral emissivity of terrain. Dept. of Physics, Syracuse Univ., Syracuse, N. Y. ASTIA Doc. 132 829.
8. Hovis, W. A., Jr. 1966. Infrared spectral reflectance of some common minerals. Appl. Opt. 5:245–248.
9. Hovis, W. A., Jr. 1966. Optimum wavelength intervals for surface temperature radiometry. Appl. Opt. 5:815–818.
10. Sawyer, R. A. 1963. Experimental spectroscopy. Dover, New York.
11. Lyons, R. J. P. 1966. Field infrared analysis of terrain. 1st Annual Report. NASA NGR-05-020-115. Geophysics Dept., Stanford Univ., Stanford, Calif.

8

Sampling and Statistical Problems

BRUCE W. KELLY Statistical Reporting Service,
U.S. Department of Agriculture, Washington, D. C.

INTRODUCTION

The most exciting prospect for utilization of remote sensing in agriculture is its application throughout a generation rather than for one crop season. Man's food, directly or indirectly, is derived from vegetation. The surface of the earth contains a finite number of acres suitable for food, grass, and forest production. To those neo-Malthusians concerned with and alarmed by the population explosion with respect to the world's food supply, remote sensing offers a means of monitoring and maintaining an over-all reckoning of the capacity for food production. The development of a remote-sensing device capable of classifying and measuring the vegetative cover of broad bands of the earth's surface offers a new dimension in the collection of statistics pertaining to food production. In addition to providing an overview, output from such a device has many interesting prospects for incorporation into surveys supporting statistical programs in countries in all stages of development. However, utilization of the full potential of sensor data will probably require a restructuring of sample and survey designs now used by the agricultural statistician, as well as new techniques for interpreting sensor data to provide sensitive measures of prospective food supply.

In this chapter, I shall attempt to present the background against which to view the potentials of remote sensing in agricultural statistics. To do this and to place remote sensing in perspective as a device for collecting data pertaining to crop acreages, we shall look briefly at the state of development of the concepts of sampling and then at three programs of current agricultural statistics and their systems of supporting surveys.

For the former, I shall attempt a minimal mathematical sketch of sampling as applied in agriculture. In this, I shall try to emphasize how the theory has evolved to a point at which it utilizes information from sources other than the sample and thereby has increased efficiency. For the latter, I shall describe briefly the programs and methodologies that could be or are being employed in Nepal, an undeveloped country; in Turkey, a developing country; and in the United States.

CONCEPTS OF SAMPLING

Definition of Terms

For the nonstatistician working with remote-sensing information, it will be helpful to know the meaning of some of the vocabulary associated with sampling before employing it in developing the concepts of sampling.

Sampling, in the simplest terms, is the selection of a part (sample) of an aggregate of material to represent the entire aggregate (universe). Examples are the spoonful of soup from the pot, the handful of grain from the bin, the people interviewed in taking an opinion poll, the bearings selected from the production of a manufacturing process, and a group of farmers selected for interviews about farm production. The "aggregate of material" may be thoroughly mixed, as soup and grain, or quite similar, as a well-regulated manufacturing process or the members of the same political party with respect to preference for a candidate for election. When the aggregate is thoroughly mixed or quite similar for irrelevant reasons, the process of selecting the sample need not be elaborate; almost any part selected will represent the whole. However, if the material constituting the aggregate consists of dissimilar units that are not well mixed, a small sample is unlikely to represent the aggregate. The greater the dissimilarity (variability) among units, the greater the likelihood that a sample of given size will not be representative for the characteristic of interest.

A *characteristic* is a quality or quantity associated with a sampling unit. Examples of characteristics are the saltiness of the pot of soup, the moisture content of the grain in the bin, the diameter of the bearings being produced, the fraction of voters preferring a particular candidate, or the type or location (qualities) of acres of corn planted or the number of man-days of labor employed (quantities) on the farms of a state.

The *sampling unit* is the unit listed or defined in the sampling frame. It may be an elementary unit or event such as a farm or a field of corn, or it may be a cluster, group, or subset of elementary units such as all the farms in a township or all the people living on a city block.

The *sampling frame* is a listing of the sampling units in the universe. It is used to identify the units from which a sample is drawn. A listing of all the farms in a given state would be a sampling frame, as would a listing of all the townships.

The *universe* or *population* is the aggregate of the elementary units whose characteristics are to be estimated. If the characteristic to be estimated were labor employed on the farms of a state, the universe would comprise the farms of the state, even though the sampling units might be clusters of farms.

A *statistic* is a sample estimate made of a characteristic for a universe; a *parameter* refers to a characteristic compiled from the entire universe. A statistic is thus a sample estimate of a universe parameter.

The *mean,* which is the most frequently used statistic, is often called the average. It is the sum of the values of a characteristic associated with a set of sampling units divided by the number of units.

Sampling error refers to the standard error of the mean and provides a measure of the precision and reliability of a sample estimate.

Bias is the deviation of the expected value of a statistic (i.e., sample estimate) from the universe parameter. Bias may arise from the way the sample estimate is determined or may result from nonsampling errors.

Nonsampling error is a term applied to errors not attributable either to sampling fluctuations or to the estimating procedure. These errors arise from a number of sources, such as faulty frames, incorrect source data, data-conversion errors, and processing errors.

Design of Samples

In developing the basic principles of sampling, we will utilize the concept of the infinite universe rather than the finite one. This concept is sometimes called the analysis of variance model because it parallels some of the concepts of experimental design. The use of this model clarifies the relationship between different sample designs and shows the alteration in the assumptions of simple random sampling in other designs that permit utilization of information from sources independent of the sample to increase sampling efficiency.

The concept of the infinite universe is important to the development that follows. This model postulates the existence of a hypothetical infinite universe of which a finite universe is one of an infinite number of samples. For many finite universes, this is not too difficult. If one conceives of the universe of human beings throughout eternity, then an existing population—world, nation, city, or the farmers of a given county—can be thought of as samples from the infinite universe, or at least as subsets from infinite sample space. For other finite universes, similar infinite universes must be postulated, even though this may strain credulity. For example, the assumption that the 50 states of the United States is a sample from an infinitely large number of states requires a bit of imagination.

The starting point in probability sampling is simple random sampling. A simple random sample of size n consists of n sampling units drawn from a universe in such a way that each possible combination of sampling units (or subsets) of the same size has the same probability of being drawn. It works out that this occurs when n individual sampling units are drawn with equal probabilities from the universe.

The concepts of simple random sampling are basic to sample design. A few symbols needed for clarification are defined below.

(a) $$\mu = E(X)$$

The *mean* μ of an *infinite universe* is the expected value of the characteristic X. The expected value is the average of all possible values of X.

(b) $$\bar{x} = \Sigma X/n$$

The *mean* $\bar{x}$ *of a sample* of size n is the sum of the n values of X from the sample divided by n, the size of the sample.

(c) $$\sigma^2 = E(X - \mu)^2$$

The variance of the universe σ^2 is the expected value of the squared deviations (or differences) between the values of X and μ.

(d) $$S^2 = \sum (X - \bar{x})^2/(n - 1)$$

The *variance* S^2 *of a sample* is the sum of the squared deviations of the sample X's from the sample mean divided by $n - 1$. It will be shown later that S^2 is an unbiased estimate of σ^2.

(e) $$\sigma_{\bar{x}}^2 = E(\bar{x} - \mu)^2$$

The *universe variance of the mean* is the expected value of the squared deviations of all possible means of samples of size n from the mean μ of the universe.

A number of theorems are fundamental to sampling theory. Equations 1, 2, and 3 are particularly important and are closely related to definitions (a) through (e).

$$\sigma_{\bar{x}}^2 = \sigma^2/n. \tag{1}$$

$$ES^2 = \sigma^2. \tag{2}$$

$$\sigma_{\bar{x}}^2 = \frac{\sigma^2}{n}\left(\frac{N - n}{N}\right) \tag{3}$$

In Equation 3, the sample of size n is drawn from a population of size N.

The three relationships above are fundamental to sampling: Equation 1 is the relationship between the variance and the variance of a mean, the latter varying inversely with the size of the sample; Equation 2 provides in S^2 an unbiased estimate of σ^2, and Equation 3 gives the finite population correction, $(N - n)/N$. The finite population correction is the bridge between the infinite universe of theory and the finite universe from which many samples are drawn. These equations make it possible to estimate the mean, variance, and variance of the mean of an infinite universe from a single sample drawn from a population that may be finite. The variance of the mean is required for making a probability statement about how closely the sample estimate approximates the population mean. Such a statement is called statistical inference.

A theorem fundamental to statistical inference is the central limit theorem. This theorem, in essence, proves that the distribution of sample means rapidly approaches the normal distribution as sample size increases. For all practical purposes, means of a sample of 30 or larger may be treated as being normally distributed. The normal distribution is the familiar bell-shaped error curve developed by Gauss.

The parameters of the normal distribution are the mean and variance of the universe. Relative frequencies (i.e., probabilities) have been computed and tabulated in terms of deviations from the mean, with the deviations measured in units of standard error. The central limit theorem and the normal distribution make possible a probability statement about the precision and reliability of a sample. This statement is usually expressed as a confidence limit.

A confidence limit is an upper and lower limit encompassing a sample mean. For a given confidence limit expressed in units of standard error of the mean, the probability of the universe mean's being within the confidence limits can be determined from the normal distribution. A confidence limit is expressed in terms of both precision and reliability. Precision refers to the size of the confidence interval, and reliability refers to the probability that the universe mean is bounded by the limits of the interval.

The two measures of reliability used most often are the one- and two-standard-error limits, which have the probability of about 67 percent and 95 percent, respectively, of including the true mean.

For example, if a sample mean were 100 units, and an unbiased estimate of the standard error of the mean were 2 units, we could establish a one-standard-error limit of 100 ± 2 and state that the probability of the true mean's lying between 98 and 102 is 67 percent. Similarly, for the two-standard-error limit of 100 ± 4, we can say the probability that the interval contains the true mean is about 95 percent.

We are now equipped with the fundamental concepts and relationships to do simple random sampling. We can draw a sample, compute the sample mean, and make a probability statement about the universe mean. The major problem in designing a simple random sample is determining the size of the sample to obtain the desired precision and reliability. For this, we need knowledge of the universe variance; and by using Equation 1, we can determine the sample size for the proper sized standard error (square root of the variance) of the mean.

Stratified Sampling

Let us now alter one of the restrictions required for simple random sampling, that of equal probability for all possible combinations of the same number of sampling units in the universe. This will permit us to partition our population into strata and to place in each stratum sampling units that tend to be alike. We do this to gain homogeneity and to reduce variance. But the partitioning requires prior knowledge (control data) of the sampling units in the universe that is known independently of the sample.

The concept of stratified sampling is not too different from that of simple random sampling. The difference is that sample space is partitioned into strata. But each stratum is treated as a separate subuniverse in which we use simple random sampling and compute means and variances for each stratum. These means and variances of the different strata are then weighted together. For means, stratum weights are the number of sampling units in each stratum relative to the total number in the universe. The weights for variances of the mean are the squared weights for the means. These relationships are expressed symbolically as follows:

$$\bar{x} = \sum W_i \bar{x}_i, \tag{4}$$

where $W_i = N_i/N$ and the subscript refers to the ith stratum, and

$$\sigma_{\bar{x}}^2 = \sum W_i^2 \sigma_{\bar{x}_i}^2, \tag{5}$$

where $\sigma_{\bar{x}_i}^2$ is the variance of the mean of the ith stratum.

When adequate control data are available, stratification is a powerful means of reducing sampling error. Combinations of sampling units are restricted to those within the same stratum. Variances within the strata are the squared deviations of the characteristics of the sampling units in each stratum from the stratum mean. This contrasts with simple random sampling where the variance is the squared deviation from the universe mean.

The design of a stratified sample involves selection of strata criteria and the allocation of a sample to the different strata. The most efficient is the optimum allocation, which results in the minimum variance of the universe mean. The expression for an optimum allocation turns out to be

$$n_i = \left(\frac{W_i \sigma_i}{\sum W_i \sigma_i} \right) n, \tag{6}$$

where n is the size of the entire sample and n_i is the allocation to the ith stratum.

Useful stratification criteria for agricultural data include size and geographic location. Distributions for agricultural items tend to be positively skewed, so that stratification by size permits gains in efficiency by sampling the smaller-sized strata at lower rates than the larger ones. Geographic stratification usually produces gains in efficiency because agricultural production is influenced by environment, and hence tends toward homogeneity within a contiguous area.

Subsampling

For subsampling (or two-stage sampling), we remove another restriction: that of sampling within all strata. The universe is partitioned as in stratified sampling, but the objective is different. For subsampling, we want clusters of sampling units that are not homogeneous but are like the universe (microcosms). These clusters are called primary sampling units (psu's). In subsampling, samples of primary units are drawn, and each of these is then sampled as a subuniverse. Subsampling can be extended to more than one level, in which case it is called multistage sampling.

Subsampling and multistage sampling have many ramifications and great flexibility and are employed in many sample surveys. Here, we can only touch upon the basic concepts of two-stage sampling. Continuing with the notions of the infinite universe, we will postulate an infinite universe containing an infinite number of primary sampling units, each containing an infinite number of elementary sampling units.

Consider the ith primary sampling unit. Under conditions of simple random sampling, the variance of the mean of samples of size n_i drawn from the ith psu may be expressed as

$$\sigma_{\bar{x}_i}^2 = \frac{\sigma_i^2}{n_i}. \qquad (7)$$

Now, consider the means of each of the postulated infinite supply of primary sampling units. By definition, the variance of these means is

$$\sigma_b^2 = \mathrm{E}(\mu_i - \mu)^2, \qquad (8)$$

where μ_i is the population mean of the ith psu.

The variance of sample means under conditions of subsampling can now be written for samples of k psu's with subsamples of size n in each sample psu. Each of the resulting samples of kn units is an estimate of the universe mean, regardless of the size of k and n. But the variance of the sample means from the ith psu only is not an estimate of variance of means from the other psu's in the universe. However, it can become so by adding the variance between psu means (Equation 8). In symbols,

$$\sigma_{\bar{x}_{(i)}}^2 = \sigma_{\bar{x}_i}^2 + \sigma_b^2,$$

where $\sigma_{\bar{x}_{(i)}}^2$ denotes the universe variance of the mean as estimated from a sample from the ith psu. The variance of the universe mean estimated from k psu's and n samples within each sample psu is, by Equation 1,

$$\sigma_{\bar{x}}^2 = \frac{\sum^{k} \sigma_{\bar{x}_i}^2 + \sigma_b^2}{k} = \frac{\sigma_w^2}{nk} + \frac{\sigma_b^2}{k}, \tag{9}$$

where σ_w^2 is the within-psu component of variance, the expected value of Equation 7. As n and k are constants, the estimate of the universe mean by definition (b) is

$$X = \frac{\sum^{n}\sum^{b} x_{ij}}{nk}.$$

Stratified sampling and subsampling may be combined by grouping the primary sampling units into suitable strata, drawing a sample of primary units from each stratum, and then drawing a sample of secondary units from the sample primary units. With this design, the strata are considered to be subuniverses. Sampling within strata is done independently, and the within-strata estimates are weighted together in the usual way.

The requirements of equal-sized subsamples and equal probabilities may also be relaxed. When this is done, some of the more advanced designs commonly used in modern sampling become possible. But since this is only a treatment of fundamental concepts and not a treatise on sampling, we shall describe briefly only two such designs used extensively in sampling agriculture.

In one of these designs, primary sampling units are drawn with probabilities proportional to the number of secondary units they contain, and the same number of secondary units is drawn from each primary sample unit. This results in a sample that is self-weighting for the secondary unit. For example, if a sample of counties is drawn with probabilities proportional to the number of farms they contain, and a subsample consisting of the same number of farms is drawn from each sample county, it turns out that the sample of farms is self-weighting. This means that unbiased estimates of per-farm averages can be obtained by simply aggregating the characteristics for the individual farms over the entire sample and then dividing by the aggregate number of sample farms.

Another useful application of unequal probabilities is to select a subsample of fields planted to a particular crop with probabilities proportional to the acreage of that crop. When this is done, the ratios of any characteristic to acreage are self-weighting. As yield is production per acre, average yield is computed by simply summing the sample yields and dividing by the number of samples. This design is used extensively in sampling to measure crop yield.

Sampling for attributes also has applications in agriculture. An *attribute* is a quality associated with a sampling unit rather than a quantity. Variation is restricted to the presence or absence of the quality; sampling units possessing it are counted, while those that do not have it are not counted. The attribute variable X can thus have only one of two values, one or zero. The concepts and equations previously developed also hold for attributes, although simpler formulas are ordinarily used. These simpler formulas are expressed in terms of p, which is equal to $\Sigma X/n$ and is the fraction of the sample possessing the attribute, and q, which is $1 - p$, the fraction without the attribute. Examples of applications of attributes in agriculture are point samples and consumers' preference samples.

A sample of points can give an estimate of the fraction of land area devoted to different uses. This may be done by examining the utilization of land for a random (or random-systematic) sample of points, and the standard error of the estimated fraction may be computed. Sample points along random line segments may similarly be utilized with the lines serving as psu's and the points as subsamples of the lines. Attributes in terms of the preferences of a random sample of testers can provide an estimate and standard error of the relative number preferring the different products being examined.

Estimators in General Use

As outlined in the preceding section, the relaxation of the requirements for simple random sampling have made possible marked increases in sampling efficiency through the use of control data from sources independent of the sample itself. Another and different means of increasing efficiency by the use of ancillary information in estimating procedures (estimators) has been developed. The choice of estimator is part of the design of a sample, since sample size and allocation are affected. We will describe briefly those in general use: the ratio, regression, and difference estimates. These estimators utilize supplementary information about characteristics associated with the sampling units that is known from sources other than the sample. These estimators are similar in that all use relationships between sample data to provide a measure of change to be applied to the data from independent sources. Source data are obtained from the sampling units for both the characteristic for which an estimate is desired and the characteristic for which information independent of the sample is available. Common uses of these estimators are helpful both in measuring change from an earlier survey and in double sampling.

The ratio estimate presumes proportionality between the two characteristics from the sample. The ratio of the mean (or total) between the two is multiplied by the independent data. As will be shown later, the ratio is analogous to the coefficients b and k in regression and difference estimates.

The regression estimate assumes linearity between the two sample variables, but not proportionality. Computations are more burdensome because a least-squares regression line fitted to sample data is required.

The difference estimate is similar in form to the regression estimate. However, unlike the regression estimate, the coefficient k is determined independently. The efficiency of this estimator depends in part on how nearly this coefficient approaches the slope of the regression line for the entire universe.

The ratio estimate is of the form

$$Y = \frac{\bar{y}}{\bar{x}} X = R\bar{X}, \tag{10}$$

where Y is the estimated mean, $\bar{y}$ and $\bar{x}$ are sample means, and $\bar{X}$ is a mean known independently from a source other than the sample.

The ratio $\bar{R} = \bar{y}/\bar{x}$ computed from the sample is an estimate of the relative magnitudes of the two variables.

The regression estimate is of the form

$$\bar{Y}_r = a + b\bar{X}, \tag{11}$$

where coefficients a and b are the intercept and slope, respectively, of a least-squares line determined from sample values of Y and X. By shifting the origin of the mean $\bar{x}$ and $\bar{y}$, the equation becomes

$$\bar{Y}_r = \bar{y} + b(\bar{X} - \bar{x}). \quad (12)$$

The difference estimate is of the same form as the regression estimate except that the slope is not determined from the sample but from independent sources. The equation can be written

$$\bar{Y}_d = \bar{y} + k(\bar{X} - \bar{x}).$$

Variance formulas for the three estimators are as follows:

$$\sigma_{\bar{Y}}{}^2 = (\sigma_y{}^2 + R^2\sigma_x{}^2 - 2R\rho\sigma_x\sigma_y)\,\frac{N-n}{N}, \quad (13)$$

$$\sigma_{\bar{Y}_r}^2 = \sigma_{\bar{y}}^2\,(1 - \rho^2)\frac{(N-n)}{N}, \quad (14)$$

$$\sigma_{\bar{Y}_d}{}^2 = (\sigma_y{}^2 + k^2\sigma_x{}^2 - 2k\rho\sigma_{\bar{x}}\sigma_{\bar{y}})\,\frac{N-n}{N}, \quad (15)$$

where ρ is the correlation coefficient of y and x.

From the three preceding formulas, it is evident that all three of the above estimators depend for efficiency upon a high correlation between the x and y variables. The efficiency of the difference estimate depends not only upon correlation but also upon how closely k approximates the slope of the regression in the entire universe. Both the ratio and regression estimates are, in the general case, biased, but the bias is trivial for samples larger than about 30 units. The difference estimate is unbiased.

The ratio estimate is used more frequently than the other two, despite the fact that its variance is equal to or larger than that of the regression estimate. The reason for this is that the ratio estimate is easier to compute than the regression and it is difficult to estimate an efficient coefficient k for the difference estimate.

Ratio, difference, and regression estimators have many applications in agriculture. Current crop acreage may be estimated in terms of changes from a census or other basis. Stocks of grain may be estimated utilizing the relationship between present stocks and capacity of sample storage facilities together with known total capacity. Regression and ratio estimators are employed in double sampling. Double sampling, as mentioned earlier, is often applied to agricultural universes and may prove valuable in applications of remote sensing. Double sampling presumes a low-cost large primary sample that provides information pertaining to an ancillary variable correlated with the variable we want to estimate. This large sample provides control data for the desired but higher cost characteristic. These control data may be used in either a ratio or regression estimate or as a basis for stratification. Double sampling is useful only when a high degree of correlation exists between the control variable and the variable to be estimated. The choice of estimators depends upon whether the relationship is linear, proportional, or nonlinear. If proportional, the ratio estimator is indicated; if linear but not proportional, regression is the choice; and if neither, stratification and sampling within the strata is the best bet.

An example of an application of double sampling may be found in crop cutting to estimate yield. It is obviously much cheaper to harvest and weigh small sample plots than an entire field. A large sample of fields may be drawn, and one or more small plots in each sample field can be harvested and weighed. Then a subsample of these fields can be completely harvested as a double sample.

Multiple frame sampling is also being applied in the collection of agricultural statistics. However, we will defer its description until we have examined the sampling frame, which is essential to probability sampling of any kind.

Sampling Frame

The theory of sampling postulates a completely and explicitly defined sample space from which subsets may be drawn in accordance with probabilities associated with individual sample points. A listing of the points of the sample space constitutes a sampling frame. It provides a means of drawing samples and of associating the probabilities with the particular subset of points drawn.

An ideal sampling frame is a list of distinct, clearly defined, mutually exclusive sampling units containing all the elements of a specified universe. The individual sampling units may be natural units, artificially constructed units, or some convenient reporting or working unit. It is not necessary to have a complete listing of individual sampling units. Clusters of units may be used as sampling units if cluster sizes are known and if procedures are developed for making an unambiguous identification of the individual sampling units within clusters.

There are two basic types of sampling frames: area and list. Each has its place in sampling and has problems associated with its use.

Area Frame The sampling units of an area frame consist of units of land area (segments) that in the aggregate comprise the total land area of a geographic area or region. The individual sampling units are defined by maps or other cartographic material, and are numbered or otherwise identified to provide a listing of the frame units. Since agriculture and many other characteristics may be uniquely associated with these units, an area frame is a useful, complete frame. Source data, however, must be associated with the area sampling units. Useful concepts for associating data are those of the closed and open segments, and the hybrid open–closed, or weighted, segment.

The concept of the closed segment is the land area of the segment; reported data must pertain to and be associated with the land area inside the segment. The closed segment is useful for characteristics easily associated with land area, such as crop acreage or the number of livestock in a field. The closed segment has the advantages of clarity of definition and ease of identification as well as lower variances than the open segment. But for characteristics not related to land area such as farm income or farm population, the open or the weighted segment is more useful.

The open segment considers only the location of the residence of the farm operator. The entire farm is associated with the segment in which the operator resides, and all characteristics to be estimated pertain to farms with headquarters in the segment.

The open–closed, or weighted, segment associates with each segment that fraction of the total farm contained within the segment boundary. Data pertaining to the farms whose operator resides within the segment are prorated to each sample segment according to the fraction of the farm inside the segment. Variances of open-segment estimates tend to be considerably larger than for either the closed or the weighted segment, whereas those of the closed segment tend to be slightly larger than those of the weighted one.

List Frame A list of all the farms in a state should be an ideal frame for sampling, having distinct advantages over an area frame. One advantage is that it affords the opportunity to utilize control data to draw samples of specialized farming operations, for example, poultry, without including many nonpoultry enterprises in the sample. Another advantage is that a mailing address (or a telephone number) makes it possible to collect sample data by mail (or by telephone), which is much cheaper than by personal interview.

In practice, however, difficulties arise: Lists from any source are rarely complete, lists deteriorate through time, and list frame units may not be natural reporting units, or they may not coincide with the population of interest. Unless we are able to make unique associations of frame and sampling units, and of frame units associated with the population of interest, unbiased estimates will not be obtained even though a random selection of frame units is used. It is easy to see why this is so. Under conditions of simple random sampling, the probability of any sampling unit is a function of the number of these units in the universe. Also, the expected value of the mean of a sample of any size is the mean of the universe. If the frame is not a listing of sampling units of the universe, the expected value of estimates made from samples drawn from the frame will not be those of universe parameters (and hence will be biased) unless adjustments are made.

Many of the defects of lists as sampling frames can be remedied by collation of names and addresses of farm operators, by adjustment of probabilities, or by association of frame unit with universe unit. The defect of frame incompleteness, however, can be overcome only by obtaining an independent measure of the magnitude of the incompleteness or by using the list in conjunction with one or more other frames that collectively include the universe.

Theory now exists for utilizing more than one sampling frame in the same sample survey. In agriculture, one frame is usually an area frame and the other or others are list frames. Multiple-frame samples utilize the strengths and minimize the weaknesses of both kinds of frames. For the theory to hold, however, two conditions must be satisfied: Every unit in the population of interest must belong to at least one of the frames, and it must be possible to identify for each sampling unit in a sample the frame or frames to which it belongs.

Conceptually, when the sampled units are assigned to domains according to the frames to which they belong, independent estimates may be made for each domain. The different estimates for each domain may be weighted together optimally into an estimate for that domain. When these domain estimates are aggregated, they provide an estimate of the entire universe. Domain weights may be selected so as to minimize sampling error for the characteristics being estimated.

Possibilities are readily apparent for economies and gains in efficiency. By utilizing more than one frame, combinations of list frames of large or specialized farms with an area frame are quite efficient. This is because even an incomplete list covers part of a population of interest and makes possible cost efficiencies by a reduction in over-all sample size. When the list frame includes current mailing addresses so that data may be collected by a cheap means, multiple frame sampling becomes particularly attractive.

Cost Constraints

The preceding discussion has attempted to depict some of the devices by which sample size may be reduced. These include sample designs, estimators, sampling frames, and sampling units. The discussion of the first three of these devices emphasized how independent information may be utilized to increase sampling efficiency and reduce costs, but without specific reference to a cost function. We will assume the independent information is without cost (i.e., not chargeable to the survey being planned). This assumption is usually realistic and simplifies the problem of optimization by making variances and costs the determining factors.

Cost is a very important consideration in sampling. In most cases, total survey costs must not exceed available or specified resources, and in any case, prudent stewardship dictates maximizing efficiency in terms of precision per dollar cost. The mechanism employed in sampling theory is the constrained minima in which the variance of the mean is minimized with respect to a constraining cost equation. For simple random sampling, the total cost is the product of the number of sampling units and the unit cost. In agriculture, there are rarely significant differences in costs between strata so that design is usually optimized with respect to stratum variances only. Sizable cost differences are found in subsampling designs between primary and secondary units, and optimization with respect to both variances are required for efficient allocation. The form of the cost function for a subsample design is

$$C = kC_p + nkC_s, \tag{16}$$

where C_p and C_s are average costs associated with primary and secondary units, respectively. The general form of the function (F) for which minimum value is sought is

$$F = \sigma_{\bar{x}}^2 + \lambda C, \tag{17}$$

in which λ is the LaGrange multiplier and C is the related cost equation.

As was the case with the normal distribution and the central limit theorem, the solution of equations of this type will not be given here.

Solutions for the different designs in terms of the n_i and k_i give the optimum allocation of sampling units. These solutions may be found in any textbook on sampling. We shall give only the solution for subsamples, which is expressed in terms of k, nk, and λ.

$$k = \frac{\sigma_b}{\sqrt{\lambda C_p}}, \qquad nk = \frac{\sigma_w}{\sqrt{\lambda C_s}}, \qquad \lambda = \frac{\sigma_b \sqrt{C_p} + \sigma_w \sqrt{C_s}}{C}, \tag{18}$$

where σ_w and σ_b are the within and between psu components of variance (see Equation 9).

Equation 17 offers an interesting possibility sometimes utilized—that of determining the optimum design. Required are estimates of variance components and estimates of costs within and between the hierarchy of sampling units at the different likely levels of stratification or multistage sampling. The solution determines the optimum number of sampling units at each level (and hence the design), but is a bit messy to obtain.

A very powerful method of allocating a general-purpose survey employs nonlinear programming to optimize the allocation of a number of characteristics with respect to costs and variances so that a preselected level of precision for each characteristic is equaled or exceeded. This method, as yet unpublished, was developed by Hartley and Hocking of Texas Agricultural and Mechanical University. It has been utilized by the Statistical Reporting Service of the U.S. Department of Agriculture in allocating several general-purpose samples and has proved quite efficient.

The optimization of the size of the sampling unit is somewhat similar to that of an optimum allocation. Costs and variances are used. For optimization of size, the most useful concept of sampling unit size is the number of elementary units in a cluster. A technique frequently employed is based upon experimentally determined variances and costs associated with clusters of different sizes. Variance is adjusted for intracluster correlation, equations of variance and cost are fitted to the associated cluster sizes, and then by substituting into Equation 17 and solving for n (cluster size), variance is minimized with respect to cost to obtain the optimum-sized sampling unit. This procedure is employed routinely to determine plot size for objective yield and segment size in an area frame.

Except in determining size of sampling unit, cost functions are not employed in the sampling frame construction. Cost functions are important, however, in multiple-frame sampling. Here the costs associated with the sampling units of the different frames are likely to differ significantly as there may be an opportunity to use different means of eliciting source data (e.g., mail, telephone, interview). Although the expressions for optimizing allocations to different frames are tedious to derive, they, like single-frame optimization, are based upon costs and variances.

Programs of Agricultural Statistics

In looking at program and practice, we should keep in mind that agricultural statistics are part of economic statistics, and although production and inventory statistics are important, so are price, income, debt, cost of production, cultural practices, labor, machinery, fertilizer, and land and water utilization data. Some of these data are associated with land and may be detected and associated by observation. Other data cannot be collected by observation, so other sources and other means of collection are required.

Program and practice of countries in three stages of development are examined below: Nepal, as yet undeveloped; Turkey, developing rapidly; and the United States, by present standards, a developed nation.

Agricultural Statistics in Nepal

There is no program of current agricultural statistics in Nepal. The Central Bureau of Statistics, however, conducted an agricultural census in 1962 and plans another in 1970 or 1971. Although summaries of the 1962 census were not completed by February 1968, we suspect sizable nonsampling errors induced by imprecise units of measurement, incomplete coverage, and perhaps deliberate underreporting by a suspicious peasantry.

The country of Nepal is about 400 miles long and 100 miles wide. A 20-mile-wide strip along the southern border, running the length of the country, is an extension of the Terai Plain of India. In this 20-mile strip live about one third of the 10 million people of Nepal. This segment of the population produces about two thirds of the agricultural output of the country.

North of the Terai Plain lie the hills and mountains of the Himalayan Range. In the hill country, transportation of goods is by pack animals and human porters, except along the newly constructed highway to Tibet. Communication systems include a sporadic postal service and a government-operated radio net. This part of the country, although farmed intensively, is a deficit food area, requiring importation of grain from the Terai area or from India. The agriculture of Nepal is quite primitive. The principal food crops are wheat, rice, millet, and maize. Most of the rice is produced in the Terai area, with wheat as a second major crop. Here, animal power is used extensively in plowing and in transport. In the hill country, rice is also grown, as are wheat, maize, millet, and mustard. Steep slopes have been terraced to permit cultivation, resulting in extremely small fields—often wide enough for only a single row of maize. Here animal power is not used: tilling is done with hand tools.

Although primitive, Nepalese agriculture is important, comprising about 90 percent of the economy. This fact, together with the necessity for moving food from surplus areas to deficit areas and for adjusting imports from other countries, makes a means of measuring agricultural production highly desirable, but not easy to do. The rugged terrain, lack of transportation and communication facilities, illiterate rural population, and small, inaccessible, irregularly shaped fields make assessment by conventional means very difficult. Some means of direct measurement of areas harvested and of yields is needed. A likely means would be application of remote-sensing techniques, either directly or indirectly. These possibilities are discussed later in this chapter (page 350).

The Program of Turkey

In Turkey, the program of current agricultural statistics consists mainly of crop area, yield, and production, livestock, price, and equipment statistics. For ten cereals, area-sown and expected-production data are collected twice each year, and data on harvested area and production are collected at the end of the year. Data on prices paid and received are collected at bimonthly intervals. Some of 115 items are priced, but the number of items varies from survey to survey. Agricultural equipment and tools (49 items) are surveyed at the end of the year, as are poultry, honey, and livestock. The production of meat, milk, wool, and mohair is derived. Fruit production and number of fruit trees, the number of olive trees, and the production of olives for oil and for eating are also reported at the end of the year. The principal needs not now being met are costs of production, stocks, recent marketing channel and consumption surveys, farm slaughter, and livestock death rates from natural causes.

Current agricultural statistics are collected by district agricultural technicians and veterinarians of the Ministry of Agriculture. The sources of information are village head men, individual villagers, and other informed people, together with the personal observation, general knowledge, and judgment of the technicians and veterinarians. These men are responsible for gathering data for the villages in their districts, and their reports are summaries for their respective districts. The district reports are summarized at the province level and sent to the Ministry of Agriculture for internal use and to the State Institue of Statistics for publication. Except for the quadrant method of estimating yield, described later, the entire program of current statistics is based upon these reports.

Both the State Institute of Statistics and the Ministry of Agriculture are aware that these reports contain reporting errors or biases, both mathematical and personal. The agricultural technicians and veterinarians, in the course of their normal work, learn a great deal about the agriculture of their villages. But these men have many responsibilities and are able to devote only a part of their time to data collection. Under these circumstances, they appear to be doing a reasonably good job, but they seem to be guided too much by reports previously submitted and are reluctant to show decreases. As a result, the reports tend to have an upward bias. However, office procedures have been established to reduce this bias by finding relationships between past reported data and the estimate adopted and published, and adjusting current reported data by these average relationships. The estimates that are adopted and published result from the combined judgment of officials from a number of agencies. But the procedure makes no provision for adjusting the levels of current estimates to check data in the form of independent, unbiased estimates such as may be produced by a probability sample or complete enumeration.

The simplest way to reduce further the bias in the reports is to introduce adjustment to independent check data such as the decennial census of agriculture. To the extent that the present biases persist through time, ratios or regressions between reported data and check data can be made to remove this bias. This procedure differs little in principle from present procedure except for the substitution of independent check data as the basis for adjusting reported data. The only difficulty, however, is the lack of a reliable agricultural census or other measure for check data. Obviously, such a measure is critical for the program of current statistics.

The Census of Agriculture, which has been a sample survey, has been severely inhibited by lack of a suitable sampling frame and by the poor quality of reported data.

The best available sampling frame appears to be a list of villages compiled in taking the population census. However, village and field boundaries and areas are not generally available. The land survey office has surveyed a portion of the villages, but this project is not expected to be completed for several years. Aerial photography and some cartography for at least part of the country exist, but these materials are not available for nonmilitary use. If farm operator reports were reasonably accurate, it would be possible to sample villages and to list the households in the sample villages and subsample them. However, this is not the case. Areas planted are usually estimated rather crudely, and production may or may not be measured. There is evidence of widespread deliberate underreporting because of fear of taxation.

Samples from an area frame would undoubtedly produce better statistics. Farm operators seem to be less hesitant to report accurately about a particular field and are less prone to forget or deliberately fail to report on part of their farming operation. However, lack of cartography and photography effectively precludes construction of an area frame. Perhaps the land survey may in time be completed and its maps and measurements of land holdings can be utilized.

Little has been said about crop yield, which, even though it is quite variable, is far easier to estimate and far better known than area planted. For certain crops important for export, yield estimates are made independently of the agricultural technician's reports. Wheat, barley, oats, and rye are estimated by the quadrant method. Four-man teams composed of one man each from the Institute of Statistics, the Ministry of Agriculture, the Soil Products Office, and the Agricultural Bank collect the data. These teams ride along roads passing through the grain-producing areas of Turkey, stop at 40-km intervals, lay out 3 one-meter-square plots according to prescribed procedures, and count the number of grain heads in each plot and the number of grains in every sixteenth head. The sample data provide a reasonably good measure of yield relative to that of former seasons, but obviously, no direct measure of current yield.

United States Program

For many years, the U.S. Department of Agriculture has collected sample data upon which to base its crop and livestock estimates, which constitute the more than 700 reports released each year. For crops, the annual cycle begins with intentions to plant, continues during the growing season with acreages planted and yield forecasts, and ends with harvested acreage, production, and final utilization of the crop. Many other items are estimated during the year. Among these are livestock inventories; numbers born or raised; slaughter; animals on feed; milk and milk products; wool and mohair; poultry statistics; honey; acreage, yield, and production of field, vegetable and tree crops; stocks of grain; cold-storage holdings; naval stores; prices paid and received by farmers; and farm labor, wage rates, population, and income.

Until the last few years, crop and livestock estimates were based upon nonprobability surveys conducted by mail. In this kind of survey, questionnaires are mailed to crop reporters and to general farm and special commodity lists. The recipients of these questionnaires are asked to fill out and return them. Because the mailing lists are small fractions of the universe and because of the selectivity caused by data from only those who are willing to complete and return the questionnaire, the aggregated reported information contains serious biases. A simple methodology has been developed to convert the voluntary reports into estimates and forecasts. This methodology is based on regression charts that remove persistent biases from the reported data. Survey data are plotted against previous estimates for a dozen or more years. Sometimes a function of the reported data, such as the ratio to the previous year's acreage or production, is plotted against previous estimates. Trend adjustments are incorporated into the charted relationships when needed. Then, when a new survey is conducted, the charts are read by using an argument derived from current survey data, and persistent bias is removed by the charted relationships. Historically, revisions are made if indicated by check data such as Census of Agriculture, State Farm Censuses, marketings, and the like. Forecasts of yield are prepared by a similar technique. Farm operators are asked to report crop conditions or expected yield per acre. Regression charts are prepared by plotting early-season estimates by growers against final yield, and these charts are used to convert grower appraisals or estimates into forecasts of yield.

Over the years the estimates and forecasts resulting from the nonprobability sample surveys have been generally good. The principal advantage of surveys of this kind is the low cost associated with collecting data by mail. Disadvantages are lack of independence, some residual bias, and no current measure of precision.

In the last quarter century, there have been a number of developments that have resulted in modifying the methodology used in making crop and livestock estimates. First, improved cultural practices, fertilizers, pesticides, and farm machinery have changed the character of the farming enterprise. Farms have become much larger, fewer in number, and more specialized. Greater demands are placed upon farm management, which, in turn, requires more reliable information for decision-making. Yet, because of changes in the universe of farms, data from questionnaires returned by farm operators who are willing to report are not nearly so stable as when farms were more nearly alike. Another important development is the theory of probability sampling, briefly summarized earlier. Probability sampling permits unbiased estimates with known precision and reliability, provided suitable sampling frames exist. A third development is the high-speed computer, which is capable of doing the complex computations required by modern statistical methods so that the estimates, variances, and other statistics from probability samples can be computed in a minimum of time. These developments have led the Department of Agriculture to base spring and fall general-purpose surveys and yield surveys for certain crops upon probability samples. Both list and area sampling frames are used.

The original area frame for drawing the master sample of agriculture was prepared in the early 1940's at Iowa State University. In preparing this frame, three types of land areas were identified on the basis of density of population and delineated on county highway maps as strata. These were: (1) open country, (2) urban places, and (3) rural places. The open-country stratum was then divided into count units. The word "count" was used because a count of farms and dwellings on the maps was made for each unit. For each count unit, a number of sampling units, or segments, was assigned, based on the farm count. These three strata plus geographic stratification were adequate for the relatively homogeneous agriculture of the early 1940's, because of the general or diversified nature of the family farm. Since that time rapid changes have occurred that make the degree of stratification inadequate, especially in the western states. This inadequacy was recognized early by the designers of the master sample materials, and efforts were started in the late 1940's to stratify those states west of the 100th meridian. However, this project was never completed.

In the late 1950's the U.S. Department of Agriculture began to construct a new area frame for the 11 western states and the 13 eastern seaboard states. In the West, the problem was essentially that of separating cultivated land from land with other utilizations; in the East it was urbanization and the decline in intensity and density of agriculture that made a new frame necessary. Maps and aerial photography were used to delineate the utilization strata. In the western states, the primary strata are cultivated land, cities, and towns, grazing land, and nonagricultural land. In the eastern states, cities were stratified into industrial, residential, and agri-urban areas. The latter stratum contains a mixture of small farms, dairy and poultry farms, and nonfarm residents of subdivisions. It is an area of rapid change. In the open country, strata were constructed according to three levels of intensity of agricultural activity. The strata in both urban and nonurban areas were determined from maps, local sources, census data, and aerial photography.

At present, two general-purpose probability sample surveys are conducted each year. The purpose of the first, conducted about June 1, is to obtain planted acreages, farm numbers, livestock inventory, poultry and dairy statistics, farm labor, and various items of economic information. This survey provides an early season base of planted acreages and farm and livestock numbers, but it was designed as a vehicle to collect any desired on-farm statistic. The second general-purpose survey, which is conducted December 1, is similar to the June survey, but its emphasis is on fall-seeded grains and livestock. During the growing season, usually at monthly intervals, objective yield surveys are also conducted for corn, cotton, wheat, and soybeans, and for a number of tree crops, including oranges, lemons, peaches, pears, sour cherries, walnuts, filberts, and almonds.

For the June survey, the present design is a three-frame sample drawn from the area frame, a list of large livestock operators, and a list of large employers of hired farm labor. The list samples are stratified by size, and the area samples are broken into geographic strata. Since state as well as regional and national estimates are important, the sample is allocated to provide precision for state estimates as well as for those at the regional and national levels. The size of the area sample for a state averages about 400 segments, each of which contains approximately 1.2 farms. For the 48 contiguous states, the size of the area sample is about 17,000 segments containing approximately 23,000 farms, and the size of the list sample is about 6,000 farms. Estimates from the area and list samples of the domain of overlap of the lists with the area are computed. These estimates are weighted together by inverse variance weights, and the resulting estimate for the domain is added to the estimate from the area-only domain. This sum is the estimate for the universe.

Five estimators are used in expanding the area sample data: the direct expansion, ratio to previous survey, ratio to an optional base, a difference estimate, and a censored estimate. The censored estimate truncates the distribution within a state at about the upper one percent of the sampling units and substitutes means and variances computed from similar data over several seasons for the rejected values. When distributions are relatively stable between seasons, the bias of this procedure is small and results in a smaller sampling error than when the truncated portion of the distribution is included in the estimate.

The December survey is conducted as a subsample of the June survey. The sample is about one sixth the size of the June survey sample, but is supplemented by a larger sample of livestock farms from the list frame. The correlation between planted and harvested acreages is quite high, so the acreage estimates are almost as precise as those for June. For livestock items, correlations are lower, and relative standard errors from the area sample are roughly twice those obtained in June, hence the necessity for a larger list sample.

Techniques of estimating and forecasting yields have been developed that are based on counts and measurements rather than on grower appraisals. It is apparent that when crops are mature and ready for harvest, estimating yield is nothing more than a sampling problem. Crop-cutting techniques based on a well-designed sample of suitably sized plots can produce estimates of yield that are as precise as need be, provided biases associated with small plots can be controlled. Forecasting yield before the crop is mature and even before the fruit has been set by the plant is more difficult. Since yield has not materialized, it is necessary to use as components of yield the number of plants, the number of fruit, and the weight or size of the fruit. Each of these can be measured while the crop is immature, and by means of relationships determined empirically, each can be projected reasonably well to magnitude at harvest.

The estimates and forecasts generated by about 1,300 supporting surveys per year and published in some 700 reports constitute the current program of agricultural statistics in the United States.

Prospective Utilization of Sensor Data

The U.S. and Turkish programs of agricultural statistics and supporting methodology show many similarities. The scope of the Turkish program is rather limited, and many users of the statistics think both content and accuracy are inadequate. However, a similar situation exists in the United States. The methodology employed in Turkey is primitive, but improvement would require a much greater expenditure to develop the requisites for sounder survey techniques. Methodology in the United States is somewhat more advanced and is in the process of rapid change, but here, too, progress requires additional resources. In both countries, it is apparent that the programs require a system of adequate supporting surveys and that the accuracy of the statistics is a function of survey methodology within the constraints of applicable theory and available resources.

It is evident from the concepts described earlier that the theory of sampling in its present development is flexible enough to utilize many different inputs of auxiliary information for increasing efficiency and reducing survey costs. Remote sensing as a possible input into a system of sample surveys is a very attractive prospect.

Utilization of an input from a remote sensor would, I think, depend on two conditions. The first is the level of discrimination attained by means of the sensor, of the difference between classes of objects on or near the earth's surface, together with the probability of a correct identification. The second is the degree of accuracy and associated error function (not derived here) with which the identified object can be located on the earth's surface and with which a homogeneous area can be measured.

For the purpose of speculation, assume that precise locations and measurements of areas are possible from a sensing device on some sort of remote platform, and that the platform can be programmed for any desired path over the earth's surface. This seems a reasonable assumption in view of modern developments in navigation and space technology. Also assume that in interpreting sensor output (by tonal signature, pattern recognition, or whatever the technique might be), a probability statement may be attached to the identification at each level of discrimination. Now, let us speculate about two levels of discrimination as follows: the sensor can distinguish between (1) water and land, bare soil and vegetation; and (2) between plants with different growth characteristics, e.g., pasture grasses, small grains, vegetables, corn and sorghum, and orchards; and provide two or three gradations of plant vigor, reflecting the incidence of disease or insect infestations and available moisture. The reason for choosing these two levels is that technological developments making possible the attainment of both seem reasonably imminent.

We now examine possibilities in three settings: an undeveloped country, a developing nation, and the United States.

In Nepal, the agricultural sector of the economy is by far the largest and most important. Inventory and production statistics are derived largely from population and consumption statistics, which are themselves of questionable accuracy. Little is known of the economics of production or marketing. The primitive culture makes data collection of any kind difficult and uncertain. The lack of survey control measures and difficulty of quantitative assessment make the application of conventional methodology hazardous. In this setting, it seems likely that sensor data, at level one, could be utilized either to stratify the land area of the country or to subsample it. These strata might be refined by whatever is known about climate, soil, and agricultural practice. Then, point or line samples from air or ground could be used to establish ground truth, to provide estimates of the fraction of non-forested land planted to the different crops. In this case, the sensor output is used as a frame, stratified by non-cropland (desert, forest, water) and land subject to cultivation.

Utilization of sensor output in this way requires only a simple application of existing theory. The sensor output in the form of conventional imagery showing the boundaries of contiguous areas might well be the starting point. Scale could be rectified as necessary. Isotropic areas delineated on the imagery together with the sensor delineation of land subject to cultivation would provide the stratification. Strata would simply consist of the land subject to cultivation within each of the areas likely to have a degree of similarity in agriculture. The land area of strata could be determined by measurement. Then, a suitable sample of points (which may be clustered) could provide a sample estimate (and its standard error) of the fraction of the area of each stratum planted to the different crops and an estimate of total area within the stratum (and the standard error of that total) planted to each crop obtained by multiplying each fraction (and standard error) by the total land area of the stratum. Strata totals (and variances) are additive to national totals (from Equations 4 and 5 multiplied by N and N^2, respectively). Point sampling is suggested to avoid the problem of small fields, the areas of which are not precisely known and that would otherwise require measurement.

Subsampling could be done by selecting samples of parcels of cultivated land within strata, by sampling strata (unlikely), or by securing imagery or other sensor output of parts (samples) of the land area of the country. The latter, which would require cartography as a basic sampling frame for the definition and selection of sampling units to be covered by the sensor, is more likely to be the cost-minimizing design. Subsamples of points, probably clustered, could provide sample estimates by methods described in the section on subsampling, if the subsample is self-weighting by either of the methods described; if not, weights are required.

The sampling units, sample design, and estimators to be employed should be determined by the kind and amount of pertinent information from sources independent of the survey and by variances and costs. Utilization of independent information (as stressed in the section on sampling) is the means by which the sampling statistician reduces sampling errors; application of cost constraints is the means by which he minimizes sampling errors with respect to the expenditure of resources. These principles apply in this case.

Now assume level two. For the undeveloped country, areas planted to the different types of crops, together with a crude measure of yield or yield potential, would likely indicate food production (in absolute or relative terms) accurately enough for planning purposes. Sensor data alone would provide rough measures both of total production and of production by locality. In a country like Nepal, where cereals and vegetables constitute the diet, where practically all cereals are for human consumption, and where much of the food produced is consumed by the producer, a measure of production by type of crop by locality should be adequate. In case the production of each crop should be needed, sensor data for areas planted to the different crop types could be sampled to provide estimates of areas and yields of individual crops.

This could be done by sampling within type strata if sensor data covered the entire country, or by subsampling if the data covered primary samples of the country. Techniques of stratified sampling and subsampling would be similar to those employed under level one, except for the additional information (crop type) that would make increased efficiency possible. If cartography suitable for an area frame were available, double sampling could be employed to adjust area measurements from sensor data for distortions of scale. Sensor data would provide the primary sample. Area sampling units from sensor data should be highly correlated with similar units from cartography, making double sampling highly efficient. A small subsample of sensor area units measured on high-quality cartography should provide an adjustment factor for sensor data. As discussed in the section on sampling concepts, either the ratio or regression estimators would be used, depending upon the relationship between the sampling units as measured from sensor data and from cartography.

As is usual, the sampling statistician will be concerned with both variances and costs. The design and allocation of the sample must be done with a view to obtaining the most precision per unit of resources expended. In this case, optimization of the sample requires applying cost constraints to sensor data as well as to the dual-purpose area sample to obtain ground truth for graduation and adjustment.

In a developing country such as Turkey, level-one discrimination seems to have little to offer. The land areas of political subdivisions down to the district are known, and there is little forestland in the major crop areas of the country. Level-one sensor output would probably contribute little useful information. Level two, however, is a different story. For most crops, areas planted could probably be determined by adjusting sensor estimates of areas devoted to crop types by the relative fractions within types as reported by district technicians. For the important grain crops, sensor estimates of areas could be subsampled to determine the fraction of total planted to each grain and to estimate yield by crop-cutting at the same time.

For the animal and animal-products estimates, sensor data would likely be of marginal value. The same would be true for stocks, expenditures, marketing channels, and consumption statistics. At level two, none of the raw data is discernible to the sensor. Statistics of this kind would have to be generated by means other than remote sensing. However, economies would undoubtedly result from a general-purpose sample survey utilizing sensor and ground samples designed and allocated to optimize crop and other estimates with respect to variances and costs. It should be possible to extend the theory of nonlinear programming (mentioned earlier) to solve this very complex problem.

In the United States, with an extensive program of current statistics based on a complex system of sample surveys utilizing different frames and combinations of frames, speculation about the utilization of sensor data is more difficult. It is clear, however, that level-one sensor information could be utilized effectively in constructing and maintaining sampling frames. The growth of towns and cities, alteration of forest boundaries, new irrigation projects, and other changes in land utilization would be discernible to the sensor. The ability to maintain sharply delineated land-utilization strata in the area frame should reduce within-strata variability and should increase efficiency. This capability should be of particular value for crop statistics because of the greater reliance on the closed segment area sample in the generation of crop-production statistics and of the effect on variance of incorrect stratification. For livestock found on pasture, range, and forestland, gains in efficiency resulting from a better stratification of the area frame would likely be more modest.

Level-two information should be of much greater value as input into the data-collection system. As in less-developed countries, for crop statistics the area sample could become a subsample or double sample of sensor data. However, in general-purpose surveys, the area sample would have other functions. These include providing area sample (only) estimates for characteristics most easily associated with land and, in conjunction with the different lists, providing multiple-frame estimates for those characteristics that may be associated with an individual, a farm, or an establishment.

In the complex data-collection system required to support a broad program of current agricultural statistics, the extent to which inputs of sensor data could be utilized would again be subject to considerations of cost. The relative contribution of sensor data to the sampling precision of the estimates being generated per dollar of cost would have to be assessed with respect to similar contributions from the other sources of data. However, it is reasonable to presume a multipurpose sensor that provides many kinds of information for many purposes other than agricultural statistics. The aggregate of data collected by a multipurpose sensor is likely to make the agricultural statistics share of sensor costs a relatively small fraction of total costs. On a multiple-use, shared-cost basis, the economics of sampling would greatly increase the optimum input of sensor data over the input on a nonshared basis.

For a given state of technology and economic conditions, monitoring land utilization and assessing food supply could probably be accomplished reasonably well by level-two discrimination, together with inputs from other data sources. But over the longer time span, it is also reasonable to presume advancement in remote-sensing technology far beyond our rather modest speculations, which could increase the ability of a sensor to discriminate and measure far more precisely than we can possibly imagine at present. Also possible is a concomitant development in the theory of statistics that would enable full utilization of the ability of such a device to collect agricultural and other data. Both developments lie within the realm of the possible. If they are deemed sufficiently important and given adequate resources, both might well materialize.

9

Research Needs: The Influence of Discrimination, Data Processing, and System Design

MARVIN R. HOLTER The University of Michigan, Ann Arbor, Michigan
Contributing Authors: H. W. COURTNEY, T. LIMPERIS

INTRODUCTION

There exist in the United States and in many other countries of the world, organizations charged with the responsibility for generating, analyzing, and employing information concerning agriculture and forestry (see Chapter 8). The conduct of these activities tends to be more intensive and more highly organized in the United States than in most other countries of the world. It is becoming apparent that even in the United States significant improvements in the performance of these functions is becoming necessary, for the many reasons discussed in Chapter 1. The quality of the types of information now generated has to be improved, additional types of information have to be provided, and the timeliness with which information and operating decisions are provided has to be much improved. These are the reasons for the interest on the part of the agriculture and forestry communities in remote sensing, because it appears that this means of data collection has the potential for making these types of improvements possible.

The existing systems in the United States, and elsewhere, tend to be based on information inputs generated directly by humans (county agents, direct mailing to farm operators, etc.). Much of this information is in the form of written reports, which are not directly amenable to automatic processing. A typical description of present methods is contained in *Statistical Reporting Service of the U.S. Department of Agriculture—Scope—Methods*.[1] The Statistical Reporting Service does not encompass all the survey and analysis operations of the Department of Agriculture, but is the largest single element thus engaged, and its methods are typical. The consequences of these largely "manual" methods are that (1) the precision of the information is not as great as will be needed in the future because of the low density of sample points and the long interpolations required between them, (2) many classes of information of increased importance are not generated, and (3) the degree of timeliness of the decisions based on that information is becoming increasingly inadequate.

There exists today a wealth of means to sense information relative to agriculture and forestry: several types of photography, ultraviolet sensors, nonphotographic visible region sensors, infrared sensors, and passive-microwave and radar sensors. The characteristics of these sensors and their outputs as they presently exist in operational form are described in Chapters 2 and 3. These sensors make it possible to generate much additional information, although to exactly what extent remains to be determined by additional research and experimentation in applying them to agricultural and forestry problems. Although they can make great amounts of additional information available, they are presently employed in the form of imagery, and the direct intercomparison and interpretation of such data by human beings is demonstrably too slow by several orders of magnitude. To illustrate the size of this problem, consider that almost any of the existing single-channel sensors can produce information at a rate between 10^6 and 10^9 elemental items of information per second. In comparison, the firing of an individual human neuron has a maximum rate of approximately 10^3 per second; i.e., there are many orders of magnitude difference in the speed of performance of a human interpreter and the speed of even a single-channel modern sensor. For an alternative way of viewing some of the problems, see Table 1, which shows communications and film requirements for covering various portions of the earth's surface at a number of different scales. It is interesting to observe in this context that the present Department of Agriculture Statistical Reporting Service[1] appears to rely very little on even ordinary aerial photography as an information source.

Although modern sensors, in themselves, will not solve the problems mentioned above, when considered in conjunction with a number of other more-or-less modern developments, the concept of a remote-sensing system emerges; and such a system does hold the promise of solving those problems. A remote-sensing system like the one illustrated

TABLE 1 Large-Area Coverage[2]

A. *Film Requirements*						
Photographic scale	1 : 60,000		1 : 800,000		1 : 2,400,000	
Area covered by 9 × 9-in. film (sq mi)	73		13,000		116,000	
Number of 9 × 9-in. frames (assuming no overlap)						
Continental U.S. (3,022,000 sq mi)	41,500		232		26	
Earth's surface (197,000,000 sq mi)	2,700,000		15,200		1,700	
Oceans and seas (139,000,000 sq mi)	1,900,000		10,700		1,200	
Land areas (58,000,000 sq mi)	800,000		4,500		500	
Weight of film (lb)[a] (assuming no overlap)						
Continental U.S.	2,560		14		1.6	
Earth's surface	167,000		940		105	
Oceans and seas	117,000		660		74	
Land areas	50,000		280		31	
B. *Data-Link Requirements*[b]						
Resolution (ft)[c]	2		25		100	
Bandwidth (mc)	10	1,000	10	1,000	10	1,000
Square miles/second	0.29	29	45.3	4,530	725	72,500
Time for area coverage						
Continental U.S.	121 days	1.21 days	18.5 hr	0.185 hr	1.16 hr	41.7 sec
Earth's surface	7860 days	78.6 days	50.4 days	12.1 hr	3.14 days	0.75 hr
Oceans and seas	5550 days	55.5 days	35.5 days	8.5 hr	2.22 days	0.53 hr
Land areas	2310 days	23.1 days	14.9 days	1.6 hr	0.92 days	0.22 hr
Swath width for real-time transmission (mi)	0.062	6.2	9.8	980	156	1,000

[a] Does not include weight of film packaging.
[b] Continuous time required to transmit imagery by data link, assuming 5 bits per picture element, 1 bit transmitted per cps of bandwidth.
[c] Dimension of barely distinguishable picture element. Corresponds to one half the distance between center lines of two resolvable lines.

in Figure 1 is defined as the synergistic result of employing, in a coordinated fashion, *a priori* knowledge of common agricultural practices and ecological relationships, modern sensors (both contact and elevated), modern data-processing equipment, modern information and decision-making theory and processing methodology, modern communications and navigation theory and devices, space-borne and airborne vehicles, and the entire structure of functions optimized by use of large-scale systems theory and practice. All the elements of such a system appear to be available or achievable with a realistic amount of research work.

In the type of remote-sensing system for which a need is foreseen, there is a continuous spectrum of activity ranging from the initial sensing operation through crop and species identification, acreage assessment, health assessment, yield forecasting and so on, down to ultimate acreage allotments, price control decisions, planting decisions,

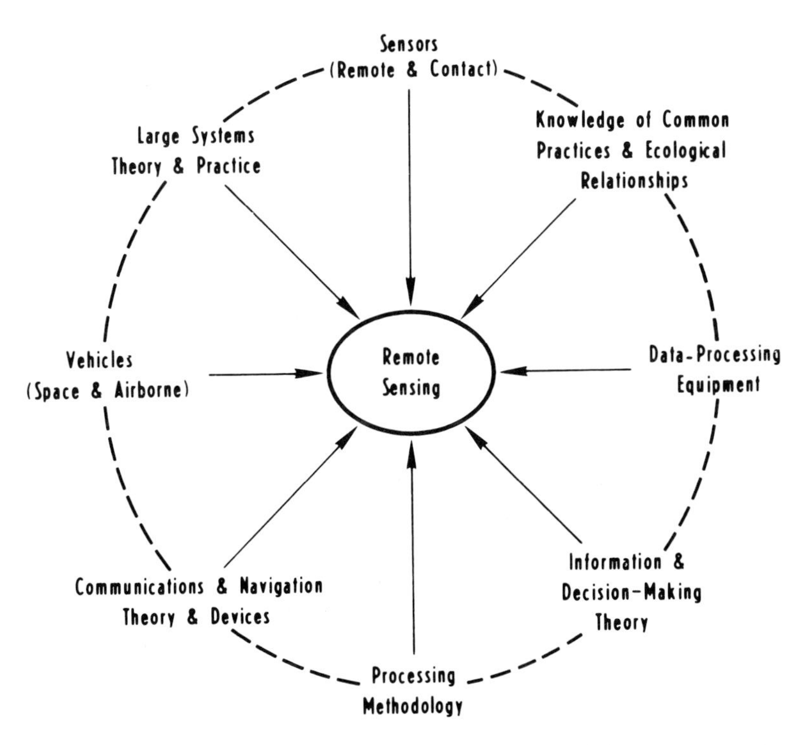

FIGURE 1 Diagram showing interrelated technologies of the remote-sensing system.

disease treatment, etc. The authors are aware that research and development are needed over this full range. However, all subsequent operations depend directly on the initial steps of sensing, species determination, and health determinations, and these functions must be at least partially automated.

The present capabilities of sensor technology appear to exceed our knowledge of how to employ the sensor outputs. Therefore, this chapter emphasizes the research needed for learning how to employ the sensor outputs to identify automatically or semiautomatically species types and their states of health and immediately related matters such as soil type and soil moisture. This will be termed the general discrimination problem. The chapter ends with a brief sketch of the over-all structure of a representative operational system and a listing of some of the more important problems downstream from the discrimination problem that will need consideration.

Discrimination

As employed here, the term "discrimination" denotes the successive classification of larger classes into smaller, more finely defined subclasses. The crudest phase of discrimination is the simple recognition that different materials are, indeed, different. This appears to be what is meant by the often-heard term "contrast enhancement." A second and more diagnostic phase is coarse identification of the materials that have been differentiated, e.g., identifying the differentiated materials as being animal, plant, mineral, or free water. A third step is to distinguish subclasses within the above (identify botanical species, soil types, etc.). A fourth step is to identify varying conditions within a species (determine age and or health within a botanical species, identify varying degrees of soil moisture and compaction, etc.). Additional and finer subclassifications can be readily imagined. Similar classifications can be made among conditions such as microclimate as well as among material things. The concern here will be with the extent to which these successively finer classifications can be made, based upon information sensed from a distance as opposed to *in situ* contact observations and measurements. The economy and convenience of the survey system will vary directly with the degree to which these classifications can be made, based on remotely sensed data with minimal recourse to contact measurements.

Basic to this process of classification or discrimination is the concept of a "signature." A signature is any collection of observable features of a material or condition that can be employed to characterize or classify the material or condition to some degree of fineness. The features that make up a signature may be observable either remotely or by contact measurements. The authors recognize that a number of features are likely to continue to require contact measurements within the foreseeable future; however, this discussion will be focused on signatures made up of remotely observable features. The features that constitute a signature may all be observed simultaneously at a single instant, or they may be sensed in a sequence of observations spread over a considerable length of time. For example, in the state of Michigan, consider a field with regular boundaries that is observed as plowed in July or August, has green vegetation obscuring most of the soil by the first snow in November, is a bright green when the snow first disappears in March, and turns yellow in June. Almost certainly, the crop to which this field is planted is a winter cereal grain, and that sequence of observables can be said to constitute a signature for winter cereal grains for that part of the world and for that climate. It is anticipated that such sequences distributed in time will, in many cases, have to be employed in any operational remote-sensing system. It is, however, obviously more rapid and economical if a collection of features is discovered that can be observed in a single instant to constitute a diagnostic signature. Therefore, the initial effort should be to search for such instantaneous signatures and to resort to time-sequential observations only when this is forced. Because of that reasoning, the discussion here is directed primarily at instantaneous signatures, recognizing the likely ultimate need to employ some time-sequentially distributed signatures.

There are a number of means of sensing more-or-less remotely: gravity sensors, static magnetic-field sensors, static electric-field sensors, acoustic sensors, gamma-ray sensors, x-ray sensors, high-energy particle sensors of many types, chemical-gas sensors, seismic sensors, and sensors of electromagnetic waves, i.e., ultraviolet, visible, infrared, microwave, and radio-frequency radiant power. For several reasons, principally, very limited range, the inability to focus or to get fine resolution, and the lack of adequate sources of natural radiation, most of these means of remote sensing are unsuitable for use from aircraft or spacecraft in applications to agriculture and forestry. For all practical purposes, the usable techniques are restricted to the sensors of electromagnetic radiation.

Four principal characteristics of electromagnetic radiation can be employed to effect remote discrimination by means of either instantaneous or time-sequential signature features. They are (1) spectral variations (i.e., variations in the sensed electromagnetic radiation as a function of wavelength); (2) spatial variations (i.e., shape and texture); (3) polarization variations (i.e., differing and perhaps characteristic amounts of polarization introduced by the materials in the radiation reflected or emitted); and (4) time variations, which are of two types. The first class of time variation consists of rapid changes—rapid enough to cause a Doppler shift in radiation reflected—that can be detected by the moving target indication (MTI) type of radar. The second class of time variation is much slower and consists of diurnal and seasonal changes observed and studied by the familiar device of comparing time sequences of observations.

Each of these four features of radiation that may be employed for discrimination interacts with the others (time sequences may be different in different spectral intervals, shape may be different at different times and in different spectral intervals, etc.). Each of these phenomena available for discrimination is discussed below in a separate section.

Spectral Discrimination

Anyone with normal eyesight is familiar with one form of spectral discrimination: using the colors of the visible spectrum as perceived by the human eye either directly or on color film (both ordinary color film and Infrared Ektachrome film). Many investigators have explored the application of color film for remote sensing (see, for example, Heller, Chapter 2; Colwell, Chapter 4; and for early detection of cereal grain diseases, references 9 and 4; for detection of potato blight, Manzer[5] and Brenchley and Dodd[6] and for detection of poor vigor in cotton arising from high soil salinity, Thomas and Myers[7] and Chapter 6). However, color films are restricted in at least two ways. First, data in photographic form are not readily adaptable to automatic processing. Automatic processing has, of course, been done with photographic data but is subject to a number of difficulties for which there is no convenient, satisfactory solution. For one thing, the registration of several photographs is difficult and never as good as can be achieved with some nonphotographic sensors, so fine resolution, the principal advantage of photography, is compromised. For another, film and film processing are more difficult to control and to keep uniform than the processes in electronic-optical sensors, so that maintaining uniform performance in a quantitative sense is more difficult. Also, it is more difficult to calibrate the records continuously with photography than with other types of sensors. Finally, it is more difficult in photography to adjust automatically for very wide signal dynamic ranges than with some other sensors. The net result is that, although automatic processing of photographic data can be done, it is more difficult and less satisfactory than with data in electronic form.

The second restriction on color film is that the fineness of the distinctions and subclassifications that can be made are limited both by the restricted spectral resolution of the film and also by the unavailability of films that function more than slightly outside the visible spectral range.

The belief that superior performance is attainable was motivated originally by spectrometer measurements in the ultraviolet, visible, and infrared; these showed that finer spectral resolution in the visible and spectral structure in the ultraviolet and infrared could be obtained. Following that, the multichannel optical-mechanical scanning sensor described in Chapter 3 and by Lowe *et al.*[8-10] was conceived and implemented at The University of Michigan. The multiband imagery (shown in Chapter 3) of Indiana farm terrain from that instrument illustrates the type of data that can be produced. It also illustrates the difficulty of human interpretation of such data when presented in pictorial form to human interpreters. Some early attempts at "manual" interpretation of such data are discussed by Legault and Polcyn.[11] However, the data, as initially produced in electronic form, can be easily synchronized exactly in both space and time by proper sensor design, can be calibrated, and if kept in electronic form, are quite amenable to automatic processing. This provides the first necessary ingredient for automatic spectral discrimination. Similar remarks apply to radar systems, although their implementation lags the optical system.

A second ingredient in spectral discrimination is the necessity for realizing that spectral signatures cannot be completely deterministic. If spectral reflectivity and emissivity measurements are made of a large number of wheat fields, for example, all at the same stage of growth, the measurements between one part of a field and another part of the same field and also between different fields will not agree exactly. At each wavelength there will be some mean value and some dispersion around it; i.e., spectral signatures are statistical in character. This should not be surprising, because it is well known that taxonomy based on any characteristic or feature demonstrates dispersion.

A third element at the foundation of discrimination theory is the necessity to realize that optimum discrimination techniques require not only that the procedures be tailored to recognize the item or material of interest, but also, and simultaneously, that they be tailored to reject other items or materials that lie in the vicinity of the desired materials but that are not of interest, i.e., the backgrounds in which the items of interest are embedded. This is simply a statement of the communications engineers' "matched-filter" concept;[12] the filter must be both matched to the signal and "matched against" the noise. To illustrate: consider the problem of detecting wheat at some stage of growth. Two types of error are possible: failure to classify all of the wheat actually present as wheat and misclassification of other crops or soils as wheat. Photo interpreters commonly call these errors of omission and commission, respectively. In the extreme, errors of the first class can be reduced to zero by classifying everything as wheat, or, more sensibly, matching the decision process as well as possible to wheat. Either of these procedures will reduce errors of class one to very small or zero values. This is not very useful, however, because the errors of class two will be very large; i.e., many things will be misclassified as wheat, and total yield estimates or whatever information is to be extracted will be grossly in error. It can be shown that in all but trivial cases it will always be found that class-two errors are large when the discrimination technique is matched only to the item of interest. To do any better requires simultaneous tailoring of the process to discriminate for the item of interest and to reject the items not of interest. It is this need that gives rise to the central importance of signatures of both items of interest and the backgrounds in which they may be imbedded. Any effective discrimination process or procedure must begin with an understanding of the signatures (in this instance spectral reflectivities and emissivities and their statistical dispersions) of all materials viewed by the sensing systems. The same factors are pertinent in spatial discrimination and, in fact, for discrimination based on any of the features of the sensed radiation.

A fourth element in discrimination is a general description of the types of decision procedures employed to take account of the statistical character of the signatures and the use of signatures of both items of interest and backgrounds. Consider a one-dimensional case, the discrimination of wheat using a single spectral band, $\Delta\lambda_1$. Some set of measurements on known wheat is used to generate a wheat signature. This is shown in Figure 2, plotted as hypothetical frequencies of occurrence of measurements versus radiation intensity for wheat and two background crops. The frequencies of occurrence can be considered as probability density functions. Assume first that only the hypothetical wheat and background No. 1 are present. Now suppose a new measurement (the signal from a single-resolution element of a scanning sensor) of a field is available and a decision must be made as to whether the crop is wheat or background No. 1. If the measurement falls to the left

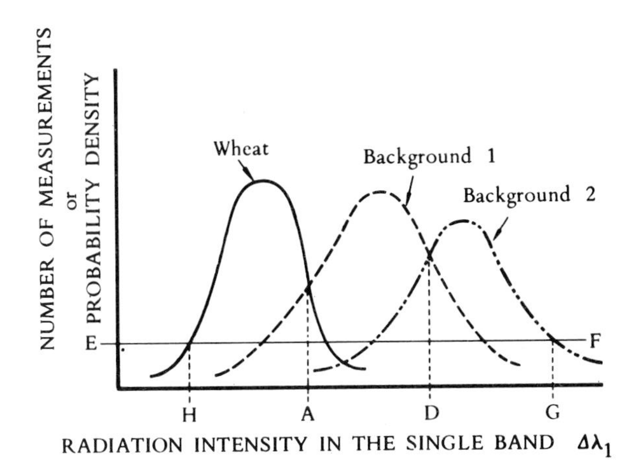

FIGURE 2 Plots of signatures of wheat and two backgrounds.

of point A, the probability is higher that the crop is wheat. A method of expressing this is as follows:

If $\dfrac{P_w}{P_{B_1}} \geqslant 1,$ decide wheat.

If $\dfrac{P_w}{P_{B_1}} < 1,$ decide background No. 1.

In both equations P_w is the probability that wheat is present, obtained by entering the graph with the observed radiation intensity from the field to be determined, and reading the corresponding number of measurements from the measurements set on known wheat used to obtain this signature. Likewise, P_{B_1} is the probability that the new measurement sample indicates the presence of background No. 1. The term P_w/P_{B_1} is the simplest form of the *likelihood ratio*, and this form assumes that the costs of making errors of the two types described above are equal. If the value V of classifying a high proportion of actual wheat correctly is different from the cost C of misclassifying background No. 1 elements as wheat, the decision line at A in Figure 2 should be shifted to the right or to the left, based on the relative magnitudes of V and C. This can be accomplished by rewriting the decision rules as:

$$\text{If} \quad \frac{VP_w}{CP_{B_1}} > 1, \quad \text{decide wheat.}$$

$$\text{If} \quad \frac{VP_w}{CP_{B_1}} < 1, \quad \text{decide background No. 1.}$$

The nature of the procedure is this: given a measurement, the probability or likelihood that wheat is present is based on the distribution of previous measurements of known wheat alone. Likewise, the probability or likelihood that background No. 1 is present is based on the distribution of previous measurements of known background No. 1 alone. On the other hand, the decision to classify the new measurement as indicating wheat or background is made on the basis of the joint effects of the two likelihoods, one for wheat and one for the background.

To extend this somewhat, assume two backgrounds present, as indicated in Figure 2. A number of representative alternative but similar decision procedures follow.

First, given a new measurement, it is possible to obtain the likelihood of wheat and each background and select the largest; i.e.,

$$\text{If} \quad P_w \geqslant P_{B_1},$$

and

$$P_w \geqslant P_{B_2}, \quad \text{decide wheat;}$$

or

$$\text{if} \quad P_{B_1} > P_w,$$

and

$$P_{B_1} \geqslant P_{B_2}, \quad \text{decide } B_1;$$

or

$$\text{if} \quad P_{B_2} > P_w,$$

and

$$P_{B_2} > P_{B_1}, \quad \text{decide } B_2.$$

In terms of Figure 2, this means that all radiation levels to the left of A lead to a wheat decision, all radiation levels between A and D lead to a B_1 decision, and all levels to the right of D lead to a B_2 decision. These decision lines can obviously be shifted by values and costs as was done above. P_w, P_{B_1}, and P_{B_2} can have values below which it is not sensible to decide that any of the materials is present. This can be implemented by establishing line EF in Figure 2, based on the sensitivity of the system and the costs and values of correct decisions and errors. Then radiation levels to the left of H and to the right of G would not lead to a decision.

Alternatively, given a new measurement, a second similar decision procedure is as follows:

$$\text{If} \quad \frac{P_w}{P_{B_1} + P_{B_2}} > K, \quad \text{decide wheat.}$$

$$\text{If} \quad \frac{P_{B_1}}{P_w + P_{B_2}} > K, \quad \text{decide } B_1.$$

$$\text{If} \quad \frac{P_{B_2}}{P_w + P_{B_1}} > K, \quad \text{decide } B_2.$$

In these equations K is a constant selected to suit the problem. Obviously, these probabilities can be weighted with costs and values as was done above. Other similar decision rules are possible but will not be described here because the purpose is to illustrate the methods without being exhaustive.

This methodology developed for measurements in a single wavelength interval $\Delta\lambda_1$ can obviously be extended to sets of simultaneous measurements in two or more wavelength intervals $\Delta\lambda_1, \Delta\lambda_2, \Delta\lambda_3, \ldots, \Delta\lambda_n$. The analytic expressions become progressively more lengthy and intricate, and the visualization gets more difficult as the number of $\Delta\lambda_i$ increases, but the principles remain the same. A review of the present status of these pattern-recognition methods is given by Nagy,[13] and a development of a branch of the field suitable for these applications is given by Sebestyn.[14]

Many investigators are attempting to extend and improve the methods of pattern recognition described in a very general way above. These efforts are directed at extending the methods so that they become "adaptive" or "self-correcting." That is, after a signature has been determined on the basis of a limited number of measurements and put into use, computational and decision procedures are sought that will examine the decisions resulting from their use and modify the signatures continuously to reduce errors. A further extension also being investigated is aimed at eliminating the need for the original measurements of known materials to generate the initial signatures. If successful, this will lead to a class of so-called "self-teaching" or "learning" procedures.

The class of techniques, with the possible extensions described very briefly above, represents the most powerful and most sophisticated techniques available and under development. A wide variety of simpler techniques of lesser but significant power have been investigated and employed. A few of them are described briefly here.

A set of n measurements of the spectral reflectivities and emissivities on a given material in n wavelength bands $\Delta\lambda_i$, $(i = 1, 2, \ldots, n)$ constitutes a sampled reflectivity and/or emissivity spectrum of the material. A hypothetical example is shown in Figure 3. The n measurements referred to here represent equally either sampled laboratory measurements or the instantaneous signal levels in the several channels of a multispectral optical-mechanical scanning sensor of the type described in Chapter 3. In the latter instance there would be n such signal levels for each resolution element or instantaneous field of view of the instrument; i.e., such a scanner can generate a sampled spectrum for each

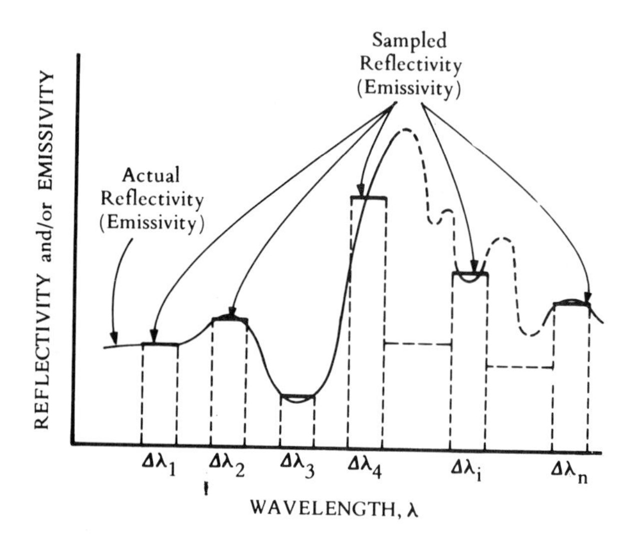

FIGURE 3 Diagram of actual versus sampled reflectivity (emissivity).

resolution element. The time-varying signal from any one of those n channels can be employed to produce a single-band pictorial presentation, as was done for each of the eighteen bands shown in Plates 6 and 7, which portray typical Indiana farm countryside. However, a more useful method of employing the n signals is to combine them to produce a single picture in such a manner that materials of interest are enhanced in contrast and, if possible, can be automatically identified.

The method of plotting these n measurements or instantaneous signals (Figure 3) brings out their relationship to the familiar spectral curves but is not the most convenient portrayal for understanding the discrimination methods. It is useful to conceptualize these methods by employing geometrical notions. Consider the n measurements or instantaneous signal levels to be the components of an n-dimensional vector. Imagine this vector plotted in an n-dimensional rectangular Cartesian coordinate system with its tail at the origin. Then the head of each vector defines a point in hyperspace.

As an aid in visualizing this, consider an example where $n = 2$, i.e., measurements or instantaneous signals in only two wavelength bands, $\Delta\lambda_1$ and $\Delta\lambda_2$. Refer to Figure 4, where each small dot represents the head of such a vector, the horizontal coordinate of each point being the measurement in band $\Delta\lambda_1$, and the vertical coordinate being the measurement in $\Delta\lambda_2$. The dots are clustered into three groups labeled A, B, and C, with a boundary drawn around each. Suppose each dot in group A represents a pair of measurements on a crop such as wheat. As mentioned earlier, a series of separate measurements on a single crop such as wheat will not all agree exactly; there will be some dispersion. The cloud of points, each slightly displaced from its neighbors, is the expression of this dispersion when the points are plotted in "color space." Near the cluster labeled A is another recognizable cluster labeled B; suppose it to be oats. Similarly, suppose cluster C to be bare earth. Each of these points represents a pair of measurements or instantaneous signal levels in each of two wavelength bands $\Delta\lambda_1$ and $\Delta\lambda_2$ from a single "object," i.e., a small patch of crop or soil. Each cluster represents a large number of measurements on a large number of individual objects all belonging to the same object class, e.g., wheat patches.

Given a set of measurements in two bands of known materials, a set of signatures exists, and the signatures can be displayed as three regions (Figure 4). Now suppose a new pair of measurements is to be classified. A simple procedure is to establish a rectangular region, containing class A, for instance, and to note whether the new pair of measurements, when plotted on a diagram of the type shown in Figure 5, falls

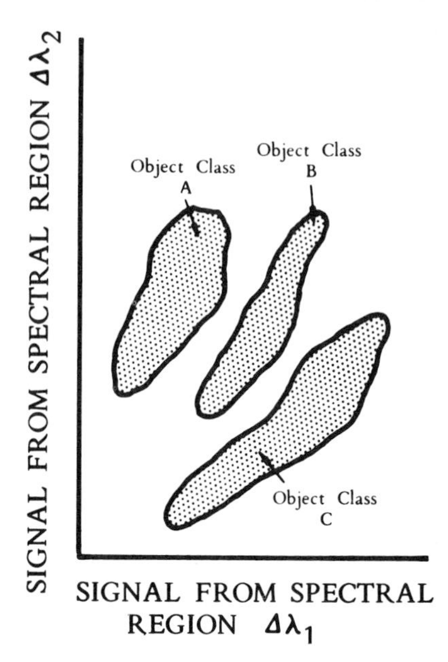

FIGURE 4 A representation of objects in two-color space.

inside or outside the rectangular "decision" region. If the pair of measurements to be classified is designated as S_1 and S_2, when object class A is wheat, this decision operation can be expressed as:

$$\text{If} \quad S_1^a \leqslant S_1 \leqslant S_1^a;$$

and

$$S_2^a \leqslant S_2 \leqslant S_2^a, \text{ decide } S_1, S_2 \text{ indicates wheat.}$$

If either of the above is not satisfied, decide S_1, S_2 does not indicate wheat. Similar expressions apply for object classes B and C. This procedure is very like that of a spectroscopist when he compares a spectral measurement to a set of known spectra. This procedure can be extended in an obvious way to measurements and signatures made up of more than two wavelength bands. For three wavelength bands, the decision region is a rectangular parallelepiped, and there are three sets of inequalities to be satisfied. For more than three wavelength bands, the situation cannot be sketched or visualized directly, but the decision methodology carries over directly to sets of measurements in any number of bands.

It is apparent from Figure 5 that rectangular regions with edges parallel to the coordinate axis will lead to some misclassifications. Witness that the lower left corner of the decision region includes some points from object class *B*. Therefore, decision regions of shapes other than rectangular might be better. Another possibility is to determine the mean of the positions of the points constituting the signature and to designate the mean as $\bar{S}^{1a}$, $\bar{S}^{2a}$. This will be the centroid of region *A*. Then the decision can be expressed as follows:

$$\text{If} \quad \sqrt{(S_1 - \bar{S}_1{}^a)^2 + (S_2 - S_2{}^a)^2} \leqslant K, \quad \text{decide wheat;}$$

and

$$\text{if} \quad \sqrt{(S_1 - \bar{S}_1{}^a)^2 + (S_2 - S_2{}^a)^2} \geqslant K, \quad \text{decide not wheat.}$$

In these expressions, *K* is a constant determined by the signature. This decision rule viewed geometrically is merely a circle of radius *K* containing region *A*. Any other analytically expressible closed regions, e.g., ellipses, can be used as a basis for similar decision procedures.

The decision regions need not be closed figures. Shown in Figure 6

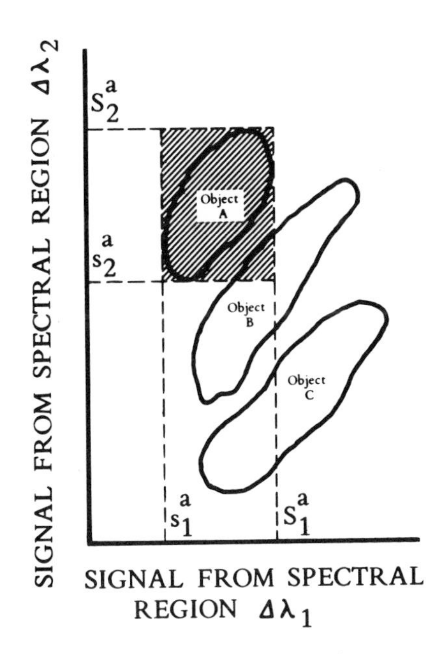

FIGURE 5 Diagram of two-color spectral matching.

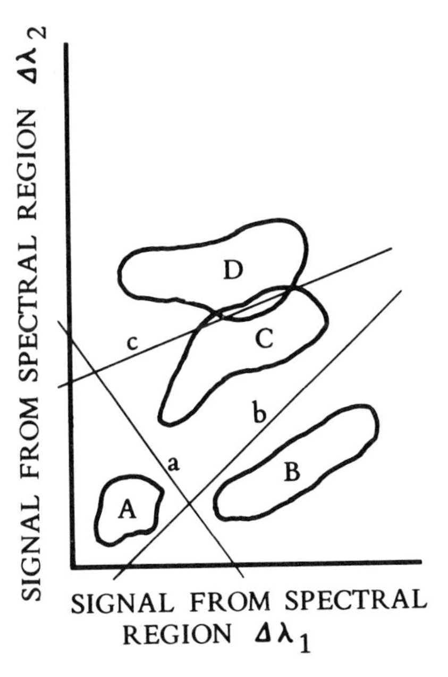

FIGURE 6 Other linear decision rules for discrimination.

is a set of decision regions made up of straight lines. In a manner similar to the above, a decision that a new measurement indicates C can be made when the new point is above line b and below line c and to the right of line a, etc. Nor do the lines have to be straight. They can be hyperbolas, parabolas, or any other analytically expressible curves. These methods are also extendible, without complication, to measurement and signature sets made up using any number of wavelength bands greater than 2.

The likelihood-ratio methods discussed first above, which were alleged to be the most powerful, are extensions of these simpler methods. The dispersion of the points representing each object class is employed to construct a decision region without sharply defined boundaries. Rather, the boundaries are bands within which the frequency of occurrence of points decreases toward zero; i.e., the probability that a new point at each position represents that object class decreases as the distance from the center of the cluster increases. Figure 2 shows a sketch of this for measurements and signatures using a single wavelength band, and the method is extendible to any number of bands. The statistical approach provides a more satisfactory decision than the simpler methods for those regions where the object classes overlap, as they do in a portion of Figure 6. It is common practice to describe the statistical distributions of points making up a signature as multivariate Gaussian distributions. This is an approximation invoked to simplify the analytical expressions and the electronic implementations. The cost, in increased errors, of making this approximation for ease of computation and implementation is not known with any degree of precision. This is an area requiring investigation.

Employing reasoning similar to the above, it is possible to show an important limitation of photographic methods. Assume the use of panchromatic film and a filter passing light in two wavelength bands, $\Delta\lambda_1$ and $\Delta\lambda_2$, as representative of all the "best film–filter combinations." Much of the information in a photograph resides in the differing densities of the photographic images of different objects. The film density, d, is proportional to the sum of the radiant signal S_1 in band $\Delta\lambda_1$ and S_2 in band $\Delta\lambda_2$; i.e.,

$$d \approx T_1 S_1 + T_2 S_2,$$

where T_1 and T_2 are the filter transmissions and T_1, T_2, S_1, and S_2 are all positive quantities. Referring to Figure 6, this means that all the different possible combinations of S_1 and S_2 that can occur giving rise to the same density will represent points lying along a straight line that, by the above equation, must intersect both the horizontal and the vertical axis on the positive side of the origin, i.e., a line much like line *a* in the figure. This means that the film–filter combinations cannot discriminate between objects whose two band measurements plot along such a line because they will all produce the same photographic density. The film will distinguish between points lying to the right and left of line *a* because their photographic densities will differ. The significance of this is that photographic film–filter combinations can be found that will differentiate, by density, object class *A* from the other object classes *B*, *C*, and *D*; but no film–filter combination can possibly differentiate, by density, the object classes *B*, *C*, and *D*. (An algorithm for calculating optimum filter characteristics for photographic sensors is developed by Legault and Riordan.[15])

Another more general way of stating the general limitation of photographic systems is that the discrimination processes that can be implemented by selecting best film–filter combinations are restricted to those processes that can be represented by linear first-order equations with all positive terms. That is a severely restricted class of processes. The nonphotographic multispectral sensing technique is an enormously more powerful tool for spectral discrimination. To illustrate this, consider an example: Suppose that the visible region can be divided into 12 wavelength bands by filters and that 10 levels of intensity can be discerned in each band. The maximum photographic film density will occur when the maximum intensity is present in each band. Below that there will be $10 \times 12 = 120$ discernible density gradations possible. Each of the intermediate densities can come about in more than one way by having more or less intensity in each band as long as the sum for all the bands is the same. It is for this reason that different objects can produce the same film densities. A nonphotographic multispectral sensor does not have this redundancy or lack of discriminating ability. The intensity in each band is sensed independently. Therefore, the total number of different input conditions that can be recognized in the same bands as the photographic system is 10^{12}. This is a much larger number than the 120 for the photographic system. Some improvement is obtainable by using color film in the photographic system, but it still falls far short of the nonphotographic multispectral system. As a simplified model, assume that each color-film emulsion is sensitive to one third of the twelve bands; then each emulsion will have a maximum of $10 \times 12/3 = 40$ possible densities. Three of these are independently recognizable. Thus the color film will discern 40^3 different input conditions. For comparison, then, the various methods have the following "information capacities":

Panchromatic film	1.2×10^2 states
Color film	6.4×10^4 states
Nonphotographic multispectral sensor	10^{12} states

One of the pattern-recognition methods of intermediate capability will be of interest later in discussing some processing results. Therefore, the method of implementing the method is described. As mentioned in Chapter 3, a commonly employed method of producing imagery from the electronic signal resulting from the use of a scanning sensor is to modulate the intensity (z-axis) of a cathode-ray tube (CRT) while moving the CRT beam horizontally (along the x-axis) in a manner determined by the scanning of the sensor and by advancing the photographic film so that each scanned "line" is displaced from but contiguous with the one immediately preceding it. The result is a pictorial presentation on the film. This is not the only way the electronic signal can be displayed on a CRT. Instead of the above method, suppose the signal from one wavelength band, $\Delta\lambda_1$, is employed to deflect the CRT beam horizontally (along the x-axis) and the signal from a second wavelength band, $\Delta\lambda_2$, is employed to deflect the CRT beam vertically (along the y-axis), and that the intensity of the CRT beam is held constant. The result, on the face of the CRT, is a "color-space" plot of the data such as was shown for a hypothetical case in Figures 4–6. Photographs of actual CRT plots of scanner data obtained over an agricultural area near Davis, California, in June of 1966 are shown in Figure 7. Only two types of materials are present: immature rice and bare earth. The part of the area from which the data were taken to produce this "color-space" plot is indicated in Figure 7. In part *A* of the figure, the signals from the two fields are geometrically separated for the two wavelength bands used. Therefore these two bands will be useful for two-color discrimination. In part *B* of the figure, another pair of wavelength bands is used, and the fields are not geometrically separated. These two bands will not be useful for discriminating between these two materials.

Suppose that for the CRT presentation shown in part *A* of Figure 7 a piece of cardboard covered the entire face of the CRT. Then, suppose that a hole was cut in this cardboard that conformed to the shape of the blob on the CRT face that represents material type *A*. This is illustrated in part *C* of the figure. Now a photodiode exposed to the entire area of the cardboard mask will be activated and have an elec-

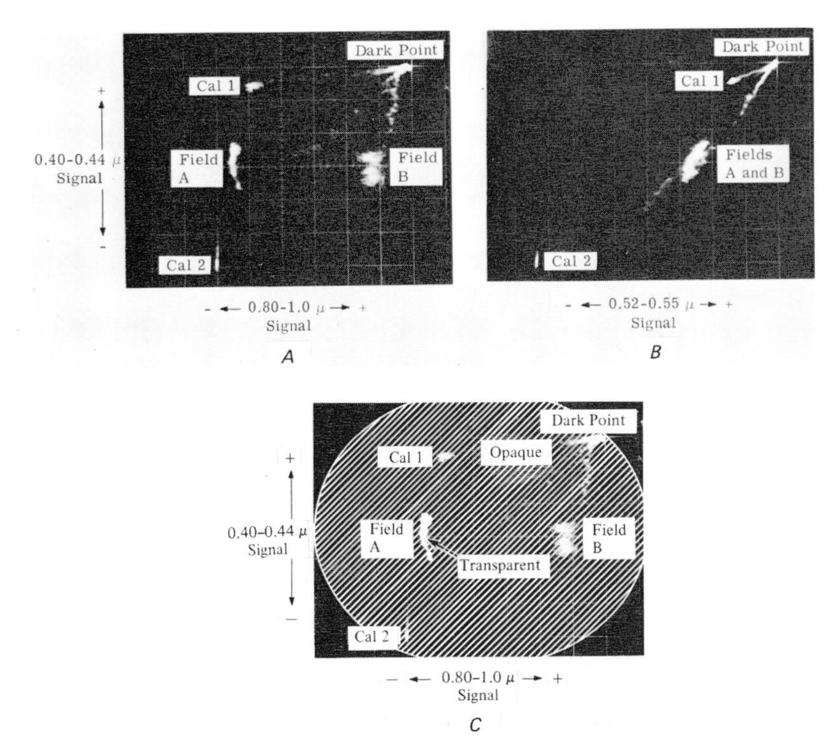

FIGURE 7 Cathode-ray-tube plots of scanner data for California fields, May 26, 1966, 2:00 P.M. *A*: typical *x-y* scope picture of fields not geometrically separated; *B*: typical *x-y* scope picture of fields not geometrically separated; *C*: cathode-ray-tube presentation masked. (*University of Michigan.*)

trical output only when the beam of the CRT is positioned under the hole in the mask, i.e., only when material of type *A* is causing the pair of electrical signals being displayed on the CRT. If, in conjunction with this type of display, the type of display and photographic recording described above and in Chapter 3 that results in a pictorial type of presentation on the film is employed simultaneously, and if the intensity (*z*-axis) of the CRT used for the pictorial display is suppressed except when the photodiode viewing the cardboard mask is activated, then there will appear on the pictorial display an image of material type *A* only. Alternatively, materials of types *A* can be presented but at reduced contrast or in colors different from a color assigned to material *A*. In short, a pictorial display is produced in which material *A* is automatically recognized and designated by one of several possible methods. Examples of this are shown and discussed below.

The implementation of this technique is deceptively simple. The technique is actually a very powerful discrimination tool. The freedom to cut holes of any shape in the cardboard mask and to give them edges that are not sharp by constructing neutral variable-density masks makes the method in some, but not all, respects more powerful than the likelihood-ratio methods, because the shapes of the decision regions that can be implemented by the likelihood-ratio methods are somewhat restricted by the amount of electronic circuitry or digital computer time that can be afforded.

This discrimination method has been called the *light-pencil technique* by scientists at The University of Michigan, where the technique was conceived and implemented. It is described more fully by Polcyn[16] and Hasell,[17] who also describe other discrimination techniques. The light-pencil technique is not restricted to two wavelength bands, as might appear at first glance. Any number of CRT displays with masks may be employed, each CRT displaying two bands. The outputs of the photodiodes, one for each CRT, may then be combined with computer-type "and" or "or" gates or other decision-making components. When this is done, in terms of the previous geometrical notations, it is equivalent to effecting the discriminations based on the projections on the coordinate planes of the complete signature distribution plotted in a many-dimensioned rectangular Cartesian coordinate system. (Actually any orthogonal coordinate set may be employed.) It is in this respect that the light-pencil technique is not as powerful a tool as the likelihood-ratio methods, since the latter employ the full hyperdimensional distributions rather than merely their projections on the coordinate planes and therefore contain more information.

Any of the discrimination methods described above can be implemented with digital, analog, or hybrid computational circuitry. Which may be optimum depends on the specifics of the application and is decided on the bases of convenience and cost as determined by those specifics.

With regard to the future research that is required to bring these methods to operational fruition, it is readily apparent that development, evaluation, and employment of these discrimination techniques all rest directly on a knowledge of the signatures. This knowledge is quite inadequate, and its acquisition and analysis represents the pacing item in the development of adequate discrimination means. Although research will be needed on sensors and on the theory of discrimination, these two areas are presently in a somewhat more advanced state than knowledge and understanding of the signatures.

Finally, a word concerning the more distant future. There is reason to believe that the discrimination techniques now being developed and applied to the relatively concrete problems of materials classification may be extendible to carry out some of the more abstract but still fairly routine decision-making functions downstream from the materials-classification function. It is not inconceivable that ultimately some of the decisions leading to acreage allotments, price supports, and other operating decisions may be automated by further developments in discrimination and pattern-classification methods. Learning how to classify the materials represents a significant step toward some elementary form of artificial intelligence about which there has been much conjecture in recent years.

Examples of Spectral Discrimination Imagery and discrimination results for data obtained at four geographical areas are presented and discussed in this section. One of the sites is near Davis, California; the other three are in the vicinity of Lafayette, Indiana. In all instances the aerial photography and scanner data were obtained by scientists at The University of Michigan. Three of the sets of discrimination results were produced by scientists at The University of Michigan. The fourth set was produced by scientists at Purdue University. Ground data and measurements were provided by scientists of the University of California, Berkeley, for the Davis site, and by scientists of Purdue University for the Lafayette sites.

The results presented demonstrate great progress in discrimination over current operational aerial survey techniques. Lest they be misinterpreted, however, it is well to point out here with equal emphasis the limitations of what they demonstrate. The discrimination results shown have been achieved by employing the data from a single flight over each site to derive signatures and by then applying those signatures to data from the same flights used to generate the signatures. Therefore, what is demonstrated is that signatures can be generated and that a high order of discrimination can be achieved by applying the signatures in pattern-recognition processes to data taken on the same date, at approximately the same time, with the same weather and sun angles, and at geographical areas in the immediate vicinity of the areas used in deriving the signatures. The extent to which those signatures and those decision processes can be employed to effect satisfactory levels of discrimination at different times of day, at other times of the year, under different weather conditions, at geographically distinct locations, etc. remains to be determined. However, the level of performance achieved under the restricted conditions outlined above does provide encouragement and motivation for continuing the research to

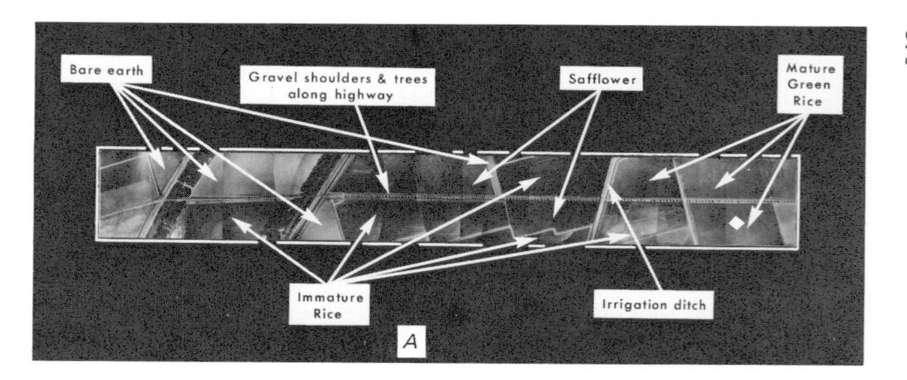

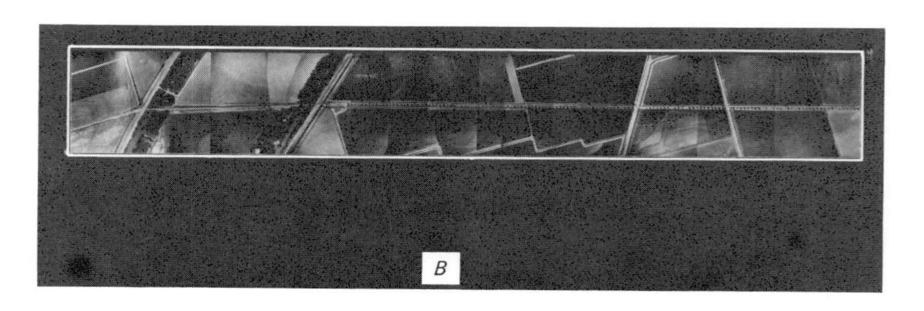

FIGURE 8 Panchromatic mosaic of agricultural area near Davis, California, May 26, 1966, 4:00 P.M.; altitude 2,000 ft; sky condition, clear and bright, 10 percent cloud cover at 30,000 ft; surface temperature 27° C. *A:* with annotations; *B:* without annotations. *(University of Michigan.)*

investigate the extension of the methods in time, space, and a variety of environmental conditions. Based on the results that have been obtained, the prognosis for achieving marked improvements under the wider variety of conditions is favorable.

Data are presented first for an agricultural area in the vicinity of Davis, California. A number of display and recognition results are shown. Figure 8*A* is a photographic mosaic of the area produced from panchromatic film exposed with a K-17 aerial camera, with a K-2 filter used under conditions described in the figure caption. As indicated there, only four types of surface material are present in significant amounts: mature green rice, immature rice, safflower, and bare soil. Figure 8*B* shows the same panchromatic photographic mosaic without annotations. Figures 9*A* and 9*B* show 18 images of the same area, each in a different narrow-wavelength band, produced by a nonphotographic optical-mechanical-multispectral scanning sensor of the type described in Chapter 3. Two things are immediately apparent: by the simple expedient of making the observations in narrow spectral intervals, greater contrasts among the materials present are frequently achieved; and the relative contrasts tend to be different in the different bands. The latter fact suggests that spectral discrimination may be profitable.

Before we proceed to the automatic discrimination results, it is worth pointing out several other features and uses of such data. Having data available in many narrow spectral bands makes it unnecessary, except for purposes of proving the point, to employ photographic sensors. Results identical to those obtainable with photography (except the very high geometrical resolution and fidelity of photography) can be obtained by combining and printing data from the subset of the total scanner channels that cover the same spectral regions covered by the photography. Plate 22*A* is an image reconstituted from scanner imagery to appear as panchromatic photography. The reconstituted image is an improvement over the mosaic in that it has greater fidelity with respect to the real situation. In the panchromatic mosaic, note that the four large fields at the right of the image, three identified as mature green rice and one as immature rice, are lighter in tone than the adjacent fields to their left, whereas in the reconstituted panchromatic image they are not. Examination of the original mosaic, where the individual frames overlap along the irrigation ditch that bounds the two groups of fields, reveals that the lighter fields are artifacts, the exposure of the lighter-toned prints having been different than the darker-toned ones. In this respect, the reconstituted image is a more faithful reproduction of nature than the camera-produced mosaic.

Plate 22*B* shows a reconstituted image representing the way an infra-

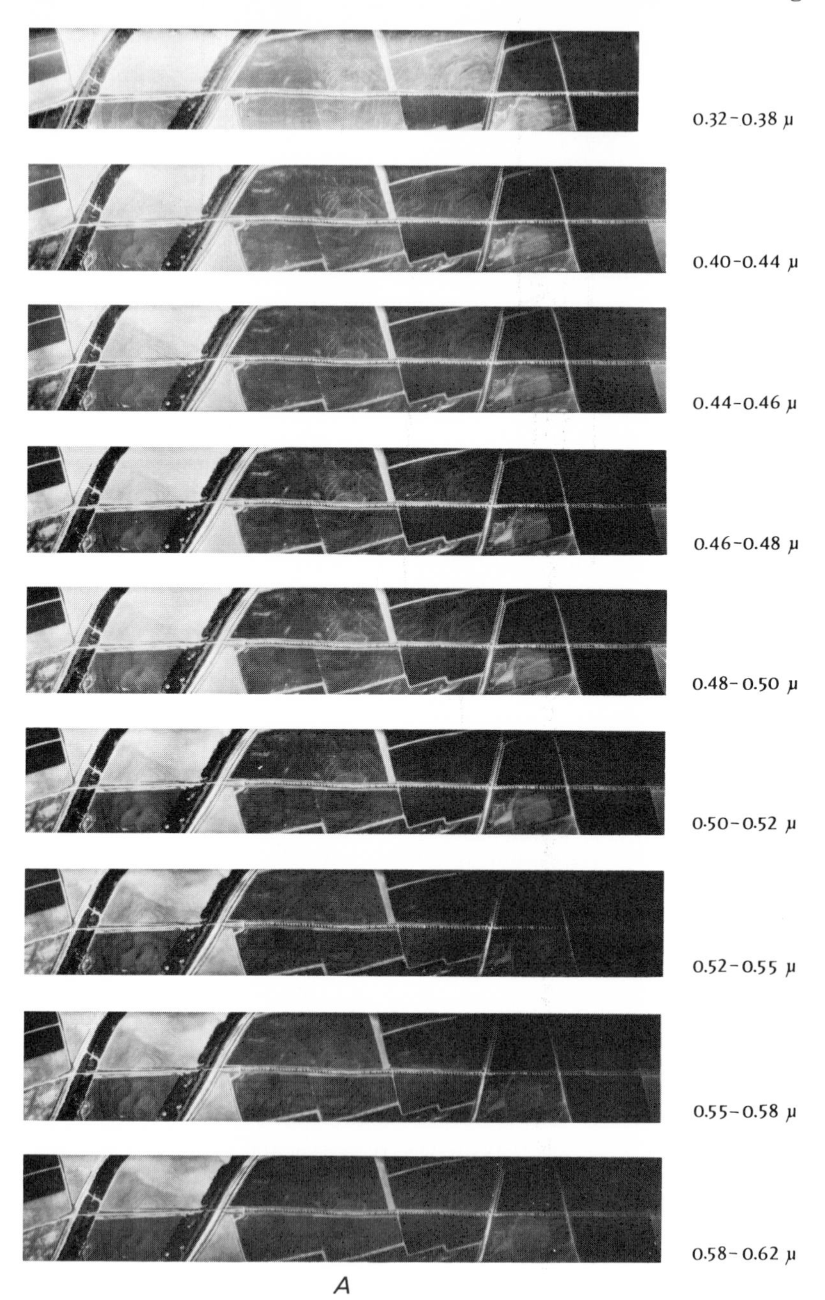

FIGURE 9 Multispectral imagery of area shown in Figure 8, same conditions. *A:* nine wavelength bands from 0.32–0.38 μ to 0.58–0.62 μ.

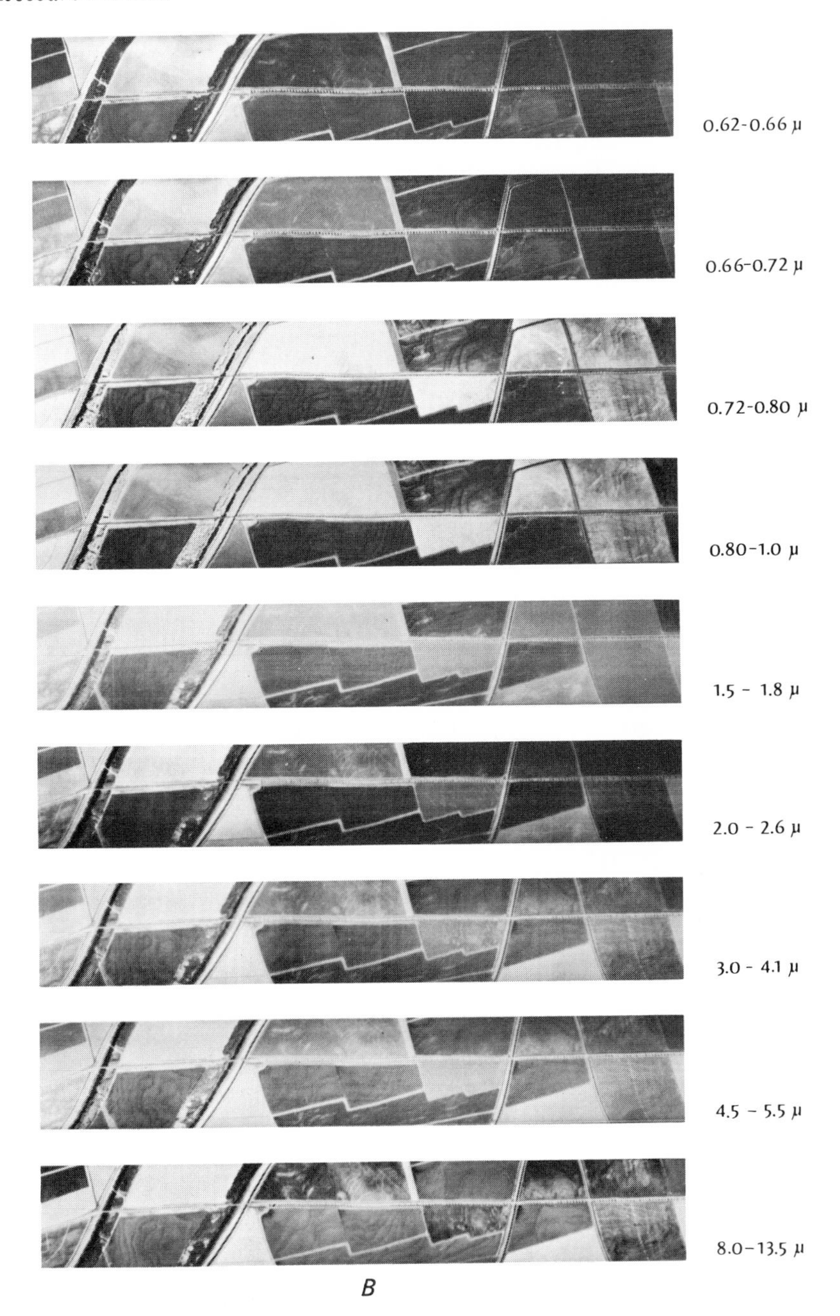

B: Nine wavelength bands from 0.62–0.66 μ to 8.0–13.5 μ. (*University of Michigan.*)

red aerographic film would appear. Plate 22*C* shows a reconstruction of the way normal color film would appear. This differs from color film in that it is more faithful in reproducing the surface colors. On all color films exposed at altitude the images have more or less of a bluish cast, a result of differential scattering of the visible wavelengths. This blue "haze" is absent from the reconstituted color image because it is possible to remove most of it from the electronic data before the image is printed. Such removal is not easily possible with the photographic sensor. Plate 22*D* shows a reconstructed Infrared Ektachrome (sometimes called infrared-camouflage) film image.

In the imagery presented here, three distinctly different uses of color are used. Ordinary color and Infrared Ektachrome film and reconstituted images represent the first of these three uses of color, namely, the coding of wavelength in color. Ordinary color film is not usually regarded as a coding of wavelength by color, because the coding is very much as the human eye sees the colors, but it is no less a color coding than any false-color film such as Infrared Ektachrome. There is no technical reason why any wavelength, visible or nonvisible, cannot be coded in visible colors, especially when the data are available in many narrow bands as represented in Figure 9.

A few of the many other possible color-coded wavelength presentations are shown. Plate 22*E* is, in a sense, the symmetric twin of Infrared Ektachrome photography in that two visible bands and one ultraviolet band are coded in color. There are tonal distinctions, although of a somewhat subtle character, present in this image that are not apparent in the images previously discussed; presumably, additional information can be obtained from Plate 22*E*. Exactly what the added tonal distinctions represent and to what use they may be put are not known because the details of the ground situation are not available and little is known regarding the spectral signatures of these crops and soils. The yellow band along the bottom of this image is a correctable defect of the existing sensor system. (One channel has a slightly different ground field of view than the others, and the yellow band represents the region out of the field of view of that channel. It was deemed preferable to present this image to include the same ground region as the other imagery rather than to remove the defect by cropping this image.)

Plate 22*F* is a color coding of three infrared bands as indicated. This is how the area might appear to infrared-sensitive eyes. There are some tonal distinctions that can be made here that are not apparent on any of the previous images. The strongest of these is the pair indicated by the arrows on the figure. The reason for this tonal difference between two fields containing the same crop, immature rice, will be clarified later. Plate 22*G* is a color coding of one ultraviolet, one visible, and one infrared band. Again, there are differences not noticeable in the reconstituted photographic imagery. Here too, the strongest difference is seen in the pair of fields containing the same crop noted on Plate 22*F*.

These examples represent only a few of the many possible codings of wavelength by color. (Many thousands of combinations are possible.) Their value and utility remain to be established by further research. They represent the simplest of all discrimination techniques. They accomplish the translation of invisible wavelengths to the visible and then rely on the human interpreter to use tonal distinctions to draw conclusions regarding crop type, condition, etc. Such presentations may provide a useful bridge for the traditional photo interpreter between conventional imagery and highly processed imagery. They have obvious resemblances to conventional imagery, but represent a step in the direction of using nonvisible wavelengths and refined processing. On the other hand, they provide further experimental evidence that simple additive techniques are far less powerful than the more refined processing techniques described above and subsequently illustrated.

Plate 23*A*, *B*, and *C* portrays three different ways of encoding the amplitude of the signal within a single band. The 8–14-μ wavelength band has been shown because it represents apparent temperature. The radiant power in this band is, for all practical purposes, entirely due to self-emission during both darkness and daylight. The temperature has been termed apparent because of true emissiveness of the surface materials and the effects of the atmospheric transmission have not been taken into account.

In Plate 23*A* the signal has been quantized into eight levels, and only the boundaries at which it changes from one level to another are printed. Thus, the image consists of contours between regions having temperatures uniform within the ranges of the quantizing levels. The center image of the figure gives the same information but is presented in a different fashion: each quantized level has been assigned a density level, and the entire region within the limits of that quantized level is presented at the same density. The lower (colored) image of the figure shows each quantized range coded in a different color. Note that this is a distinctly different use of color than the wavelength coding discussed above. Any one of these presentations of apparent temperature may be useful in providing information leading to understanding of microclimate as one of the several uses to which they may be put.

Reference to these images of apparent temperature now make it possible to understand why, as previously mentioned, two fields of immature rice appear quite different on two of the wavelength-color-coded presentations, but appear essentially identical on the others. Reference to any of the apparent temperature presentations shows the lower, triangular field to be at a significantly higher temperature than the adjacent field of the same crop. As corroboration, note that among the images with wavelength coded in color, these fields are differentiated only in those images containing data from either the 8–14-μ band or the 4.5–5.5-μ band, which are the only two bands dominated by emitted radiation. Furthermore, in the two wavelength-color-coded images, the contrast between the two adjacent fields is strongest in the image containing a contribution from the 8–14-μ band. This is in agreement with the fact that there is a greater amount of emitted radiant power in the 8–14-μ band than in the 4.5–5.5-μ band. In the color-coded apparent-temperature image, note that the triangular field is blue, indicating an apparent temperature next to the highest in the area. The highest apparent temperature, colored violet, occurs in the fields identified as bare earth. This provides a clue to the likely reason for the differing appearance of the two fields containing young rice. The young rice plants do not completely cover the ground, so there will be significant amounts of soil visible between the plant leaves. If the triangular field had been drained while the other field was still partially flooded, with plant leaves growing above the water surface, the results observed would be expected. The drier soil of the better-drained field should reach a higher temperature at that time of day. The ground information is not sufficiently complete to verify this conjecture, but it fits the phenomena observed in the imagery, and a rational physical explanation is available.

A word of caution regarding the application of this quantizing and contouring technique to infrared scanners is in order. For the levels or contours to be meaningful on a relative basis, the scanner electronic passband must extend down to very low frequencies, essentially dc. Otherwise there will be level shifts in the signals because of the characteristics of the electronics and the signal will not be related to true temperatures on the ground in any simple manner. Under those circumstances, the meaning of the contours or levels is quite questionable. The only scanning equipment meeting these requirements, that the authors are aware of, is the instrument at The University of Michigan that produced the data shown. If, in addition to attaching significance to the relative levels or intervals, it is desired to associate quantitative meaning to the quantized intervals, the scanner must meet additional requirements: it must carry internal calibration sources on the ground adjacent to the area of interest; and also, to simplify interpolation between calibration levels, it is desirable that the scanner output be linearly related to radiation input. Again, the authors know of only one scanner meeting these conditions—the one at The University of Michigan.

Plate 23*D* through *G* shows four images, each containing only one type of surface material. Hence, these materials have been "recognized" and "identified." The process employed is the light-pencil technique. Its implementation is described above. Each recognition was effected using only two wavelength bands. The bands employed are indicated under each image. The recognition image for immature rice shows the fields not filled in solidly. This may indicate failure to classify correctly some of the rice present. It appears more likely, however, that the classification is correct and that those areas represent regions where the rice plants are less fully developed, allowing more of the soil to show through and thus altering the wavelength distribution of radiation from them. Similarly, the left ends of the two mature green rice fields not classified as rice may not be errors but regions of the fields where the crop is stunted because of poor drainage or for some other reason.

A question frequently asked by potential users of these techniques is: "What is the minimum number of wavelength bands needed to identify a crop or soil?" The set of recognition images in Plate 23*D*–*G* provides an answer of a sort and may indicate that it is not a very illuminating question to ask. Note that, although only two bands were employed in each recognition, it was a different pair of bands that worked best for each crop. This indicates that, while discrimination of a single material from a very limited set of "background" materials may be satisfactorily accomplished by the use of only two bands, the requirement to identify more than one material and one or more materials in the presence of a greater diversity of backgrounds will almost certainly require more than two bands. As the number of materials to be identified and the diversity of backgrounds present increase, the number of bands required for satisfactory performance almost certainly increases. Therefore, a more useful phrasing of the question might be something like: "For a desired probability of correct classification of a specified number of materials embedded in a specified number of backgrounds, what is the minimum number of bands for specified discrimination methodologies?"

As mentioned above, it is not necessary to devote a separate image to each material recognized. There are several ways of combining the individual classifications. One is shown in Plate 23*H*, where each surface material is coded in a different color. Note that this is a third use of color, distinct from each of the two discussed above. In this

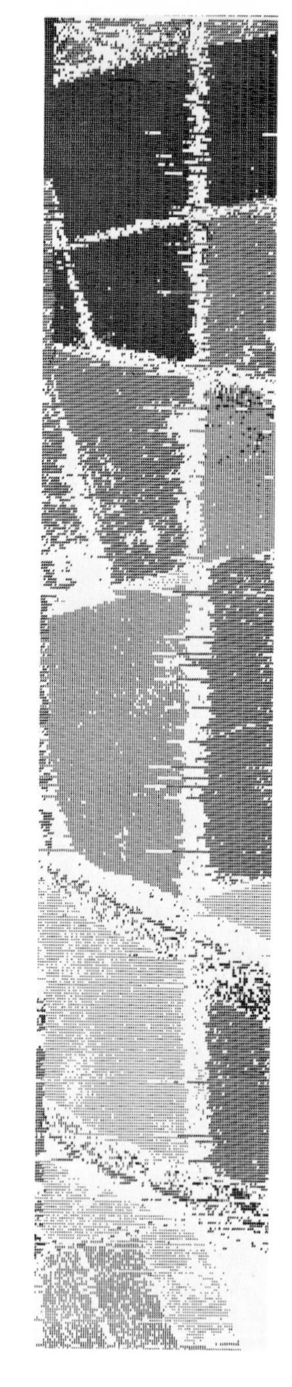

FIGURE 10 Digital recognition map of area shown in Figure 8.

instance, decisions and classifications are made by some pattern-recognition methodology implemented through some device, and then the results of the decisions are assigned colors for display purposes.

This type of display is intriguing in that it furnishes a mechanization of what are now standard practices carried out "manually." Quite frequently, agriculturalists, foresters, and a host of others working with data from geographically distributed materials and conditions will laboriously gather great quantities of data on foot, and with tremendous labor on the part of draftsmen and others, they transform it into maps that are color-coded, cross-hatched, or labeled with symbols or contours. Much past experience appears to indicate that such a presentation of the data is most comprehensible by the human user. What the image of Plate 23*H* appears to indicate is that methods are being developed to gather and present that type of data in such presentations without requiring the large amounts of manual labor; i.e., it is becoming possible to go directly from the data gathering to the final presentation in a single step. As will be indicated below, it appears possible to carry the process at least one step further to the production of summary tabular results that may be more useful than coded maps for some purposes.

It was remarked previously that the decision processes can be carried out with any of the available computational methods. Figure 10 shows recognitions of the several materials carried out by a digital computer. A pattern-recognition process of intermediate power was used. The six wavelength bands indicated on the figure were used to calculate the likelihoods of the presence of the various materials. This is less powerful than the likelihood-ratio methods, but the results are still quite striking. The prominent vertical lines of obvious misclassification are not deficiencies of the process but intermittent computer electronic malfunction.

Having classified the surface materials according to type, there is no reason why the computer cannot carry the process one step further and sum the areas of each. The result of doing this on the same data is shown in Figure 11. For some of the operations of the U.S. Department of Agriculture Statistical Reporting Service, for instance, that would be the preferred output.

The second area (near Lafayette, Indiana) for which recognition results are presented is shown in Figures 12 and 13. These are digital computer recognition printouts created by scientists at Purdue University using data from The University of Michigan multispectral sensing system. The area is the same as that shown on Plates 6 and 7 and discussed in Chapter 3, and is recognizable by the rectangular

Crop	Symbol	Acres
Bare Soil	.	190
Safflower	(	290
Immature Rice	+	410
Mature Rice	X	270
Other		440

FIGURE 11 Crop acreage (obtained by digital processing) of area shown in Figure 8.

hollow field of wheat with oats planted in the center. The imagery on Plates 6 and 7 was made on May 6, 1966. The data upon which the recognition results presented in this chapter are based was made nearly two months later, on June 28, 1966. Figure 13 shows a blowup of Figure 12 containing the rectangular hollow wheat field. In June, the wheat is near maturity.

Four spectral channels were used in producing these results. It was assumed that data in each of the four channels was distributed in Gaussian form for the wheat and for all other materials. The two mean vectors and covariance functions were then determined (one of each for wheat, and one of each for all other materials). At each sample point the ratio of the two probability functions was calculated, and a classification was made, depending on whether the ratio was greater than or

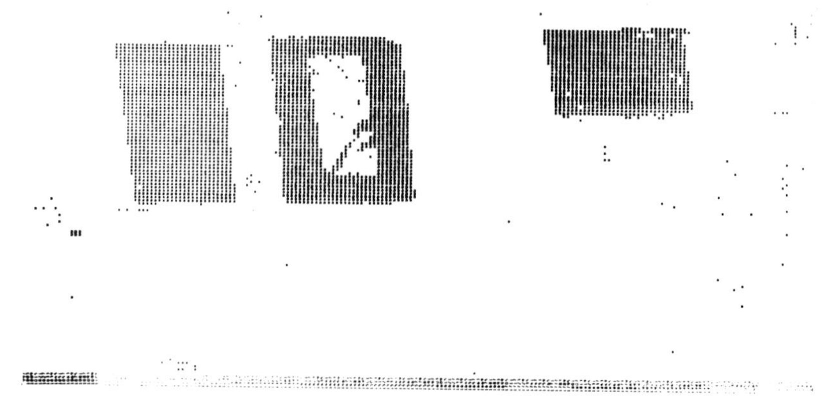

FIGURE 12 The first attempt to use truly automatic pattern-recognition techniques.[18]

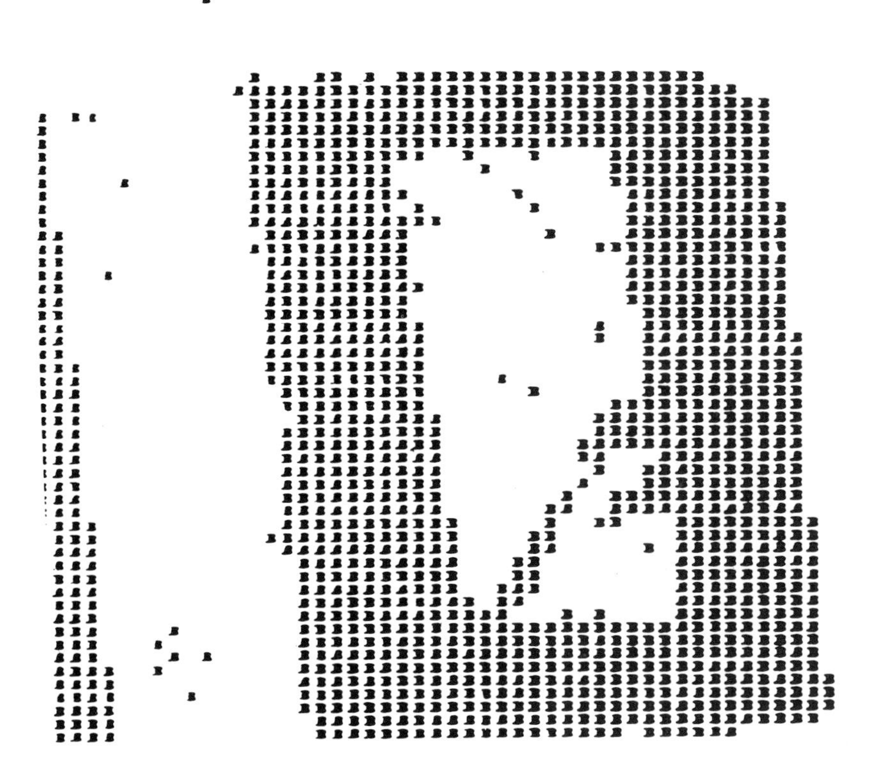

FIGURE 13 Enlarged section of Figure 12, showing the truly automatic pattern-recognition techniques.[18]

less than 1. This procedure is simpler than that employed subsequently at Purdue and elsewhere; nevertheless the results show a high degree of correct classification. Further information on this work is given by Landgrebe *et al.*[18]

The third area for which recognition results are presented is also near Lafayette, Indiana. This work was done by University of Michigan scientists, and the results are presented in Figure 14. The identifying numerals and letters are explained in Table 2. The area portrayed contains a number of small plots of crops under a variety of experimental conditions, as indicated in the table. Of primary interest are the approximately 16 long rectangular fields, some subdivided, in the center of the picture, having on the corners fields W4, O1, C16, and W13. Figure 15 shows panchromatic and infrared photographs of the area. Plate 24 shows color and Infrared Ektachrome photography of the area.

TABLE 2 Ground-Data Table, Purdue Agronomy Farm, June 28, 1966

Field Symbol	Crop	Height (in.)	Estimated Ground Cover (%)	Row Direction
O1	Oats	32	90	N–S
S2a	Soybean variety	16	35	N–S
S2b	Soybean variety	9	5	—
S3	Soybeans	4	5	E–W
W4	Wheat varieties	—	—	—
W5	Wheat	38	90	—
W6	Wheat (fertilizer experiment)	40	90	—
S7	Soybeans	—	—	N–S
C7	Corn variety	—	—	N–S
C8	Corn (planting date experiment)	—	—	N–S
C9a	Corn (fertilizer experiment)	—	—	N–S
C9b	Corn (fertilizer experiment)	—	—	N–S
O10	Oats	36	95	—
A11	Alfalfa varieties	—	—	—
M11	Legume varieties, soybeans, sorghum, oil crops	—	—	—
M12	Legume seedlings, sorghum, and sudan grass	—	—	—
W13	Wheat	40	90	—
C14	Corn	10–12	5	E–W
C15	Corn	4–6	< 5	E–W
C16	Corn	10–12	5	N–S
SmG17	Small grains	—	—	—
O18	Oats	26	90	—
W19	Wheat varieties	—	—	—
C20	Corn varieties and diseases	—	—	—
W21	Wheat	42	100	—
SmG22	Small grains	—	—	—
W23	Wheat	—	—	—
S24	Soybeans	—	—	E–W
C25	Corn	15	10	E–W
S26	Soybeans	4	< 5	E–W
E27	Erosion experiments	—	—	—

Contrast reversals between the two black-and-white photographs occur in a number of fields: O1, S2a, parts of C8, C9a, A11, and O18. These fields contain healthy green crops that provide good ground cover and produce bright images on the infrared photograph and dark images on the panchromatic film. Wheat fields W19, W4, W5, W6, W13, and W23, which have turned yellow, show no strong contrast changes in

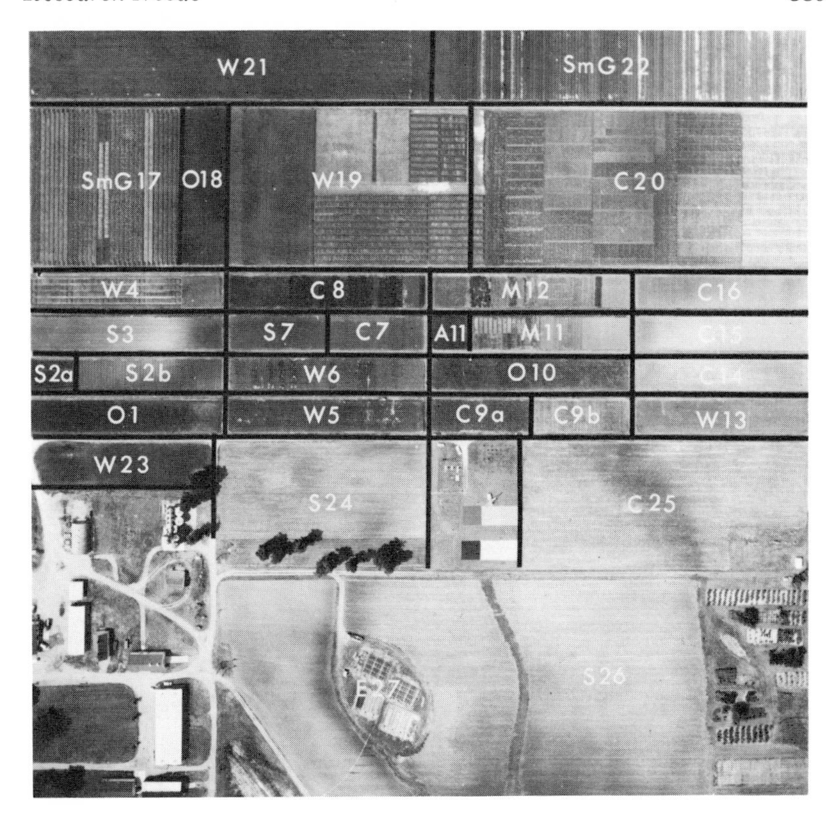

FIGURE 14 Ground data map of portion of Purdue University agronomy farm near Lafayette, Indiana, late June 1966.

W = wheat
O = oats
C = corn
S = soybeans
A = alfalfa
SmG = small grains
M = miscellaneous (legumes, soybeans, sorghum, oil crops, sudan)

going from the visible to the infrared photograph. Similar comparisons can be made between the two color photographs. The color print is, unfortunately, a poor exposure, as can be inferred from the tones of the color panels in the center of the picture. The left end of the lower series of panels is green and appears dark on all the imagery in contrast to the varying appearance of the green vegetation. Varying shades of red in the Infrared Ektachrome image reflect many differences in maturity and vigor of the vegetation. Field C8 is an experiment in corn planting date. The more reddish tones indicate earlier planting dates and hence more fully developed foliage. Field C9 is a corn fertilization

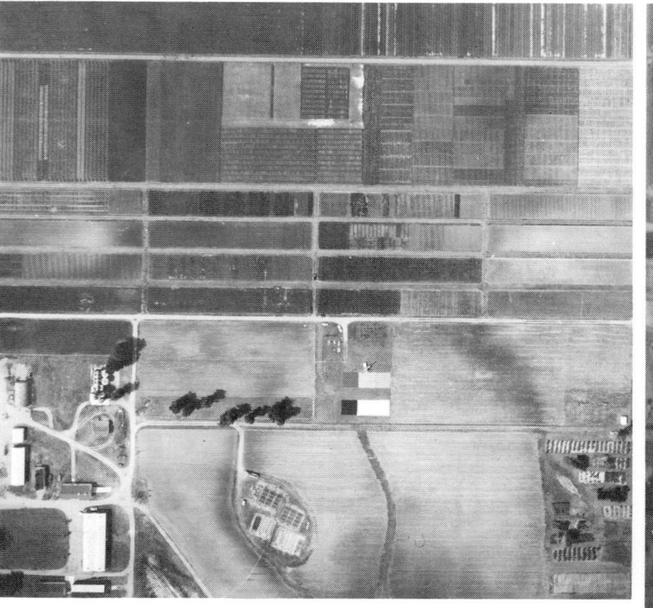

A

B

FIGURE 15 Black-and-white aerial photographs of area shown in Figure 14, taken at 2,000 ft. *A*: Panchromatic (visible), 3:14 P.M., June 30, 1966. *B*: Infrared, 1:04 P.M., June 29, 1966.

experiment, and C9a would appear to be the better fertilized field, for it has the more-luxuriant vegetation. The different varieties of soybeans in fields S2a and S2b produced different tones. (S2a is barely visible at the left edge of the picture.) Fields S24, C25, and S26 exhibit faint pink tones of sparse young foliage against the blue-gray tones of the predominating soil. The wheat fields do not appear red in the Infrared Ektachrome print because the wheat has matured and turned yellow.

Plate 25 shows a view earlier in the season in the spectral region to which Infrared Ektachrome film is sensitive. Notice that the wheat fields, still green in May, appear red. The trees at the lower part of field S24 are dark, indicating that they have not yet leafed out fully. A number of the fields appear green, indicating that vegetation has not yet appeared above the surface in significant quantities in May, and planting may not yet have occurred.

The above incomplete interpretation of the photographs reveals an important point: it is seldom possible to obtain direct diagnostic identification of crop types from photography. The various types of photography can produce contrast enhancement by proper selection, but without much cross checking and drawing on ancillary information, they can seldom produce identifications. In contrast, refer to Figure 16, a recognition map of wheat fields that was produced by a digital computer. Twelve wavelength bands were used in a likelihood-ratio decision process. All wheat fields are correctly classified, and very few false alarms are evident in the other fields. The only seriously questionable area is the field of small grains at the upper left, which might well contain significant wheat; the available ground data are not definitive on this matter. The altered shapes of the fields are partly due to aircraft crabbing during the data-taking flight and partly due to distortions deliberately introduced to force the data onto the computer printout format.

Results from the fourth and final area (also near Lafayette, Indiana) presented are of a somewhat different character. They were also obtained by scientists at The University of Michigan. Although they are not so pictorially pleasing as the previous results, they illustrate some of the ways in which the present methods are inadequate and show early solutions to some of the problems. The particular means chosen for this investigation required degrading the geometrical resolution of the basic data.

The area shown in the panchromatic photograph and labeled sketch (Figure 17) is in the C5 area of the experimental agricultural area managed and studied by the Purdue University Agricultural Experi-

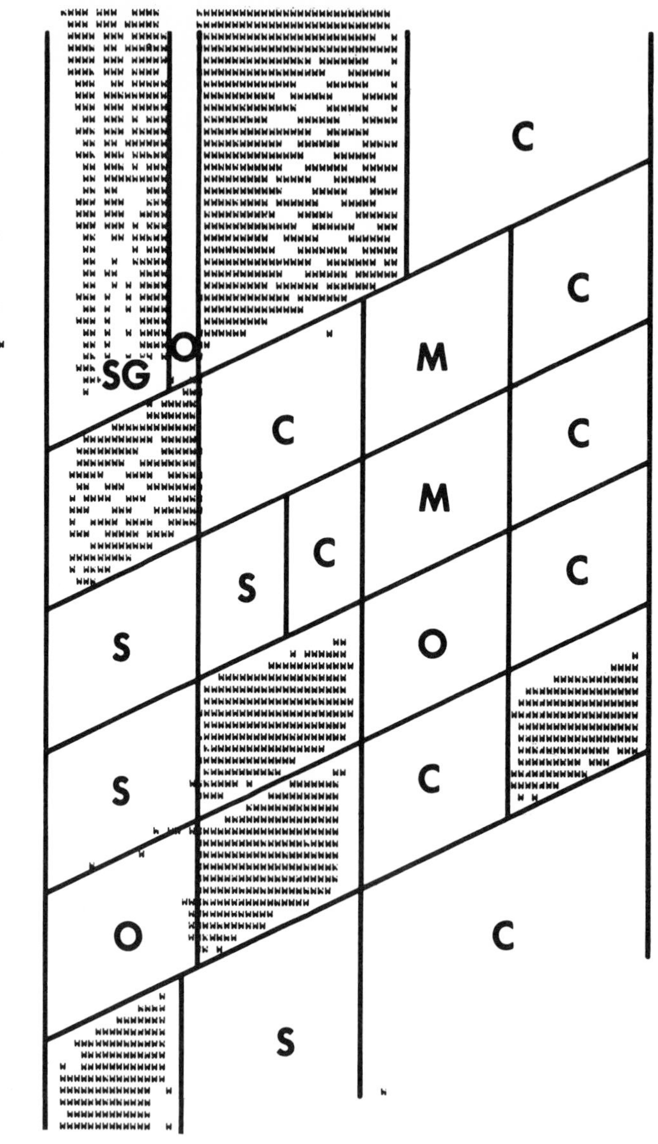

FIGURE 16 Recognition map for wheat, area described in Figure 14. Maximum-likelihood processing of digitized 12-channel scanner data. Skewness is due to crabbing of aircraft and sampling distortion. June 30, 1966, 3:14 P.M., 2,000 ft.

W = wheat
O = oats
C = corn
S = soybeans
SG = small grains
M = miscellaneous (legumes, sorghum, sudan, soybeans, oil crops, alfalfa)

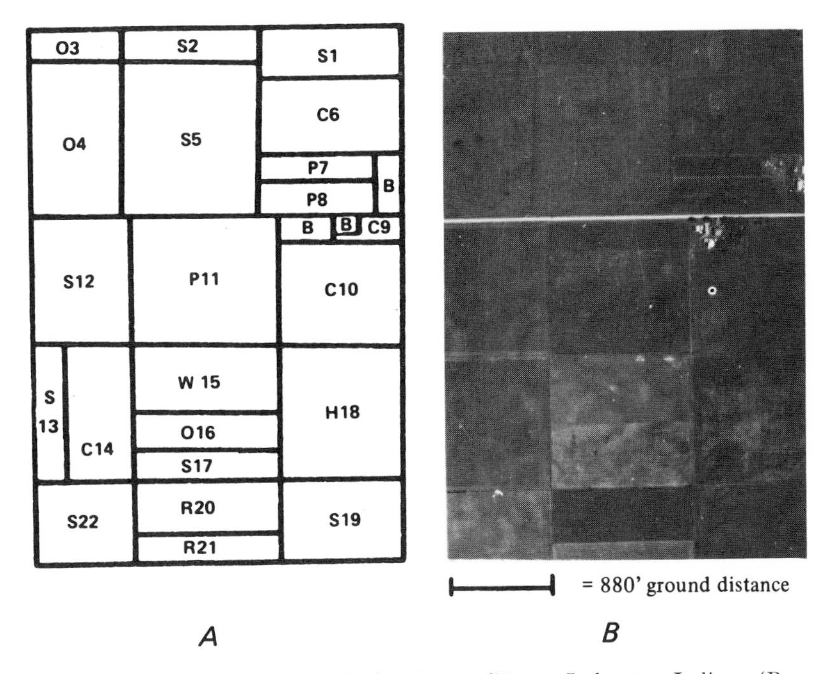

FIGURE 17 Ground data for agricultural area C5 near Lafayette, Indiana (Purdue University), June 30, 1966, 8:30 A.M. *A:* Field identification map. *B:* Panchromatic photograph with K-2 yellow filter, taken from 10,000 ft.

S = soybeans	W = wheat
C = corn	H = hay
R = red clover	P = pasture
O = oats	B = buildings

ment Station staff. In the subsequent processed imagery, the ratio of length to width of the images has been increased to show more detail.

Figure 18*B* is a processed image, obtained with a likelihood-ratio method, designed to recognize soybean field S22 and to reject red clover field R20, and soybean field S19. No explicit use of data from the other fields is made in this operation. The results show recognition of soybeans S22, partial recognition of soybeans S13, S12, and S5, recognition (misclassification) of corn C10, but complete and correct rejection of red clover R20 and soybeans S19. Corn C10 was erroneously recognized as soybeans because the illumination geometry (relative positions of the sun, the field, and the scanner) was such that C10 on the right side of the aircraft gave approximately the same spectral distribution of radiation as the S22 on the left side of the aircraft. Corn and

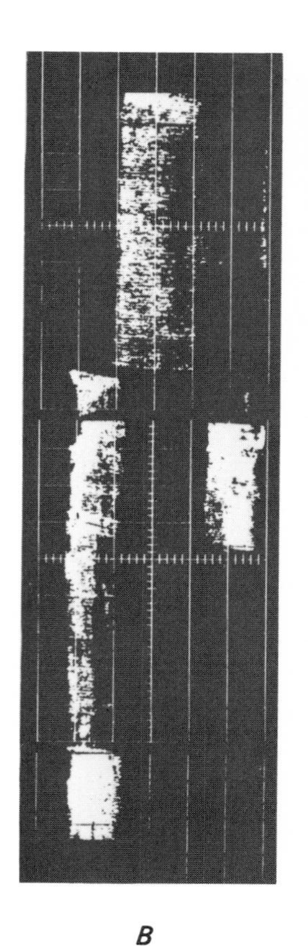

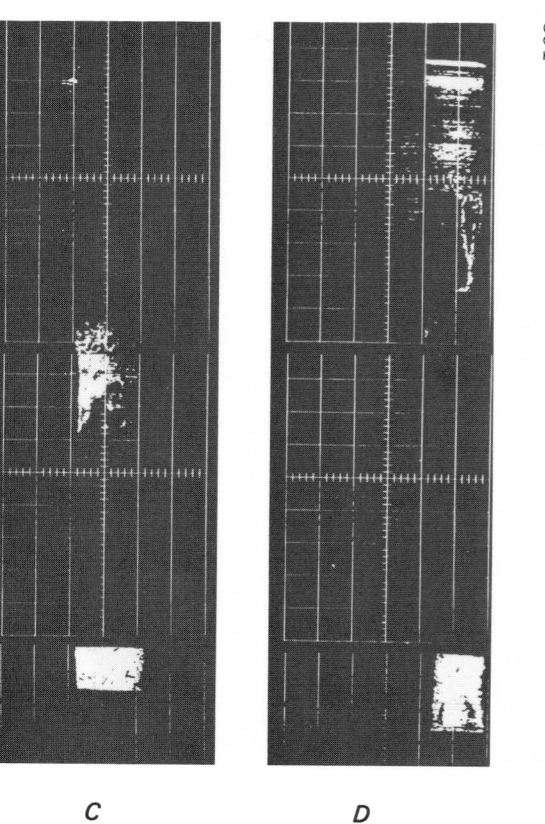

FIGURE 18. Recognition picture for Phase I obtained from the analog computer, agricultural area C5 near Lafayette, Indiana (Purdue University), June 30, 1966, 3:00 P.M. Aircraft altitude was 2,000 ft; heading was 180°. Wavelength bands used in processing were 0.40–0.44 μ, 0.46–0.48 μ, 0.50–0.52 μ, 0.58–0.62 μ, 0.66–0.72 μ. Date of processing, November 1966. Likelihood processing was based on the Bayesian decision rule. *A:* Field identification map. (See Figure 17 for key.) *B:* Soybean (S22) recognition with S19 and R20 as backgrounds. *C:* Red clover (R20) recognition with S22 and S19 as backgrounds. *D:* Soybean (S19) recognition with S22 and R20 as backgrounds.

soybeans at a certain stage of growth are the most difficult crops to differentiate yet encountered. Soybean field S19, on the right side of the aircraft, gave a response sufficiently different from S22, on the left side, that it could be rejected, while S22 was recognized, even though both fields contained soybeans. This illustrates the variation of spectral signatures with viewing and illumination angles (goniometric effects), which must be taken account of in many instances in the discrimination processing. It is shown below that in at least this instance it can be successfully coped with by more refined processing.

Figure 18*C* shows the results of a likelihood-ratio process designed to recognize red clover field R20 while rejecting soybean fields S22 and S19, no explicit account being taken of other fields. Field R20 is clearly recognized, but partial recognition takes place also for field P11. Possibly the recognized portion of this pasture did, in fact, contain red clover.

Figure 18*D* resulted from a likelihood-ratio process designed to recognize soybean field S19 and explicity reject soybean S22 and red clover R20. The imagery shows a strong response to S19, a fair response to S1, and a weak response to C6.

The imagery of Figure 19*B* resulted from a likelihood-ratio process using an acceptance signature constructed from data on all soybean fields and a rejection signature constructed from data on all corn and wheat fields. Although all soybean fields except S13 gave strong responses, fields C10 and C6 were also recognized when their rejection was desired.

Although not all materials whose rejection was desired were explicitly taken into account in the rejection processes, most of the errors in recognition and rejection of specific fields appear to have been caused by the differences in spectral radiance from the left and right sides of the scanned area, i.e., goniometric effects in the illumination and viewing geometry. When a correcting function for the average change in spectral radiance across a scan "line" from left to right was applied to the multispectral data before they were processed, and a soybean recognition process was then employed, the imagery in Figure 20*B* resulted. Here, all soybean fields are recognized, and the only significant error is partial misclassification of cornfield C6 as soybeans.

The results presented for this last geographical area illustrate some of the aspects of the discrimination problem that require further research. It is readily apparent that much additional signature data must be obtained and the manners and amounts of their variances studied and analyzed in order to derive fully satisfactory decision pro-

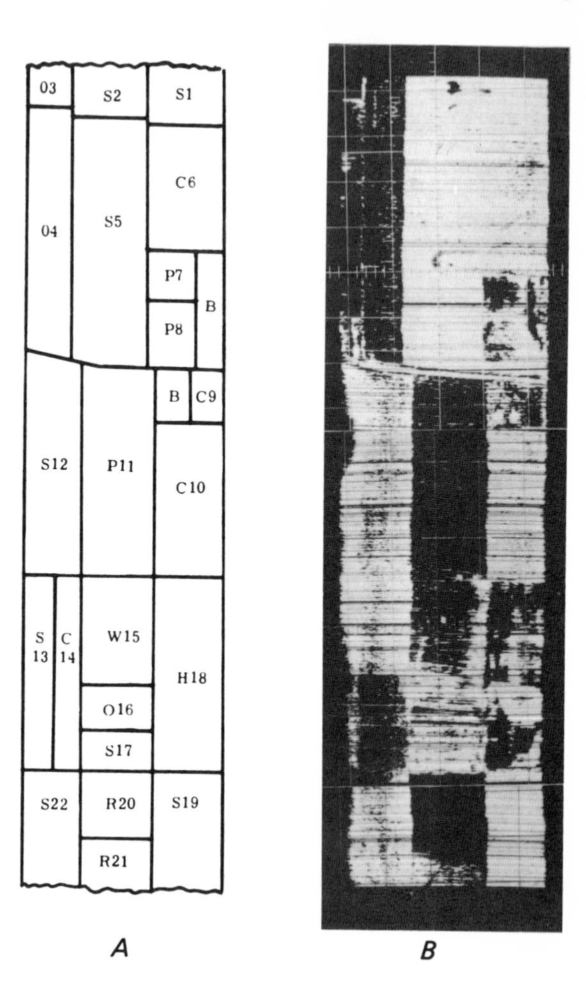

FIGURE 19 Recognition picture for Phase II obtained from the analog computer, same area, conditions, and wavelength bands, as in Figure 18. Date of processing, May 1967. Likelihood processing was based on the Bayesian decision rule. *A:* Field identification map. (See Figure 17 for key.) *B:* Soybean recognition with corn and wheat as backgrounds.

cedures. The results to date, however, give strong evidence that satisfactory methods can be developed with a realistic amount of research.

All data presented in this section on spectral discrimination have been from the optical (ultraviolet, visible, infrared) wavelengths. This does not mean that the methods are not applicable to radar wavelengths; they are. The construction and operation of multifrequency radars has, however, lagged the optical region work, so no radar results are currently available. Efforts should be made to accelerate this aspect of radar research.

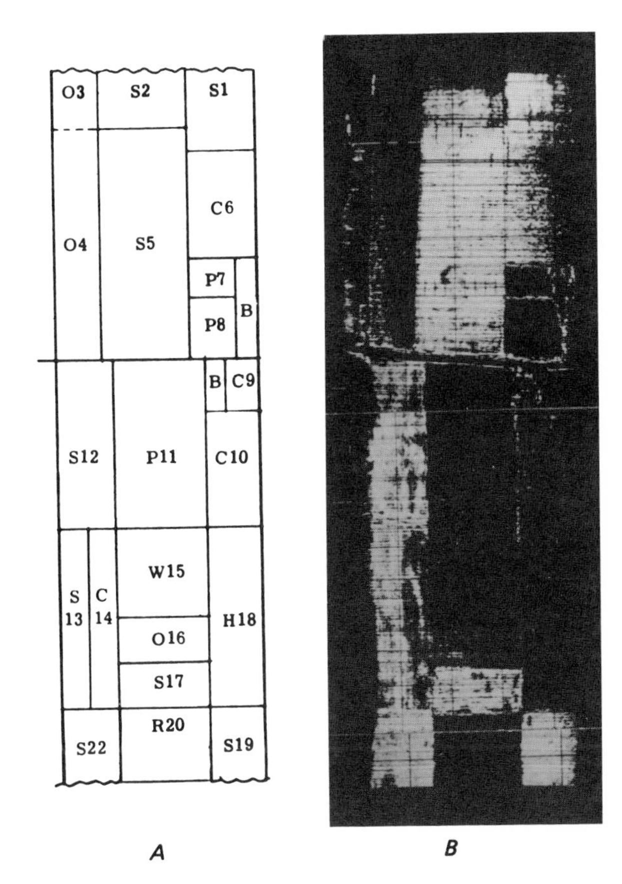

FIGURE 20 Recognition picture for Phase III obtained from the analog computer, same area and conditions as in Figure 18. Wavelength bands used in processing were 0.46–0.48 μ, 0.58–0.62 μ, 0.80–1.0 μ. Date of processing, March 1968. Preprocessing with illumination geometry functions and then likelihood processing based on the Bayesian decision rule. *A:* Field identification map. (See Figure 17 for key.) *B:* Soybean recognition with illumination geometry effect removed.

Spatial Discrimination

In principle, the very same methodologies found so effective in spectral discrimination are applicable in spatial discrimination. The observables, i.e., measured characteristics, will of course be different. In spectral discrimination the observables are straightforward and obvious, namely, radiant power levels in narrow spectral bands spanning the usable spectrum. In spatial discrimination the situation is much less clear. Just what set of observables can best, or even adequately, characterize shape or textural features is not yet known to any satisfactory degree. In addition a number of factors that will modify shape and texture are not operative or much more weakly influence spectral information. Viewing angle has a very strong influence on perceived shape and texture; therefore, spatial signatures will be modified by rotations of the viewed materials about all three possible axes. Although goniometric effects do modify spectral signatures, their influence is much weaker. Spatial signatures will be strongly influenced also by the scale of the image, whereas spectral signatures are relatively uninfluenced by scale. Shapes and textures can be seriously modified if other objects partially obscure the object of interest, whereas the spectral content of the exposed portions are quite uninfluenced by partial obscuration. Obscuration can be caused by trees partially overhanging a crop field or clouds covering part of the surface under observation. Shadowing does influence spectral signatures to some extent, but not at all as strongly as it does spatial signatures. The shadows can be due to clouds or, alternatively, a dark-toned object many times cannot be differentiated from its own shadow. Textural information on crops, even large crops such as trees, requires much finer geometrical resolutions than spectral information does. The result of these and other difficulties is that there is not a generally useful set of techniques for obtaining spatial information that will directly identify crops from aircraft and spacecraft altitudes. Larger features such as shapes and sizes of field boundaries may, by inference, indicate land-use types and thus crop types. This situation is all the more unsatisfactory because the human interpreter makes such extensive use of shape and textural information by means presently not well understood.

Methods of automatic shape and texture recognition have been studied intensively for the past two decades, although not for agricultural applications. Much of the effort has been directed at reading written texts, and no generally satisfactory solution is yet available. Several spatial and textural features have been investigated with partial success. By analogy with electrical engineering practices, frequency content and filtering have been investigated. It is well known that the Fourier transform of a time varying signal, $A = f(t)$ (where electrical signal amplitude A is function f of time t), results in the frequency spectrum of the signal. Frequency filtering can then select (classify) some parts of the signal and separate them from the remaining parts. A common implementation of this is tuning a radio or television receiver for one station and rejecting all others. In an analogous manner, spatial distribution of intensity or pictorial density can be analyzed for their frequency content by Fourier methods.[19,20] In this case the frequencies are "spatial" rather than time frequencies, i.e., cycles per unit distance or unit angle from the sensor, rather than cycles per unit time as in the electrical example. A complication in the spatial problem is that there can be variations in two independent directions, which makes it a two-dimensional problem in contrast to the one-dimensional electrical problem. However, optical methods exist that effect two-dimensional Fourier transforms.[21,22] Other methods have also been investigated: linear edge followers noting curvature rates, line crossings, and other features; scanning images with apertures of different size and shape; apertures to modify the image in a search for characteristic features; various digital-computer-sampled decompositions of the images; etc. Since none of the approaches yet pursued has produced a generally satisfactory solution to automatic spatial and textural classification, these methods will not be further described here. Uhr[23] and Nagy[13] provide an entrée into the pertinent literature; they have very extensive bibliographies.

As noted above, these methods have been investigated very little for agriculture and forestry applications. Since spectral discrimination is unlikely to provide solutions for all the problems, the spatial methods should be investigated for these applications and, where appropriate, the agricultural community should undertake research in spatial discrimination. As in the spectral studies, this will entail obtaining and analyzing spatial signature information as the first step. Following that will be the difficult step, already solved for spectral discrimination, of elucidating what spatial observables and features can characterize the information of interest to agriculturalists and foresters.

Polarization Discrimination

With the polarization discrimination method, objects can be distinguished from each other by making use of differences in the orientation of the resultant electric field vector of the reflected and/or emitted electromagnetic radiation. Radiation power arriving at a remote sensor can be considered as a superposition of many plane waves, each of which consists of a time-varying electric field and magnetic field. At every instant of time these two fields of the plane wave are at right angles to each other and at right angles to the direction of wave propagation. The geometrical relationships involved in polarization phenomena are sketched and described in Chapters 3 and 5.

The relationship between the orientation of the electric field vector and the geometry and composition of an object has been studied in somewhat more detail in the microwave portion of the electromagnetic spectrum than in the optical region. Some of the microwave data are presented and discussed in Chapter 3. These data indicate that significant effects do occur that justify further studies. In the optical spectrum, exploratory measurements have been made by Coulson[24] and by Limperis and George.[25] Figure 21 is a sample of some of the data obtained by Coulson. Here, degree of polarization is given for a

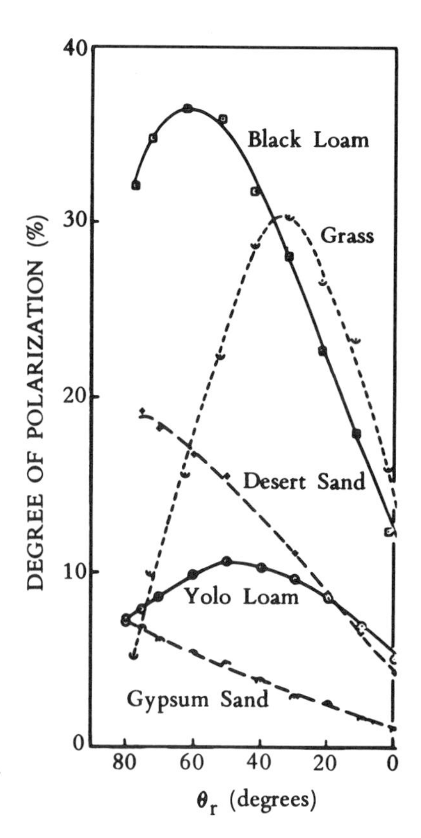

FIGURE 21 Plot of polarization characteristics of some soils.[24]

number of soil samples measured at a wavelength of 0.49 μ. The degree of polarization is defined as

$$P = (I_{\perp} - I_{\parallel}/I_{\perp} + I_{\parallel}) \times 100,$$

where $I_{\perp}$ and $I_{\parallel}$ are the intensities of radiation with electric field vector perpendicular and parallel, respectively, to the reflectance planes. Assume that the object is being irradiated with an electromagnetic source that is unpolarized (i.e., $I^{\circ}_{\parallel} = I^{\circ}_{\perp}$), and the resulting differences in $I_{\parallel}$ and $I_{\perp}$ are due to the reflectance properties of the material. By rearranging the relationship given above, the ratio of $I_{\perp}/I_{\parallel}$ is $(P + 100)/(P - 100)$. Thus if the degree of polarization is zero, then $I_{\parallel} = I_{\perp}$. For the black loam shown in Figure 21, the degree of polarization is about 35 percent at an angle θ_r of about 60°. In comparison, gypsum sand has a 5 percent degree of polarization at the angle θ_r. Therefore, when these soils are viewed at an angle of 60° by unpolarized irradiating power,

$$I_{\perp}/I_{\parallel} \approx 2 \quad \text{for black loam.}$$

$$I_{\perp}/I_{\parallel} \approx 1.1 \quad \text{for gypsum sand.}$$

Clearly, with the appropriate instrumentation one could distinguish between these two types of soils. Conventional remote sensors can be used for this with the addition of a linear polarization analyzer. The analyzer is an optical component that will transmit electromagnetic radiation that has an electric field colinear with the optical axis of the analyzer. For irradiation with noncolinear electric-field vectors, the transmission through the analyzer is the vector dot product

$$\tau = I \cos \delta,$$

where δ is the angle between the optical axis of the analyzer and the electric-field vector, and I is the radiation intensity.

Exploratory studies of polarization effects in the visible spectrum observed from an airborne platform have been performed at The University of Michigan. In this experiment, two Graflex cameras with Plus-X film and an analyzer, which operated over the 0.3–0.7-μ region, were mounted on an aircraft. The optical axes of the two analyzers were placed at 90° to each other (Figure 22); the cameras were operated simultaneously. This experiment was designed to investigate the existence of polarization in reflected solar radiation from natural and arti-

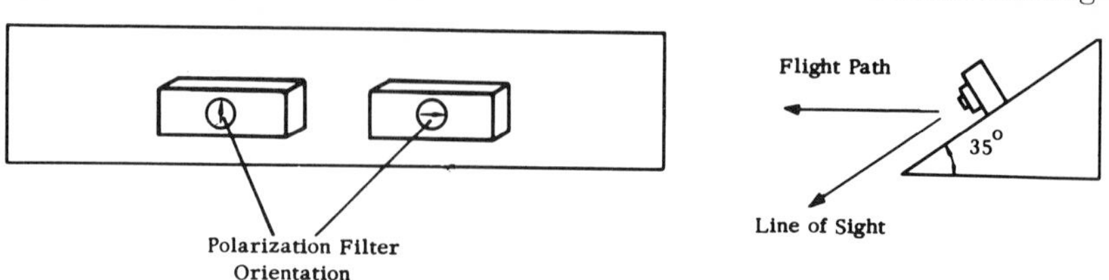

FIGURE 22 Diagram of camera orientation for aerial studies of polarization.[25]

ficial objects. Figures 23 and 24 are examples of the photographs taken. Figure 23 indicates both the texture and gross shape-discrimination effects. The water, which is a specular reflector in the visible region, has a considerably different intensity in the two photographs because most of the reflected radiation from it is polarized. On the other hand, the image of a small ice sheet on the surface of the lake (which is also flat, but has a different surface texture) appears equally bright in both photographs because the reflected radiation from it is predominantly unpolarized, and therefore is not as severely affected by the polarization filters. The white patches that appear at the bottom center of the twin photographs in Figure 23 are snow, which is also a nonpolarizing material. Note how clearly the trees along the bottom edge of the lake show up in detail against the darker background in the bottom photograph. An extremely interesting feature is the contrast inversion occurring at the tip of the island (shown in Figure 24). It implies that the radiation from the tip is polarized in the plane of incidence. The composition of that portion of the island and its state at the time at which the photographs were taken are unknown.

These photographs were taken with the aircraft flying into the sun for the greatest amount of specularly reflected light. When skies are overcast, however, such preferred flight paths are not necessary because there exists an essentially hemispherical emitting source (the illuminated clouds) that distributes the specular reflections to all angles.

In summary, polarization of microwave radiation of objects is in the early stages of study. Preliminary results show some characteristics that might be exploited. In the entire optical region, very little research has been performed. More investigations are necessary to determine the usefulness of this discrimination tool for the earth resources management problem.

Temporal Discrimination

Two classes of temporal changes can provide a basis for discrimination. The first of these, changes rapid enough to cause Doppler shifts in the radiation from the materials, has not been investigated for agricultural applications. The status of present technology is such that this technique is applicable only in the radar part of the spectrum. In the optical part of the spectrum, the shifts are too small to be isolated with the filters commonly available, and the spectral widths of most sources are wider than the Doppler shifts. It is feasible to sense the Doppler shifts at radar frequencies and, in fact, they limit the motion sensitivity of military moving target indication (MTI) radars. Thus it is well established experimentally that foliage moved by common wind speeds (3 to 5 mph) will result in detectable moving target signals. It is conceivable that the MTI signals coupled with some knowledge of wind speeds can provide crop recognition to some degree, because plants may have characteristic frequencies of motion related to stem length and other mechanical properties. The characteristic rippling of a mature cereal grain field in the wind is a familiar example. The rippling appears different for plants of differing ages. Therefore, it seems that some exploratory investigations of this technique are in order.

The second class of temporal effect occurs at a much more leisurely pace; it is the class of diurnal and seasonal changes that all plants and soils undergo. Certainly the technique of time-lapse photography is too familiar to require extensive discussion here. It is the current widely practiced form of temporal discrimination. Less familiar, however, are the temporal effects observed in other parts of the spectrum and the manner in which these may influence spectral and other automated and semiautomated discrimination techniques.

Temporal effects are due mostly to diurnal and seasonal variations that affect the material composition of objects; two examples are the material changes in plants caused by the destruction of chlorophyll and other large molecules, and the snow cover in the winter months. Also affected by seasonal and diurnal variations are temperatures of objects and atmospheric constituents.

There are data that support the use of temporal information as a possible discrimination tool. For example, Figure 7 in Chapter 5 shows the variability in the spectral reflectance of an oak leaf observed at several different times of the year as measured by Keegan *et al.*[26] at the National Bureau of Standards. The increasing numbers in each set of curves represent subsequent measurements. These data indicate that if careful selection is made of one or more spectral intervals of operation, green plants may be detected by observing the area repeatedly during the year.

In general, most plants exhibit considerable differences in reflectance and emission spectra at various stages of growth. An important factor,

FIGURE 23 Photographs taken with twin cameras as diagrammed in Figure 22.[25]

FIGURE 24 Same area as in Figure 23, also photographed with twin cameras. Note island in river.[25]

aside from the physical changes in the plant itself as indicated in Figure 7, Chapter 5, is the multiplicity of objects in the sensor's instantaneous field of view (FOV). The data used in Figure 25 were collected with a field spectrometer whose spatial resolution was 6 in.^{2}. The differences in spectra between the 3-in. and 22-in. alfalfa can largely be explained by the percentage of soil occupying the FOV. If $(y_1, \ldots, y_8)$ is the reflectance spectrum for 22-in. alfalfa $(x_1, \ldots, x_8)$, the reflectance spectrum for 3-in. alfalfa, and $(z_1, \ldots, z_8)$ the spectrum for the background, clay, then α may be computed by a least squares approximation; i.e., $\alpha(y_1, \ldots, y_8) + (1 - \alpha)(z_1, \ldots, z_8)$ is as close as possible to $(x_1, \ldots, x_8)$. In this example, $A = 0.75$; i.e., the 3-in. alfalfa comprised 75 percent of the FOV as compared with the 22-in. alfalfa. The dashed curve shows the predicted value of $(x_1, \ldots, x_8)$, which is a fairly reasonable fit for such a crude experiment. More investigation should be undertaken of this type of temporal variation.

Another relatively unexploited temporal discrimination tool is the use of an object's thermal history; however, only marginal effort has been expended in this area to date. Every material has associated with it a thermal capacity. If all the environmental factors such as wind

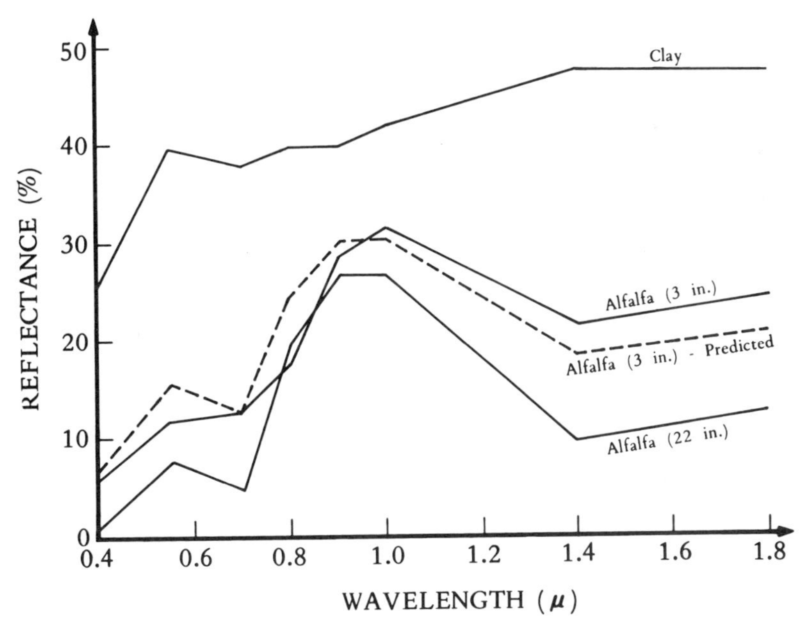

FIGURE 25 Plot of reflectance spectra for alfalfa and clay. (*University of Michigan.*)

velocity and the illuminating function are known or estimable, these thermal capacities may be determined from a remote platform. From this information the type of the material might be identified.

The temperature of an object in a changing thermal environment is regulated by surface emissivity, thermal diffusivity, total heat capacity, and surface area. The contrast between any two image points indicates the relative apparent temperature difference resulting from differing amounts of reflected and radiated energy received from those points. A photo interpreter would be aided in his task if the quantitative relationship between the actual target temperature and the photographic density of the target image were known. However, imagery is not, in general, calibrated; i.e., the transfer function of the scanning system is not known. Consequently, a given film density cannot be directly associated with an apparent temperature, much less an actual temperature. A scanner at The University of Michigan has been calibrated for the purpose, and it has been found possible to associate scanner signals with ground contact temperature measurements.

Examples of diurnal and seasonal thermal effects are shown in Figures 26 and 27. The region is Mud Lake Bog north of Ann Arbor, Michigan. The area in the center of the picture is a free-water surface. Around the edges of the water are water lilies surrounded by bog mat. Beyond the bog mat are grassy and forested areas. Lighter areas represent warmer materials. The shift in contrast between the water and the terrain throughout the morning is quite typical, as suggested by the contact temperature records shown in Plate 8. The sequence of images in Figure 26 requires clarification in one respect. The changing image tone of the water throughout the morning does not mean that the water temperature is changing radically. It is not, but rather, the terrain around the water is warming, and by midmorning the terrain has become warmer than the water. The reason the tone of the water changes is that the gain of the sensor was changed from flight to flight to keep the terrain data at the midpoint of the dynamic range of the sensor. The much smaller diurnal excursion of the water temperature as compared with terrain is a prominent feature of the data in Plate 8, and it is explained by the high emissivity and heat capacity of water.

A postmidnight seasonal sequence of images in the same spectral band is shown in Figure 27. Notice that in June and September the water is warmer than the terrain at that time of night, as would be expected. In January, however, the appearance is quite different. The water has frozen, and ice has radiation and temperature properties quite different from those of water.

It is quite clear on the basis of available evidence that there are very

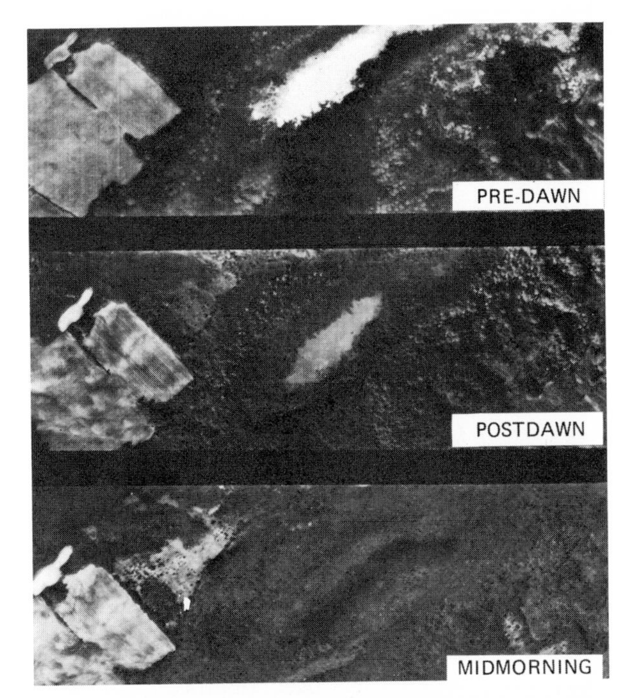

FIGURE 26 Time-of-day contrasts at Mud Lake Bog, Michigan, June 25, 1963, 4.5–5.5-μ spectral region. (*University of Michigan.*)

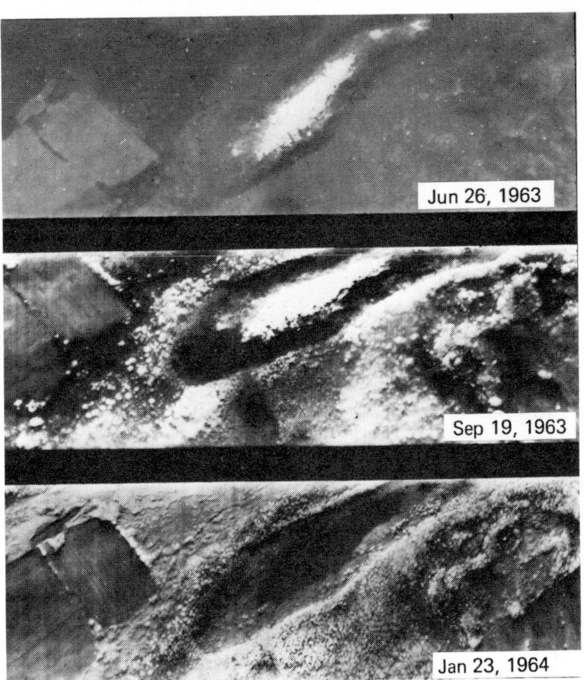

FIGURE 27 Seasonal contrasts at Mud Lake Bog, Michigan, 4.5–5.5-μ spectral region. (*University of Michigan.*)

strong diurnal and seasonal variations that can be employed when it is necessary to use time-sequential signatures. These effects have, however, been studied very little, so much work remains to be done before adequate understanding is achieved. There has been only one partly satisfactory sampling of diurnal changes successively throughout one complete annual cycle. That was carried out by scientists at The University of Michigan between June of 1963 and June of 1964. The Mud Lake Bog data referred to here were taken from the results of that investigation. There is a serious need for a number of additional investigations of that kind using the more modern instrumentation now available.

The Future of Discrimination

Each of the four preceding sections has discussed and provided some illustrations pertinent to discrimination using one of the four principal features of the radiation available. The relative amounts of material in the several sections constitute a measure of the effort that has been devoted to each, and the results and present status are dependent on the efforts that have been expended. It is apparent that research in spectral discrimination has been significant and that although the research is very far from being complete, some powerful techniques are now available. It is equally apparent that the other three classes of phenomena have received very little attention and much less of their potential has been realized. Surprisingly enough, discrimination in forestry may require finer resolution than in many types of agriculture because many agricultural applications involve quite large fields containing a single botanical species, while most forests consist of mixed stands with a high probability that adjacent trees will be of different species. A two- or three-stage sampling design may be necessary to make these difficult separations. Much more research is indicated on this problem.

It is possible to foresee ultimately, provided the requisite research is carried out, a large automated system that employs all four classes of phenomena interrelated in some fashion for optimum discrimination. This might take the form of first-stage discrimination based on spectral effects, followed by spatial discrimination on data whose contrast has been enhanced by spectral means, and that followed, for particularly difficult cases, by final discrimination based on polarization and temporal effects. Assessment of just what may be the optimum method of employing the various techniques in concert must await the completion of much research.

Research Needs

Research results clearly indicate a high probability that improved methods can be achieved to provide guidance regarding the types of research still required. It has been shown that modern sensing methods can make available greatly increased information at remote observation sites regarding plant and soil types and conditions, microclimate, etc., which can be used in decisions regarding species, yields, health, acreages, and other information of interest. It has also been shown that the vastly increased amounts of data that can be made available can be produced in such a form that they are amenable to automatic or semiautomatic processing, which makes the whole enterprise practical. The concept of a "signature" diagnostic of a material, material type or environmental condition has emerged as central to the sensing-classification process. We have seen that sensor technology and the analytic understanding of pattern classification processes appear to be in a more-advanced state of development than is the understanding of the radiation properties of plants, soils, and environmental conditions. Therefore, the pacing item, for the moment, appears to be knowledge of signatures and their amounts and modes of variation due to natural causes. Note that "signature," as employed here, implies not only a phenomenological relationship of radiation parameters to plants, soils, and environment, but also an understanding of the mechanisms involved, i.e., quantitative understanding of the influence of plant physiology on plant radiation properties, the means by which soil structure and condition influence soil radiation, etc. To elucidate these matters will require an extensive program of measurement and analysis in the laboratory, in the field from fixed platforms, from aircraft, and from spacecraft platforms.

Laboratory Programs

The factors that can cause variations in signatures are so numerous that an essential element in the research program will be carefully controlled laboratory measurements and analyses of the data. In the laboratory, control of illumination, viewing and illumination angles, atmospheric composition and path length, microclimate, soil type, and soil moisture is possible. The measuring instruments can be programmed to have only pure materials in their fields of view, avoiding the mixed signature problem, e.g., having only a single plant leaf in the instrument field of view. The measurements required will be reflectivities and emissivities at all the wavelengths of interest as functions of viewing and illuminating angles, plant type, health, stage of growth, and other parameters of interest. For nonplant materials or conditions, other similar parameters will be of interest. In the case of plants, as representative, there must be accompanying botanical studies to obtain the physiological parameters of the plants and relate them to the radiation properties. Since many investigators will, of necessity, be involved, this will pose problems of standardization of nomenclature, units, calibration, and measurement methods and instruments so that the results can be intercompared in a meaningful way. The goals and results of these laboratory programs will be quantitative understanding of the relationships between the radiation properties and the structures and compositions of the individual materials under controlled, uncontaminated conditions.

Ground Field Programs

The laboratory measurements will reduce the number of factors causing variability to manageable proportions, so understanding of the elemental processes can begin to be built. Building on those results, it will be possible to extend the work to more realistic and more variable conditions by carrying out similar series of measurements and analyses from fixed observation points, some ground-based and some elevated. In the field, additional sources of variability will be encountered, many of which cannot practically be simulated in the laboratory. There will be illumination directly from the sun, reflected from clouds, scattered by the atmosphere, and emitted and reflected from surrounding objects, and these sources of illumination will vary in both diurnal and seasonal cycles; there will be all the usual meteorological effects. It will be possible to observe larger and more varied samples; natural mixtures of soil and vegetative cover can be observed with more realistic resolution of instantaneous fields of view; more realistic atmospheric effects on the radiation transmission can be measured, etc.

Accompanying the physical radiation measurements and analyses in the field will be structural and compositional studies, similar to those carried out in the laboratory, of the materials and conditions under observation. The unifying element in this work should be a program to develop a model by which the results of field measurements can be adequately calculated on the basis of laboratory measurements, laboratory-gained understanding of the elemental processes, and understanding of additional sources of variability encountered in the field. These should all be entered into the calculation as functions of time of day, time of year, weather, etc. (See Chapter 7.) The extent to which calculations and modeling of this sort agree with actual field measurements will furnish an adequate measure of the progress being made. It appears desirable that the field measurements be conducted with instrumentation very similar (but adapted for outdoor use) to that employed in the laboratory rather than having it resemble operational sensors.

Aircraft Programs

Aircraft programs will be required. They differ from laboratory and field programs in two respects. First, added conditions and causes of variability, not easily simulated in field programs, will be encountered; and second, it seems desirable that the instrumentation employed more nearly resemble operational sensors than laboratory instrumentation.

From aircraft platforms, the illumination conditions will again be different: the longer atmospheric paths will introduce additional selective absorption and scattering, which will tend to modify signatures, and achievable ground resolutions will be coarser. It will be possible to survey larger areas with more nearly the same viewing angles. Certain areas, e.g., forests, that cannot be viewed practically from the ground, can be viewed from above; and very large areas can be surveyed in short times, giving statistically more significant samples.

Aircraft vibration and other practical difficulties, coupled with lack of precise control and knowledge of aircraft position, make the use of laboratory-type precision instruments in aircraft somewhat questionable. Furthermore, something positive is to be gained in coverage of larger areas and by experimenting with more nearly operational conditions achieved by using aircraft instruments more nearly resembling operational sensors but having, perhaps, somewhat finer resolutions and better calibration for which the price of slower speed of operation can be afforded in experimental programs.

The aircraft programs should, whenever possible, be carried out over sites where ground field programs are conducted, and the two sets of programs should be conducted simultaneously. Once again, the unifying element should be an analytic modeling effort aimed at calculating aircraft results from laboratory and field results, suitably modified for the altered conditions of measurement.

If instrumentation in the aircraft resembles operational rather than laboratory instrumentation, this, coupled with the large areas that aircraft can cover, makes it possible to conduct research on aspects of the over-all problem, downstream from the specific discrimination problem. The information generated will be quasi-operational in form and quantity, so problems of very large volume data handling, use of *a priori* information, integration of information from other sources (Tiros and ground-sampling sites) can be investigated and should be a significant part of any aircraft program. An essential element at this stage should be well-designed comparisons, with surveys made at the same time by current methods to obtain measures of improvement.

Spacecraft Programs

Spacecraft experimental programs are required because, once again, environmental effects that will modify signatures and that are not easily or practically simulated will modify signatures. The principal effects will be added atmospheric degradation, viewing even larger areas at nearly the same angles (from 500 nautical miles a 10° field of view covers a swath width of 100 miles), and ability to view very widely separated areas within very short times, so that it becomes pertinent to inquire about the stability of signatures over large regions with the concomitant differences in sun illumination angle.

Here again, the instrumentation should resemble operational instruments more than laboratory instruments, and for the same reasons as cited in the preceding section. In addition, the spacecraft program presents opportunities and motivation for research on operational hardware, data retrieval, communications versus on-board processing, and other systems problems. Once again, the unifying element in the discrimination research should be the analytic modeling effort aimed at calculating space-borne measurement results from the results of the laboratory, field, and aircraft programs suitably modified for the altered conditions. This requires that the programs in all four regimes be well coordinated and frequently carried out simultaneously at and over the same sites.

Summary

As remarked earlier in this chapter, the discussion of research needs cannot be exhaustive because of the limited space available, so it has focused primarily on the general discrimination problem. A few remarks are made in the following section on systems configurations and problems. Here it should be added that the preceding sections apply to all sensors operating in the ultraviolet, visible, infrared, passive-microwave, and radar regions. Spectral discrimination is the furthest advanced of the discrimination techniques, more so in the optical regions than in the microwave regions. The research program results will find immediate application by that technique. Spatial discrimination, not so well advanced, will require research to discover the spatial parameters capable of constituting signatures. Because of that, the results of the measurement programs will not find such immediate use for spatial signatures, but rather, will be used in a search for distinctive spatial features that may be combined into signatures. The use of polarization data from these programs will be in a similarly primitive state; i.e., it will be exploratory to determine the extent to which it is worth pursuing. Although it has been stated that sensor and pattern-recognition technologies presently lead signature knowledge, this does not mean that no research should be conducted in these areas. Suitable implementations of the existing technology are known to be currently lacking, and added signature data and understanding will almost certainly create the need for technology advances in those areas.

Resource Management System Design

This chapter and this book conclude with final mention of some of the problems associated with implementing an operational global remote-sensing system.

The sensing instruments and interpretation processes are complex, and their proper design is critical to the success of remote sensing. The sensors, however, constitute but a small part of a much larger system, and they represent but a fraction of the cost of the over-all system. It is not possible to treat here in any detail the vast number of system and subsystem trade-offs that must be made to implement an operational system. However, a few questions that can be raised are typical of those that must be answered by the users, systems designers, and equipment designers before a system can be implemented.

Assume: (1) that a global remote-sensing system consists of space-borne vehicles to provide large-area routine coverage, aircraft to provide intermediate-area coverage (but with greater detail than that obtainable from spacecraft), and ground-sampling stations for *in situ* checks; (2) that the final data-processing interpretation and evaluation is centralized at a single station with data received from a number of remote stations distributed throughout the world; (3) that present meteorological networks and meteorological satellites such as Tiros and Nimbus are tied into the system. A system based on these assumptions is shown in Figure 28.

The central data-processing station receives data from a number of

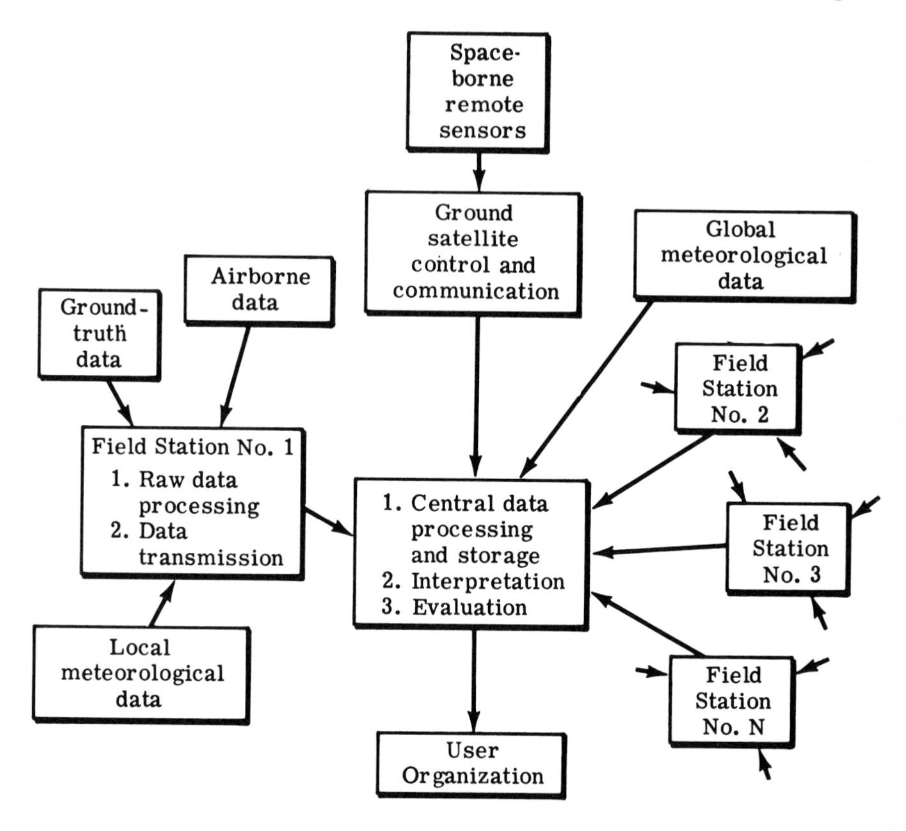

FIGURE 28 Global remote-sensing system configuration.

points. It is linked directly with the satellite system through ground-tracking and telemetry stations and receives data from a number of field stations that are gathering ground and airborne data. It also receives data from meteorological networks. It contains quantities of *a priori* information, e.g., common farming practices. The main function of the central data-processing center is to coordinate the data, analyze it, store it, and present it as rapidly as possible to users in forms that are easily interpreted.

The field stations gather data routinely and carry out special measurement missions as directed by the central station. These may be ground measurements *in situ* at some specific point, or airborne measurements, perhaps to confirm in greater detail and accuracy area measurements obtained from the space-borne equipment. At the field stations, preliminary data processing is carried out to ensure the quality of the data and to organize data into a suitable format for direct input to the central processing system. The number of field stations will be dictated by the total area to be surveyed, by the biological and physical complexity of the area, and by the total area that each station can handle. The location and number of stations will depend on physical, political, and economic factors.

A number of questions can be asked with respect to systems: How many satellites will be needed to provide the necessary remote-sensing coverage? Would such coverage consist of the entire earth's surface? How often would we wish to inspect a given point or area? If we do not provide complete global coverage, then which areas should be covered? How many field stations would be required, and how many aircraft would each field station need? What data-processing capability will be needed at the central station, and how much of this data processing should be carried out at the field stations? What kinds of communications systems will be required from the satellite to the central data-processing station, and what kinds of communications networks will be needed to tie the field stations into the processing station?

Studies dealing with the economic benefits of remote sensing to agriculture, hydrology, and geology will be able to answer many of these questions and will permit the objectives to be established with the system itself. It is the task of the large-scale systems designers to interpret these objectives in physical terms. They must decide how the various functions will be performed and how they will be integrated.

Figure 29 is a functional diagram of the central data-processing station. Its operation is envisioned as follows: data are received from the communications networks and are recorded in both digital and analog forms. Data will flow into the center as charts, tables, graphs, photographs, and telemetry signals. By automatic processing, data from the various sources and different sensors will be identified and related to a common geographical coordinate system. Each of the sensors may have a different spatial resolution, will cover a different area, and will obtain measurements at differing times and under different meteorological conditions. The volume of the data will be large, particularly from the standpoint of space systems. Present-day methods of human photo interpretation will simply not suffice. Hence, it will be necessary, for the most part, to feed data from the sensors directly into a data-processing system itself. A geographical referencing system for the data will be needed and, along with this, sophisticated navigational gear will be needed for both the airborne and the space-borne vehicles. Present satellite-tracking methods are able to provide very precise data about a satellite's position and will perhaps be adequate for those purposes.

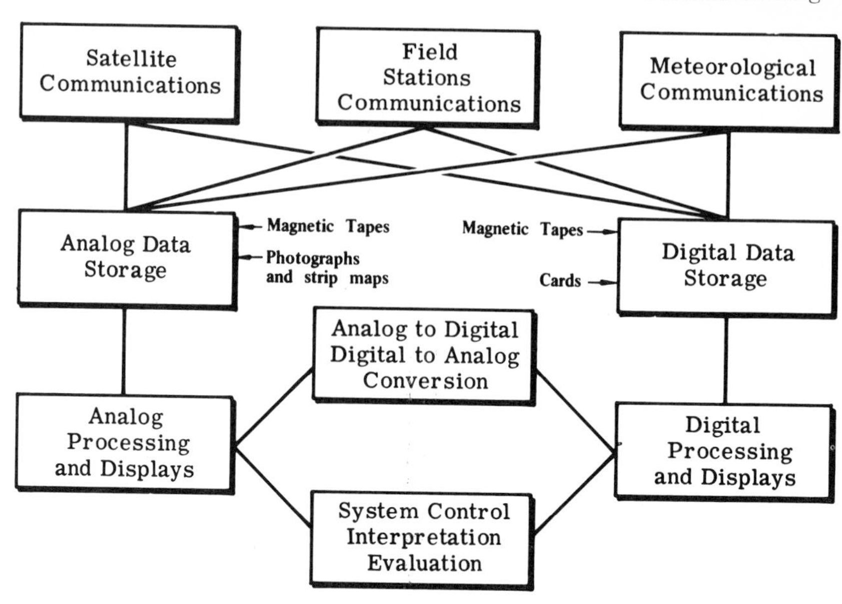

FIGURE 29 Flow chart for central data station.

However, satellite stabilization may be troublesome, particularly so if fuel must be expended to counter motion of moving parts, for the life of the satellite will be dictated by its fuel capacity. If none or passive methods are used, then pitch, roll, and yaw must be measured. Accuracy of these measurements must be examined with respect to the remote-sensor data-coordination problem itself. The airborne sensors will have another problem, for they will be operating at remote points where the ground-navigation equipment now used will probably not be available. This suggests that either new stations must be established or some type of inertial guidance system will be needed.

The system designer must also consider the various forms in which the processed data are to be presented. It will probably be necessary to have both analog and digital readout equipment. For example, consider that a user's request is for multispectral characteristics versus certain weather conditions of a certain wheat field. Since it has a single set of coordinates, this, in all probability, would be a digital printout. Or, if the request is for imaging in a certain spectral band, an analog strip map might be produced. A combination of analog and digital displays can be envisioned where it is desired to mark the strip maps with identifying markers—for example, a request such as imaging of all wheat fields receiving a certain amount of rainfall.

Figure 29 shows how the data might flow at the central data station. Broadband or video type of information received from the satellite-communication centers, from the field-communication systems, and from meteorological centers would be recorded directly onto magnetic tapes at the analog storage–processing center. Simultaneously, the data would also be converted from analog to digital form and fed to the digital storage–processing center, where the digital information would be stored on tape. Once the data were stored in both analog and digital form, they would be available for processing. From analog storage, broadband video data could be displayed on a CRT, or they could be presented in the form of strip maps recorded from the displays. The data could also be fed to an analog computer, where they could be electronically filtered for a particular characteristic prior to display. From the digital-storage center, the data would be fed to the digital-processing equipment, and similar displays could be obtained or direct printout could be made, as in the case cited above.

Figure 30 shows the flow of data at the field station. The field station consists of the airborne sensors, the ground-sensing equipment, meteorological sensing gear, and a preliminary data-processing system. In addition to the sensors that might be carried on the aircraft, airborne navigation and meteorological equipment would also be a part of the system. Data obtained from the airborne measurements would be recorded on magnetic tapes or, in the case of photography, on film. These data would then be fed into the raw-data-checking center, where as in the case of the central data-processing station, the data would be coordinated with ground measurements and local meteorological data. A number of questions can now be raised regarding the field station. How extensive should the ground-measurement program be? How many ground-measurement stations must there be? How will the data from the ground measurements be coordinated with the airborne measurements? How much data processing should be carried out at the field station? How will photographic information from the field stations be relayed back to the central data station?

Thus far, we have alluded to some of the problems the system designer will face. The equipment designer is also faced with some formidable tasks. For example, let us consider the satellite-borne multispectral sensor. It will be desirable from the user standpoint to have a system of high spatial resolution, high spectral resolution, high sensitivity, and wide coverage. These factors all work against each other and together contribute to increased demands on scanning rates, electronic bandwidth detector design, optics, and cooling system requirements. Ultimately, they reflect themselves in the design of the vehicle itself.

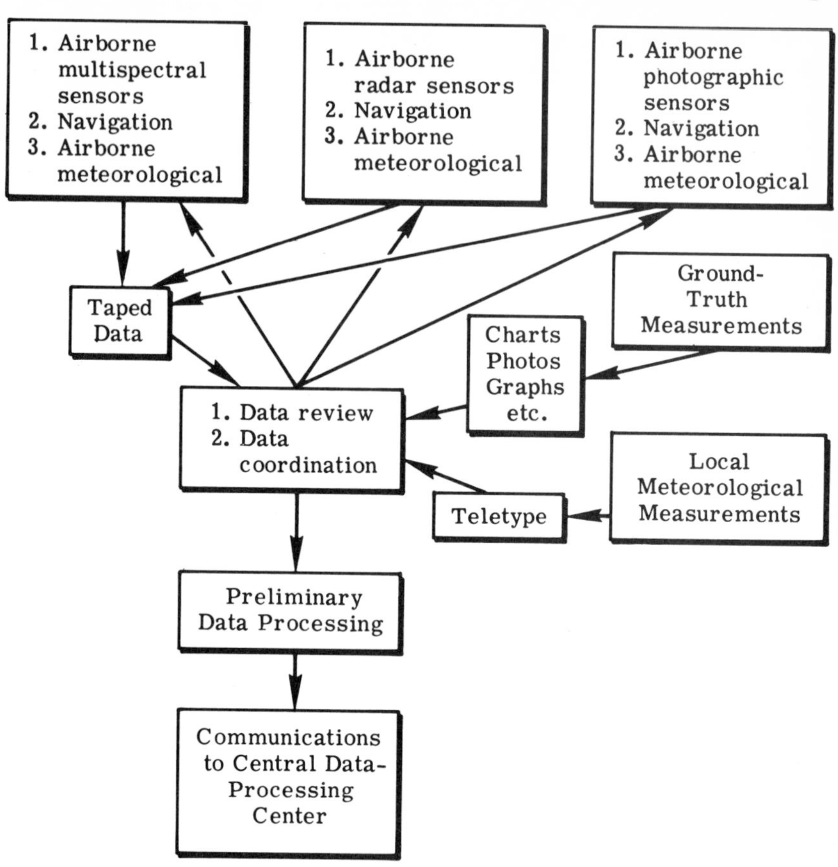

FIGURE 30 Chart of data flow at field stations.

It has been assumed here that the multispectral sensor would include a long-wavelength infrared capability; the detectors for this region require cooling. The present methods of cooling detectors in airborne scanners use either closed-loop cooling systems or cryogenic fluids (liquid nitrogen, liquid helium, etc.). Direct detector cooling methods by the use of cryogenic fluids or solids have the obvious disadvantage that the coolant must be replenished. Closed-loop cooling systems recirculate the refrigerant gases, and thus the gases are not exhausted; but unfortunately this is at the expense of electrical power. The direct cooling method has the obvious advantage that it is simple, requires no power, and is highly reliable. But if one expects to have a long satellite-use time, let us say more than a year, then these methods cannot be used. Systems have been proposed, however, that could cool an array of detectors for as long as six months to a year. For a closed-loop system, large amounts of electrical power are required. Recently developed closed-loop systems (Sterling cycle), although fairly efficient as compared with earlier systems (Joule-Thomson), still require large amounts of power. The most promising solution is a newly designed two-stage passive cooler that uses the space environment itself (i.e., radiation cooling) to achieve temperatures on the order of 70° K, which is sufficient for one new detector operating in the 8–14-μ wavelength region.

The electrical bandwidth requirements of the multispectral sensor will depend on the number of spectral channels, spatial resolution, area coverage, and satellite orbital velocity and altitude. For a 30-channel system with reasonably high resolution, the requirements are modest compared with those for television bandwidths. However, if one allows for growth in the state of the art of sensor performance and other satellite-control and telemetry requirements, television bandwidths are not excessive. This implies then, that satellite-tracking and receiving stations must have the capability to transmit and receive televisionlike bandwidths. Obviously, such a system on a global basis would be very costly. An alternative to this would be to have a single station that would have these bandwidth capabilities. It would require, however, that the satellite store its data and play it back at a specific point in its orbit. To do this, the recording equipment within a satellite would have to record at a 6-Mc rate and play the data back at an even faster rate as it passed over the receiving station. The speeded-up video transmission bandwidth would depend on its playback period, which in turn is a function of the orbit or the range to the ground-receiving station. But, to say the least, this period would be manyfold that of the real-time readout. A possible solution to this problem may be to command program the sensor to gather information over preselected sites, limiting the total data, and to then use slowed-down video at its receiving station.

An organized creative technology will be needed to solve these problems. The initial step will be basic research and mathematical modeling of the system itself. These steps must then be followed by simulation of the system on a computer and utilizing aircraft to simulate the satellite. During this period, spacecraft experimentation, measurements, and analyses should be carried out to learn about the environmental effects on signatures and also to develop space-borne instrumentation techniques. Subsequently, as the last step before implementation of an operational system, a single-thread system should be operated. A single-thread system is one having the minimum number of components necessary to exhibit all the functions to be performed by the system. The system should contain space-borne, airborne, and ground-based elements functioning in concert.

References

1. U.S. Dep. Agr. 1964. Statistical Reporting Service of the United States Department of Agriculture—Scope—Methods. USDA Miscellaneous Publication 967. U.S. Government Printing Office, Washington, D. C.
2. The Univ. of Michigan. February 1966. Peaceful uses of earth-observation spacecraft. Vol. III: Sensor requirements and experiments. Willow Run Labs., Inst. Sci. Tech. Rep. No. 7219-1-F(III), NASA Contract NASq-1084.
3. Colwell, R. N. 1961. Some practical applications of multiband spectral reconnaissance. Amer. Sci. 49(1).
4. Colwell, R. N. 1956. Determining the prevalence of certain cereal crop diseases by means of aerial color photography. Hilgardia. 26:(5).
5. Manzer, F. E., and G. R. Cooper. 1967. Aerial photographic methods for potato disease detection. Maine Agr. Exp. Sta. Bull. 646.
6. Brenchley, G. H., and C. V. Dodel. 1962. Potato blight recordings by aerial photography. NAAS Quart. Rev. Vol. 57.
7. Thomas, J. R., and V. I. Myers. April 1966. Factors affecting light reflectance of cotton. Proc. Symp. Remote Sensing of Environment, Ann Arbor, Mich.
8. Lowe, D. S., and J. G. N. Braithwaite. 1966. A spectrum matching technique for enhancing image contrast. Appl. Opt. 5(6).
9. Lowe, D. S., J. Braithwaite, and V. L. Larrowe. October 1966. An investigative study of a spectrum-matching imaging system. Univ. of Mich. Inst. Sci. Tech. Rep. 8201-1-F.
10. Lowe, D. S., F. C. Polcyn, and J. R. Shay. 1964. Multispectral data collection program. Proc. 3rd Symp. Remote Sensing of Environment. Ann Arbor, Mich.
11. Legault, R. R., and F. C. Polcyn. 1964. Investigations of multispectral image interpretation. Proc. 3rd Symp. Remote Sensing of Environment. Ann Arbor, Mich.
12. Middleton, D. 1960. An introduction to statistical communication theory. McGraw-Hill, New York.
13. Nagy, George. 1968. State of the art in pattern recognition. Proc. IEEE. 56(5): 836.
14. Sebestyn, George S. 1962. Decision-making processes in pattern recognition. Macmillan, New York.
15. Legault, R. R., and J. F. Riordan. 1964. An algorithm for optimizing a spectral filter. Appl. Opt. 3(6).
16. Polcyn, F. C. November 1967. Investigations of spectrum matching sensing in agriculture Final report, May 1964 through March 1967, Vol. I. Univ. of Mich., Inst. Sci. Tech., Willow Run Labs., Rep. 6590-9-F(I). NASA Grant NsG 715/23-05-071.
17. Hasell, P. G. January 1968. Investigation of spectrum matching techniques for remote sensing in agriculture. Interim Report, March 1967 through December

1967. Univ. of Mich., Inst. Sci. Tech., Willow Run Labs., Rep. 8725-12-P. Prepared under USDA Contract No. 12-14-100-8923(20). Published under USDA Contract No. 12-14-100-9503(20).

18. Landgrebe, D. A., *et al.* March 1967. Automatic identification and classification of wheat by remote sensing. Purdue Univ., Agr. Exp. Sta., Lab. for Agr. Remote Sensing. Res. Progress Rep. 279. LARS Information Note 21567.
19. Linfoot, E. H. 1964. Fourier methods in optical image evaluation. Focal Press, London. p. 75*ff*.
20. O'Neill, E. L. 1963. Introduction to statistical optics. Addison-Wesley, Reading, Mass. p. 160.
21. Cutrona, L. J., E. N. Leith, L. J. Porcello, and W. E. Vivian. 1966. On the application of coherent optical processing techniques to synthetic-aperture radar. Proc. IEEE. 54(8):1026–1032.
22. VanderLugt, A. B. July 1963. Signal detection by complex spatial filtering. Univ. of Mich. Inst. Sci. Tech. Rep. No. 2900-394-T.
23. Uhr, Leonard. 1966. Pattern recognition. John Wiley, New York.
24. Coulson, K. L. September 1965. Effect of surface properties on planetary albedo. Thermophysics Specialists Conf., Monterey, Calif.
25. Limperis, T., and D. George. 1965. Electromagnetic field signatures in the optical spectrum. Ann. N. Y. Acad. Sci., 140, Art. 1, Planetary and space mission planning.
26. Nat. Bur. Standards. 1964. Infrared optical measurements. Semi-ann. tech. rep., 1 July through 31 December 1964. Rep. No. 8626. Meteorology Div., Washington, D. C.

APPENDIX

Committee on Remote Sensing for Agricultural Purposes

In 1957, certain representatives of the chemical industries brought to the Agricultural Board of the National Research Council the following general question: "How can we obtain improved timeliness and accuracy of information on detection, incidence, and assessment of loss from crop and forest pests—diseases, insects, weeds, and wildlife—in order to improve development, processing, distribution, and marketing of pesticides?" From an economic standpoint, the question was valid, as evidenced by a 1954 report of the U.S. Department of Agriculture setting the annual production loss from these agents to crops and forests at some $7 billion.

After preliminary surveys were made, the Board responded in 1960 by recommending that a committee be formed to investigate the potential of aerial surveys to answer the question posed. To capitalize on current technologies, the committee included members from many disciplines—physics, engineering, statistics, plant pathology, forestry, economics. From the beginning, the committee considered not only aerial photography in the visible portion of the electromagnetic spectrum but sensing in all spectral regions where the atmosphere permits—ultraviolet; visible, reflective, and emissive infrared; and active (radar) and passive microwave. The committee also extended its efforts beyond crop-pest damage and loss information to information on crop and forest production management and marketing deemed collectible by aerospace vehicles.

The Committee on Remote Sensing for Agricultural Purposes, as the group was formally designated, by late 1962 had designed experiments to gather simultaneous multispectral data from agricultural crops and soils that would pinpoint problems of data reduction and discrimination analysis. Committee members participated in these early feasibility trials, which were funded by the National Aeronautics and Space Administration and the U.S. Department of Agriculture in 1964. Since that time, members have acted as participants and advisors to those two agencies in research programs at Purdue University, at USDA installations at Weslaco, Texas, and Berkeley, California, at the University of California at Berkeley, and at The University of Michigan.

MEMBERS OF THE COMMITTEE

T. R. BROIDA (1961–1966) Spindletop Research Corporation, Lexington, Kentucky
R. N. COLWELL (1961–1969) University of California, Berkeley, California
D. M. GATES (1965–1969) Missouri Botanical Garden, St. Louis, Missouri
R. C. HELLER (1964–1969) U.S. Department of Agriculture, Forest Service, Berkeley, California
O. W. HERMANN (1961–1962) U.S. Department of Agriculture, Washington, D. C.
M. R. HOLTER (1961–1969) The University of Michigan, Ann Arbor, Michigan
*H. J. KEEGAN (1963-1968) Clemson University, Clemson, South Carolina
B. W. KELLY (1966–1969) U.S. Department of Agriculture, Washington, D. C.
F. E. MANZER (1966) University of Maine, Orono, Maine
C. L. MINARIK (1961–1966) U.S. Army, Fort Detrick, Maryland
V. I. MYERS (1966–1969) South Dakota State University, Brookings, South Dakota.
J. A. RIGNEY (1967–1969) North Carolina State University, Raleigh, North Carolina
H. A. RODENHISER (1961–1969) U.S. Department of Agriculture (retired)
J. R. SHAY (Chairman) (1961–1969) Oregon State University, Corvallis, Oregon
H. A. STEELE (1966–1967) U.S. Department of Agriculture, Washington, D. C.
H. C. TRELOGAN (1963–1966) U.S. Department of Agriculture, Washington, D. C.

*Deceased.